Cryogenic Engineering and Technologies

Principles and Applications of Cryogen-Free Systems

Cryogenic Engineering and Technologies

Principles and Applications of Cryogen-Free Systems

Edited by
Dr. Zuyu Zhao
Janis Research Company, LLC
Dr. Chao Wang
Cryomech, Inc.

CRC Press is an imprint of the
Taylor & Francis Group, an **informa** business

CRC Press
Taylor & Francis Group
6000 Broken Sound Parkway NW, Suite 300
Boca Raton, FL 33487-2742

First issued in paperback 2022

ISBN-13: 978-1-498-76576-3 (hbk)
ISBN-13: 978-1-03-233780-7 (pbk)
DOI: 10.1201/9780429194726

Publisher's Note

The publisher has gone to great lengths to ensure the quality of this reprint but points out that some imperfections in the original copies may be apparent.

Visit the Taylor & Francis Web site at
http://www.taylorandfrancis.com

and the CRC Press Web site at
http://www.crcpress.com

Contents

Preface

One of the most important achievements of cryogenic engineering in the twentieth century was the successful development of the 4 K closed-cycle cryocoolers of Gifford–McMahon (GM) and pulse tube types. Since liquid helium has been the primary cooling method at temperatures of 4 K and below for nearly a century, the breakthrough of 4 K cryocoolers is having a fundamental impact on the world of cryogenics.

Starting from the 4 K platform of the cryocoolers, different technologies can be employed to reach much lower temperatures. Cryogen-free cryogenics have been developing rapidly over the past decade, mainly due to the increasing price of liquid helium. Another reason for this popularity is the reduced necessity of cryogenics expertise in the operation of cryocooler-based systems. The cryogenics field is inescapably on a path toward replacing the use of liquid cryogens in favor of cryocoolers.

Driven by this belief and the desire to help the communities, we would like to take this opportunity and present the latest information on cryogen-free technologies. The chapters are contributions from industrial experts of five leading cryogenic companies. The book is written based on many years' experience in research and development and is meant to be a fundamental and practical resource. The book introduces the history of helium, fundamentals of 4 K regenerative cryocoolers, vibration reduction, and enabling technologies to build cryogen-free systems. Finally, cryogen-free superconducting magnets and cryogenic refrigeration systems at the temperatures of 4 K down to a few millikelvin (mK) are presented.

The audience of this book is any individual working on low-temperature research and/or technologies. All of the contributors will feel rewarded if the book is useful to many readers, and we look forward to your feedback.

The cooperation of all the contributors in preparing and discussion for the final draft is gratefully acknowledged. Thanks also goes to editors at Taylor & Francis Group publisher for the encouragement and help to prepare the book.

Editors

Zuyu Zhao, PhD, earned his BS degree from Fudan University in 1982, Class 77. He came to the United States in 1983 with the World Bank Scholarship program and graduated from Northwestern University with a PhD in physics in 1990. He then spent two and a half years working at Harvard University as a post-doc and set up a new lab pursuing Bose-Einstein condensation on spin polarized hydrogen. Dr. Zhao joined Janis Research Company in 1993 and focused on developing custom ultra-low temperature facilities for the research and science community. He currently serves on the Board of Directors and as Executive Vice President-Principal Scientist of the company. He served from July 1, 2007, through June 30, 2013, as an elected member of the American Institute of Physics (Physics Today) Advisory Committee.

Chao Wang, PhD, has been the director of research and development at Cryomech since 1998. Dr. Wang earned his BS from Shanghai Institute of Mechanical Engineering in 1985 and MS and PhD from Xian'An Jiaotong University in 1990 and 1993. Between 1993 and 1996, he was a post-doctoral researcher and later became an associate professor at Cryogenic Lab, CAS. He later received an A.V. Humboldt Research Fellowship for study at the University of Giessen. In 1998, Dr. Wang joined the Cryomech team where his main research focus is development of pulse tube cryocoolers, GM cryocoolers, and cryocooler-related cryogenic systems. During his time with Cryomech, he has developed and commercialized the world's first two-stage pulse tube cryocoolers below 4 K. He has published more than 80 scientific papers and has been awarded five patents. Dr. Wang has won a number of awards for his contributions to cryogenic refrigeration technology, including the Roger W. Boom Award (2000) from the Cryogenic Society of America.

Contributors

Adam Berryhill
Cryomagnetics, Inc
Oak Ridge, Tennessee

Michael Coffey
Cryomagnetics, Inc
Oak Ridge, Tennessee

Charlie Danaher
High Precision Devices
Boulder, Colorado

Luke Mauritsen
Montana Instruments
Bozeman, Montana

Brian Pollard
Cryomagnetics, Inc
Oak Ridge, Tennessee

Chao Wang
Cryomech Inc.
Syracuse, New York

Zuyu Zhao
Janis Research Company, LLC
Woburn, Massachusetts

1. Evolution of Cryogenic Engineering

Zuyu Zhao

Chapter 1

1.1 Early Cryogenic Engineering: From Ice to Ice Maker (Refrigerator)

1.1.1 Cryogenic Engineering Prior to the Eighteenth Century

The word ***Cryo-*** originates from the Greek ***kruos***, meaning ***icy cold***. Cryogenic temperatures are usually considered those below −150°C, which covers the normal boiling point of most gases (except for the so-called permanent gases including hydrogen, oxygen, nitrogen, carbon monoxide, and the inert gases). Cryogenic engineering covers the technologies used to produce these cryogenic temperatures. Cryogenic engineering is not fundamentally different from refrigeration, defined as "to make things cold." However, refrigeration originated much earlier and is closely related to food preservation. Bacteria grow rapidly in many foods including meat, fish, fowl, and dairy when exposed to room temperature and above. Bacterial activity is suppressed at lower temperatures, and temperatures below −20°C inactivate any microbes (bacteria, yeasts, and mold) present in food. In addition to other food preservation techniques, such as drying, fermenting, pickling, smoking, curing, canning, and the like, storing food at freezing temperatures became an effective preservation method. Ice is the most obvious and natural "cryogenic engineering instrument." In fact, the Chinese began to use crushed ice in food about 2000 BC [1].

Food preservation for sailors in earlier times posed a significant challenge, since it was not unusual for voyages to last many months. For example, eighteenth-century German immigrant ships bound from England to Philadelphia typically sailed for 60 to 80 days, and many sea voyages lasted even longer. Since sailing ships had limited capacity for storing ice, meat and fish could only be carried in small quantities, and their use was normally strictly rationed. These storage limitations became even more severe when ships traveled in warmer geographic areas because any ice used for preservation would melt quickly.

The ice industry was a significant industry for centuries, but obtaining, preparing, distributing, and storing natural ice was expensive. As the industrial age progressed, the search began for alternatives to naturally formed ice: "necessity is the mother of invention." The word "refrigeration" was first used at least as early as the seventeenth century. The search began at that time for a "refrigerator" or "ice maker" that was able to produce ice at a lower cost.

1.1.2 Eighteenth Century: Expansion May Make Materials Colder!

William Cullen (1710–1790) (Figure 1.1) was the first person to demonstrate the principle of artificial refrigeration, in 1748 [2], by allowing ethyl ether to boil into a vacuum. He employed a pump to create a partial vacuum over a container of diethyl ether ($C_4H_{10}O$), which then boiled while absorbing heat from the surrounding air. Although Cullen's experiments only created a small amount of ice and did not immediately lead to any practical application, the concept of cooling caused by the rapid expansion of gases remains the primary means of refrigeration today. It is fair to say that Cullen laid the foundation for the modern refrigerator.

FIGURE 1.1 William Cullen (1710–1790), inventor of the artificial refrigerator. (Courtesy of University of Glasgow Story, University of Glasgow.)

1.1.3 Mid-Nineteenth Century: Realization of the Vapor-Compression Refrigerator That Produced Ice!

Oliver Evans (1755–1819) [3], remembered as the "Grandfather of Refrigeration," introduced the idea of "Vapor-Compression Cycling" at the beginning of the nineteenth century (1805). He also developed the first detailed and theoretically coherent design for a vapor-compression refrigerator, which identified all the major components (e.g., compressor and condenser, cooling coil, expander, evaporator) of a **refrigeration cycle**. The vapor-compression refrigerator uses a circulating refrigerant as the medium (usually a hot, saturated vapor). The refrigerant is first compressed to a higher pressure (and higher temperature) and then passed through a condenser. The hot, compressed vapor is then cooled and condensed into a liquid phase by passing the compressed vapor through a coil or tubes that are cooled by flowing water or cool air. To complete the refrigeration cycle, the refrigerant vapor from the evaporator is routed back into the compressor.

John Gorrie (1803–1855) [4] further adapted the basic refrigeration concept and developed the first practically useful vapor-compression refrigerator capable of producing ice (it was one of the earliest *"cryogen-free"* refrigerators). He filed for a US patent with the title "Improved process for the artificial production of ice" on February 27, 1848, which was granted on May 6, 1851 as US Patent No. 8080.

The refrigerator was shown in the patent illustrations with ice collected in a wooden box near the top [5].

In his patent application, Gorrie wrote:

> It is a well-known law of nature that the condensation of air by compression is accompanied by the development of heat, while the absorption of heat from surrounding bodies, or the manifestation of the sensible effect, commonly called cold, uniformly attends the expansion of air, and this is particularly marked when it is liberated from compression. The nature of my invention consists in taking advantage of this law to convert water into ice artificially by absorbing its heat of liquefaction with expanding air.

Interested readers may refer to Appendix A for more details of the patent (Figure 1.2).

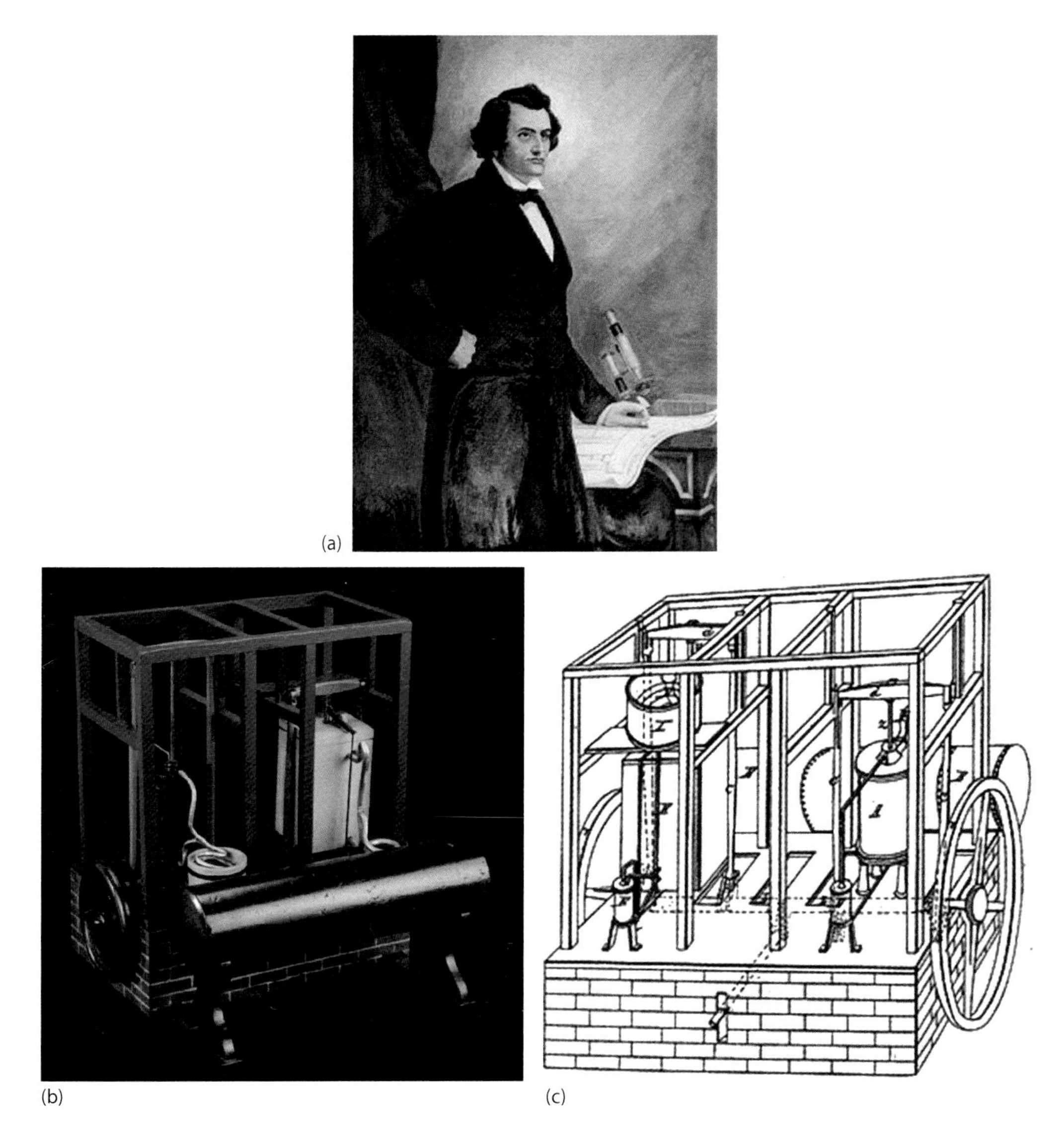

FIGURE 1.2 Dr. John Gorrie and a model of the original ice-making machine: (a) Dr. John Gorrie, (b) A model of the original ice-making machine, and (c) A sketch of the original ice-making machine. (Courtesy of Division of Recreation and Parks, Florida Department of Environmental Protection, https://floridadep.gov.)

1.2 Development of Thermodynamics

1.2.1 Understanding Heat

In the mid-nineteenth century, producing ice from water was still challenging (despite water's freezing temperature of 0°C (273 K), far higher than the "cryogenic temperature" range as defined at the beginning of this chapter. Clearly, new technologies and a better understanding of cryogenic engineering were needed in order to make a breakthrough.

While the previously mentioned pioneers focused on developing "refrigerators" that could reach lower temperatures and produce more ice during the eighteenth and nineteenth centuries, James Watt (1736–1819), a Scottish inventor, mechanical engineer, and chemist, worked on the development of a steam engine. Watt's engine was the first to make use of pressurized steam to drive the piston. Watt's invention represented a key breakthrough in the Industrial Revolution, despite the very low efficiency (**approximately 4%**). The desire to improve steam-engine efficiency naturally led to additional research into the steam engine's fundamental underlying operating principles (Figure 1.3).

Early theories of heat can be traced to the origin mythologies of Egypt circa 3,000 BC, when heat was believed to be related to the "primordial forces" from which everything was formed. The understanding of heat continued evolving, and the concept of the phlogiston was developed in the seventeenth century. The phlogiston was understood to be a substance existing in all combustible bodies, which was released during combustion. The phlogiston theory was replaced by the caloric theory in the eighteenth century, representing one of the historical markers in the transition from alchemy to chemistry. The caloric theory considered heat to be a hypothetical weightless fluid called **caloric** that could pass in and out of pores in solids and liquids, and that traveled from hotter bodies to colder bodies. Alternative theories of heat continued to be proposed, but they all viewed heat essentially as a "substance." The scientific framework of "thermodynamics" was forming shoulder-by-shoulder with that of cryogenic engineering and technologies.

A study of steam-engine efficiency was initiated by Nicolas Léonard Sadi Carnot (1796–1832), a French military engineer and physicist. Carnot emphasized the importance of heat/energy transfer, a concept not well understood at that time. In his monograph of *Reflections on the Motive Power of Fire* published in 1824, Carnot presented the first successful theory of the maximum efficiency of heat engines. Sadi Carnot has earned the name "father of thermodynamics," and the world might have understood the laws of heat much sooner had he not died of cholera at the age of 36 (Figure 1.4).

(a) (b)

FIGURE 1.3 James Watt (1736–1819) and his steam engine: (a) Watt steam engine and (b) James Watt (1736–1819). (From James Watt, *Dictionary of National Biography,* 60, London, UK, Smith, Elder & Co., 1885–1900.) (https://en.wikipedia.org/wiki/James_Watt#/media/File:James-watt-1736-1819-engineer-inventor-of-the-stea.jpg)

FIGURE 1.4 (a) Sadi Carnot's monograph of *Reflections on the Motive Power of Fire* (https://en.wikipedia.org/wiki/Reflections_on_the_Motive_Power_of_Fire) and (b) Image of Carnot and Carnot cycle. (From Carnot, S: *Reflections on the Motive Power of Fire*, Mineola, NY, Dover Publications, 1988.)

Nicolas Clément (1779–1841) [8] first gave the name "Calorie" to heat. The calorie was defined circa 1824 as the amount of heat needed to raise the temperature of one gram of water from 0°C to 1°C. However, this definition still did not clarify the relationship between heat and energy. It was James Prescott Joule (1818–1889) [9] who studied the nature of heat and concluded that energy could neither be created nor be destroyed. Instead, Joule concluded that energy could only be converted from one form into another. Joule's conclusion led to the law of conservation of energy, which in turn led to the development of the first law of thermodynamics.

One of Joule's most important experiments used a large water-filled container with a paddle wheel mounted within. The paddle wheel was connected to an axle, around which were coiled many wraps of string. The string was then looped over a pulley, and a heavy weight was affixed to the far end. When Joule released the weight, the string moved over the pulley, turned the axle, and made the paddle wheel spin. The spinning paddle wheel caused the water temperature to increase. After careful measurements and calculations, Joule concluded that the amount of potential energy lost by the falling weight was exactly equal to the amount of heat energy gained by the water. The energy conversion factor currently accepted by the international community [10] is (Figure 1.5) thermochemical calorie = 4.1858 joule.

Based on the experiments by Joule (and others), William Thomson, Baron Kelvin (1824–1907) outlined this very important view in 1851: ***Heat is not a substance, but a dynamical form of mechanical effect*** [11].

The world had finally reached a correct and scientific understanding of heat by the mid-nineteenth century, taking an important step toward the full establishment of a new branch of natural science called "**Thermodynamics**."

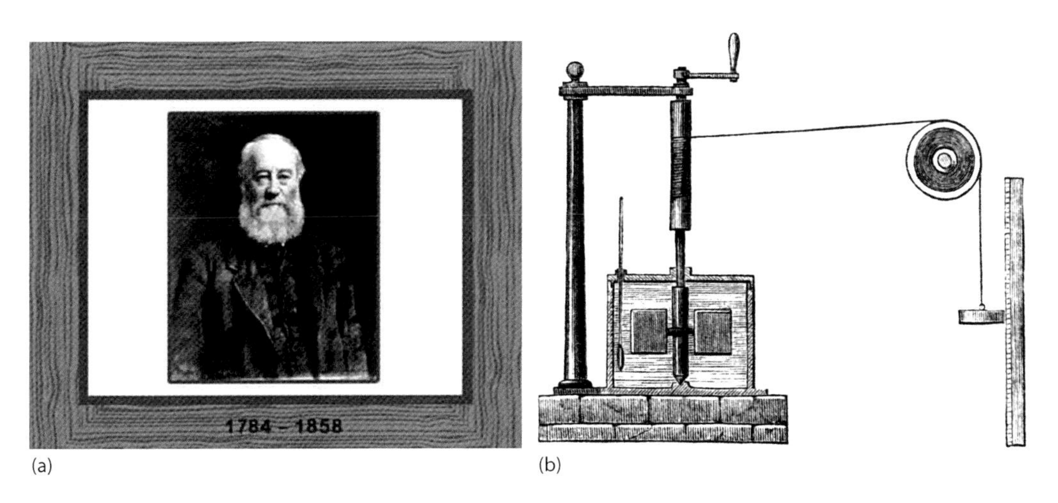

FIGURE 1.5 (a) James Prescott Joule and (b) his experiment of the "Mechanical Equivalent of Heat" that led to the correct and scientific understanding of heat (https://en.wikipedia.org/wiki/James_Prescott_Joule).

1.2.2 Summary of Basic Concepts of Thermodynamics

This section presents a brief summary of the important concepts of thermodynamics, to be used as a foundation for the subsequent discussion of cryogenic engineering.

1.2.2.1 Thermodynamics and Thermodynamic Systems

As defined by William Thomson (Lord Kelvin) in 1854: "Thermo-dynamics is the subject of the relation of heat to forces acting between contiguous parts of bodies, and the relation of heat and electrical agency" [12].

Thermodynamics studies macroscopic systems, which are composed of large numbers of "particles," such as atoms, molecules, clusters, and electrons. The macroscopic properties of the system are the average values resulting from the random motions of the constituent microscopic particles. For example, the "pressure" of a system is based on the average value of the momentum of all the microscopic particles forming the system.

The understanding of thermodynamics did not begin with physical models or assumptions. Instead, based on experience and experimentally measured macroscopic physical parameters (called thermodynamic coordinates or thermodynamic variables), statistical approaches were used to deduce the macroscopic properties of systems and to summarize their relationship with the four laws of thermodynamics.

A thermodynamic system consists of the material within a macroscopic volume in space, which can be described by several thermodynamic state variables including temperature, entropy, internal energy, and pressure. All material that is not included in the system is called the system's "surroundings."

A thermodynamic system is separated from its surroundings by certain boundaries, which, in certain cases, can be arbitrary. The system, surroundings, and boundaries must be well-defined at the start of all discussions of thermodynamics. The combination of the system and its surrounding may be called the "universe."

A few specific types of "systems" can be defined, and include:

Isolated system: A system that does not exchange energy or matter with the surroundings
Adiabatic system: A system that does not exchange heat with the surroundings
Closed system: A system that does not exchange matter with the surroundings
Open system: A system that exchanges matter with the surroundings

1.2.2.2 Thermodynamic Variables

Thermodynamic variables include both **state variables** and **path variables**.

State variables (also called **state coordinates**) characterize the (equilibrium) state or momentary condition of a thermodynamic system. They are a set of the macroscopic properties of each system including temperature, pressure, volume, and internal energy.

State variables can be divided into two categories: **intensive variables** and **extensive variables**:

Intensive variables are independent of the mass, and include temperature, pressure, etc.

Extensive variables properties vary directly with the mass, and include internal energy, heat capacity, and others.

A **state variable** depends only on the equilibrium states of a system and not on the path connecting the different states.

Path variables are variables that depend on the sequence of processes that take the system from the initial state to the final state. Heat and work are examples of path variables.

1.2.2.3 Thermodynamic States and Equation of State (EOS)

The thermodynamic state of a system is a description of the system including certain macroscopic coordinates (properties).

Thermodynamic equilibrium state:

A state that in which the macroscopic coordinates of the system does not change with time. In many cases, the following three equilibrium states exist simultaneously:

Mechanical equilibrium state: State with no unbalanced force and torque inside the system and between the system and the surroundings.

Chemical equilibrium state: A state with no spontaneous changes of internal structures or mass flow inside the system.

Thermal equilibrium: A state with the same temperature in all parts inside the system. In addition, there is no heat exchange (or heat flow) between the system and its surroundings.

Equation of state (EOS) is a functional relationship between the state variables of system in the equilibrium state, and the relationship itself is governed by the laws of thermodynamics.

Quite often, EOS represents the relationship between pressure (P), volume (V), temperature (T), and the mass **m** or number of moles **n** of a system in its equilibrium state.

For a closed system where m (or n) remains a constant, the functional relationship can be simply expressed as $f(P, V, T) = 0$

Tips:

1. EOS only exists for an equilibrium state.
2. Generally speaking, EOS applies only to pure substances.
3. EOS is based on experimental results, rather than being theoretically derived from thermodynamics.
4. It is not uncommon that EOS is a semi-empirical function based on measured data, graphs, or numerical tables. In such cases, the relationship between the variables still exists, and the function is still called an equation of state even though an explicit mathematical expression may not exist.

1.2.2.4 Thermodynamic Process

The passage of a thermodynamic system from an initial to a final state of thermodynamic equilibrium.

1. **Irreversible process, reversible process, and quasi-static transformation**

   A process is "irreversible" if the set of values of all properties cannot be restored when a system changes from the final state back to the initial state.

Examples include:

- Broken eggs
- Heat flow through a finite temperature difference
- Diffusion
- Mechanical work that changes into heat due to frictional effects
- Gas inflation toward vacuum
- The uniform mixing of two solutions of different concentrations, etc.

A process is called "reversible" if the set of values of all properties can be restored when a system changes from the final state back to the initial state, without having any impact on the surroundings. A reversible process is a somewhat idealized case (i.e., non-realistic) in thermodynamics, because "dissipation" is often involved in real systems. In order to better study reversible thermodynamic processes, an idealized concept known as quasi-static transformation was developed. Deviation from thermodynamic equilibrium during a quasi-static transformation is infinitesimal; therefore, all the states a system passes through during that process can be considered equilibrium states. Thus, a Reversible Process can be analyzed thermodynamically by quasi-static transformation without considering the effects of dissipation on the system.

2. **Some specific processes**

Isothermal process: Process in which the temperature of the thermodynamic system remains constant.

Isobaric process: Process in which the pressure of the thermodynamic system remains constant.

Isochoric process: Process in which the volume of the thermodynamic system remains constant.

Adiabatic process: Process in which there is no transfer of heat (or matter) between a thermodynamic system and its surroundings.

A spontaneous process: Process that occurs naturally in a system without continuous outside intervention from the surroundings. *A spontaneous process always proceeds in the direction of increasing entropy. During a spontaneous process, a system releases free energy and moves toward a lower and more thermodynamically stable energy state, continuing until the free energy of the system reaches a minimum at the final equilibrium state.*

The relationships between entropy, the direction of a spontaneous process, and the laws of thermodynamics will be discussed later in greater detail.

Examples of spontaneous processes include:

- Free expansion of a gas into a vacuum space;
- HEAT flowing from an object with higher temperature to an object with lower temperature when the objects are brought into thermal contact;
- The spontaneous freezing of water at one atmosphere of pressure below 0°C, and the spontaneous melting of ice above 0°C;
- The spontaneous flow of water downhill, (but never uphill); and
- The dissolving of sugar when placed into a cup of coffee (it does not subsequently reappear in its original form).

Cyclic process: A process in which the final state of a system is identical to its initial state.
Mechanical work and non-mechanical work: Mechanical work is that part of a thermodynamic process often expressed with formula of

$$dW = PdV \tag{1.1}$$

Non-mechanical work is that part of a thermodynamic process that cannot be expressed with the formula PdV. Non-mechanical work often takes place in processes such as chemical reactions and magnetic refrigeration, and it may result in changes in chemical composition, electrical energy, magnetic energy, and others.

1.2.2.5 Thermodynamic Function of State

Due to the pragmatic purpose of this book, the discussion in this section is restricted to closed systems, i.e., systems in which there is no mass flow between the system and its surroundings.

Notations:

U: Internal energy of the system
T: Temperature of the system
S: Entropy of the system
P: Pressure of the system
V: Volume of the system
F: Helmholtz (also called Helmholtz Free Energy) of the system
G: Gibbs energy (also called Gibbs Function or Gibbs Free Energy) of the system
H: Enthalpy of the system
Q: Heat transferred between the system and its surroundings

1. Internal energy: U

   Differential expression: $dU = TdS - PdV$ (1.2)

   This expression is equivalent to the first law of thermodynamics; see the discussion in the following section.

   Internal energy is the total energy contained within a macroscopic system, excluding the kinetic energy of motion and the potential energy of the system as a whole. More precisely, internal energy is associated with the kinetic energy of motion (such as rotation, vibration, and translation motion) of the microscopic items, plus the potential energy caused by interactions between the microscopic items within the macroscopic system.

   Internal energy is equal to the maximum amount of work that a closed system can perform in an adiabatic process.

   The concept of entropy will be discussed in greater detail in the next section.
2. Helmholtz energy: $F = U - TS$

   Differential expression: $dF = -SdT - PdV$ (1.3)

Helmholtz energy remains constant in isothermal and isochoric processes. *It is equal to the maximum amount of work that a closed system can perform in an isothermal process.* Helmholtz energy is an important concept in statistical physics; it is one of the parameters in the partition function in statistical physics.

3. Gibbs energy: $G = H - TS$

Differential expression: $dG = -SdT + VdP$ (1.4)

Gibbs energy represents the maximum amount of non-expansion (or non-mechanical) work that may be performed by a (closed) thermodynamic system in an isothermal and isobaric process.

Gibbs energy is also called the Gibbs function and is crucial to chemical thermodynamics (or physical chemistry). The Gibbs function is often used to describe thermodynamic systems undergoing a process at a constant temperature and pressure, for example, an isothermal and isobaric chemical process. It reaches a minimum value in a *closed system* when the temperature and pressure of the system remain unchanged.

Gibbs energy is also important for an *open system* because *it represents the increase in internal energy of an open system when the system obtains one additional particle.*

In this case, the Gibbs energy can be represented as:

$$dG = SdT + VdP + \mu dN \tag{1.5}$$

where:

dN is the change in the number of microscopic items in an open macroscopic system.

$\mu = (dU/dN)_{S,V}$ describes the *chemical potential.* It represents the change in internal energy of a macroscopic system when the number of microscopic items changes at constant temperature and pressure.

(This is the only case in which *open systems* are addressed in this book).

4. Enthalpy: $H = U + PV$

Differential expression: $dH = dQ + VdP$ (1.6)

Enthalpy is an alternative expression of energy, and *it represents the heat transferred between the system and its surrounding in an isobaric process.* Enthalpy energy remains constant in an adiabatic and isobaric process (known as a "Joule–Thomson process"), to be discussed in detail in Chapter 6. A process with no change in enthalpy is called an isenthalpic process or isoenthalpic process.

5. Summary

- When an isolated system is at equilibrium state, some thermodynamic parameters are either minimized or maximized. As examples, entropy (S) reaches a maximum value at equilibrium, while the Helmholtz energy (F) reaches its minimum value.

- The above four types of well-defined functions of state (all with units of energy), along with their corresponding differential expressions, form the complete set of functions that can be used to deduce all of the thermodynamic properties of a system.
- The above four types of state functions are alternative expressions of "energy," and all of them represent extensive properties.
- The term "maximum" as used in the preceding discussion represents a value achievable (only) when the system undergoes a reversible process.
- These state functions are not directly measurable. Rather, it is the change or difference of these state functions, as opposed to the absolute value of a system, that carries physical meaning and is of interest.

1.2.2.6 Thermodynamics and Its Governing Laws

Zeroth Law of Thermodynamics: Two systems in thermal equilibrium with a third system *simultaneously* are in thermal equilibrium with each other [13].

First Law of Thermodynamics: Conservation of energy

The amount of heat transferred to a closed system from its surroundings plus the work done on the system by its surroundings is equal to the increase of the internal energy of the system from initial state to final state.

$$\text{Mathematical expression: } U_f - U_i = Q + W \tag{1.7}$$

where:

U_f: Internal energy of the system at final state;
U_i: Internal energy of the system at initial state;
Q: Heat absorbed by the system from its surrounding; and
W: Work done to the system by its surroundings—work can either be mechanical or non-mechanical.

Refer to **Chapter 10: Cryogen-Free Adiabatic Demagnetization Refrigerator (ADR) System**.

Second Law of Thermodynamics: Defining the direction of spontaneous processes

This law has two equivalent statements:

Expression #1: Clausius statement

It is impossible to construct a refrigerator that, operating in a cycle, will produce no effect other than transfer heat from a lower temperature reservoir to a higher temperature reservoir.

In other words, heat will not spontaneously flow from a lower temperature reservoir to a higher temperature reservoir.

Expression #2: Kelvin statement

It is impossible to construct an engine that, operating in a cycle, will produce no effect other than the extraction of heat from a heat reservoir and the performance of an equivalent amount of work (Figure 1.6).

Third Law of Thermodynamics: The entropy of all systems is defined as zero at absolute zero temperature.

The third law of thermodynamics provides a reference point for the determination of entropy, that is, any entropy determined relative to this reference point has an absolute entropy.

(a) (b)

FIGURE 1.6 (a) William Thomson, Lord Kelvin (https://en.wikipedia.org/wiki/William_Thomson,_1st_Baron_Kelvin#/media/File:Lord_Kelvin_photograph.jpg). (b) Rudolf Clausius (https://en.wikipedia.org/wiki/Rudolf_Clausius#/media/File:Clausius.jpg).

The entropy of all perfect crystals approaches zero as the temperature approaches zero, but it is impossible for any process, no matter how idealized, to reduce the entropy of a system to its absolute-zero value in a finite number of operations.

1.2.2.7 Deeper Understanding of Laws of Thermodynamics

This subsection further explores the laws of thermodynamics and presents a deeper understanding of them.

1. The laws of thermodynamics are phenomenological laws based on long-term observations of physical processes, substantiated by empirical measurements. They do not describe the fundamental physical mechanism of any macroscopic thermodynamic coordinates. As an important thermodynamic variable, temperature is always included in discussions of the thermodynamic properties of a system.
2. The zeroth law of thermodynamics introduces the state function of **Temperature**. The condition "simultaneously" is important within the zeroth law of thermodynamics.
3. The first law of thermodynamics introduces the state function of **Internal Energy**. It is the law describing the conservation of energy.

   This law makes a quantitative connection between HEAT and WORK, but it does not specify the direction of the flow of energy.

The + sign ("+Q" and "+W") used in mathematical expressions is crucial since it determines the direction of both heat and work. +Q indicates that heat flows from the surroundings to the system, while +W indicates that work is being performed on the system by the surroundings.

4. The second law of thermodynamics introduces the state function of **Entropy**, which is a core principle of thermodynamics.

Entropy is defined as $\Delta S = \int dQ_{rev}/T$ (1.8)

where:

ΔS is the change of entropy of the system;
dQ_{rev} is the heat flowing into the system during a reversible process; and
T is the temperature of the system.

The total entropy of the universe is not conserved but can (and will) be created in all real (or spontaneous or irreversible) processes. This is called the Principle of Increased Entropy.

When the Principle of Increased Entropy is applied in irreversible processes, we must consider the total entropy of the "universe," including both the system and the surroundings, as shown in Eq. 1.9:

$$\Delta S_{\text{univ.}} = \Delta S_{\text{system}} + \Delta S_{\text{surroundings}} > 0 \quad (1.9)$$

where:

$\Delta S_{\text{univ.}}$: Total entropy change of the "universe";
ΔS_{system}: Entropy change of the "system"; and
$\Delta S_{\text{surroundings}}$: Entropy change of the "surroundings."

ENTROPY is a fundamental principle of thermodynamics and is sometimes referred to as ***the second E-revolution.*** (***The first E-revolution*** refers to the theory of ***Electromagnetism***, the unification of electricity and magnetism by Maxwell. Some people refer to quantum ***Entanglement*** as the ***third E-revolution***).

5. The first law of thermodynamics specifies which thermodynamic processes **may**, but not necessarily **will**, take place. In other words, any process that violates the first law of thermodynamics will not take place, but not all processes that meet the criteria of the first law of thermodynamics will take place.
6. Refrigerators and steam engines are two of the most important inventions in the history of cryogenic engineering, and both of them played important roles in the development of the second law of thermodynamics.

   Clausius made his statement of the second law based upon the performance of refrigerators; it implies that any irreversible process is one-directional, toward increasing entropy. This statement is equivalent to the Principle of Increased Entropy. Kelvin's statement was based on the performance of the steam engine and implies that all engines have a maximum limit to their efficiency. Efficiency cannot be 100%, and some **non-usable energy (HEAT)** is always involved in the process. The Carnot engine represents the highest efficiency possible in a steam engine.

While the first law of thermodynamics makes a connection between HEAT with WORK, the second law of thermodynamics reveals the fundamentals regarding the transformation of HEAT and WORK.

7. A new branch of physics called Statistical Thermodynamics, originating from the studies of cryogenics and thermodynamics, is recently being used to provide a microscopic explanation of entropy.

   Thermodynamics, in our previous discussion, takes a "macroscopic" approach to measure, tabulate, and sometimes predict the macroscopic properties of "macroscopic" systems. It also attempts to build a relationship between these the macroscopic and the microscopic systems. This branch of natural science is also called **Classical Thermodynamics;** it does not explore the physics behind these properties. Microscopic approaches that explain the thermodynamic behavior of systems that include large quantities at the molecular or atomic scale are represented by a new branch of science known as **Statistical Thermodynamics or Statistical Mechanics.**

   Statistical mechanics provides a connection between the macroscopic properties of materials and the microscopic behaviors and motions occurring within the material. It uses the properties of a material's constituent particles and the interactions between these particles to arrive at the classical thermodynamics explanation of a system at a macroscopic level. Statistical mechanics covers both equilibrium and non-equilibrium states and is beyond the scope of this book.

   Perhaps the most important contribution of **Statistical Thermodynamics** is to provide a microscopic explanation of entropy as expressed mathematically by Ludwig E. Boltzmann (Boltzmann's equation) [14]:

$$S = k_B \ln \Omega \tag{1.10}$$

where:

S = entropy;

k_B = 1.38065 × 10^{-23} J/K is the Boltzmann constant; and

Ω = is the number of different microstates an isolated macroscopic system can have at a given equilibrium state.

Boltzmann's equation explores the physics behind entropy: it represents the **measure of randomness or disorder** of the microscopic constituents of a macroscopic system at an equilibrium state. This equation builds the connection between the microscopic motion inside a system and the macroscopic properties described by classical thermodynamics.

Boltzmann's equation is a probability equation resulting in the term **entropy, S,** as a quantitative value derived from Ω. A greater number of microscopic states imply a higher disorder of the associated macroscopic system.

Boltzmann's equation also determines the direction of a spontaneous process: spontaneity of a process is associated with higher disorder as well as irreversibility (in nature) for an isolated system.

FIGURE 1.7 L. Boltzmann's grave in Zentralfriedhof, Vienna, bearing the entropy formula. (From Lebrun, P., *An Introduction to Cryogenics*, European Organization for Nuclear Research Laboratory for Particle Physics CERN, Accelerator Technology Department 18 January, CH-1211 Geneva 23 Switzerland Departmental Report CERN/AT 2007-1, 2007.) (https://www.atlasobscura.com/places/boltzmanns-grave).

$\Omega = 1$ implies that a macroscopic system is at the most ordered equilibrium state (with only one microscopic probability). This most ordered state can only correspond to temperature at absolute zero where Entropy = 0 when $\Omega = 1$,

It is consistent with the Third Law of Thermodynamics as previously stated (Figure 1.7).

8. Relationship between spontaneous processes and irreversible processes
When we say that a spontaneous process is irreversible, the surrounding conditions are implied to be unchanged during the process, and therefore the direction of the spontaneous process can be determined with certainty. For example, the transition between water and ice is a spontaneous process, and its direction depends on the temperature of the surroundings. When the surrounding temperature cools to <273 K, the direction of the spontaneous process is from liquid water to solid ice. When the surrounding temperature rises from to >273 K, the direction of the spontaneous process is from solid ice to liquid water.

9. Gibb's equation: the criteria for spontaneous processes at a constant pressure
   During an isothermal and isobaric spontaneous process (primarily chemical reactions), the total Gibbs free energy of the system always increases in negative values, that is, $\Delta G < 0$. Eventually, the Gibbs value reaches a minimum, at a system composition after which time no further net change will occur.

   Therefore $\Delta G < 0$ is a criterion that determines whether a chemical reaction will take place. $\Delta G = 0$ implies that the "process" stops, and an equilibrium state is reached.
10. Commonly used equations and relations in thermodynamics

$$\left(\frac{\partial u}{\partial v}\right)_T \left(\frac{\partial v}{\partial T}\right)_u \left(\frac{\partial T}{\partial u}\right)_v = -1 \tag{1.11}$$

Maxwell's relations (common)

$$\begin{aligned}
+\left(\frac{\partial T}{\partial V}\right)_S &= -\left(\frac{\partial P}{\partial S}\right)_V = \frac{\partial^2 U}{\partial S \partial V} \\
+\left(\frac{\partial T}{\partial P}\right)_S &= +\left(\frac{\partial V}{\partial S}\right)_P = \frac{\partial^2 H}{\partial S \partial P} \\
+\left(\frac{\partial S}{\partial V}\right)_T &= +\left(\frac{\partial P}{\partial T}\right)_V = -\frac{\partial^2 F}{\partial T \partial V} \\
-\left(\frac{\partial S}{\partial P}\right)_T &= +\left(\frac{\partial V}{\partial T}\right)_P = \frac{\partial^2 G}{\partial T \partial P}
\end{aligned} \tag{1.12}$$

1.3 Experimental Breakthrough

When the theoretical underpinnings of thermodynamics were being completed in the second half of the nineteenth century, scientists working on cryogenic engineering were focused on gas liquefaction. An important breakthrough, the liquefaction of the "permanent gas" hydrogen was achieved near the end of the nineteenth century.

1.3.1 Permanent Gases and Liquefaction of Hydrogen

Prior to the end of the nineteenth century, gas liquefaction techniques involved three basic steps:

1. Compressing the gas to a high pressure, typically several dozen times atmospheric pressure;

2. Cooling down the compressed gas to as low a temperature as possible, using one of the cooling methods previously developed; and
3. Expanding the pre-cooled compressed gas from high pressure to a lower pressure, typically at a controlled rate using a needle valve.

This third step made use of the Joule–Thomson effect for gases, which will be discussed in **Chapter 6: Enabling Technologies**.

By the mid-nineteenth century, most gases had been successfully liquefied by immersing the gas in a bath of ether and dry ice, and then pressurizing until liquefaction occurred. However, this approach failed to successfully liquefy several gases including oxygen, hydrogen, nitrogen, carbon monoxide, methane, and carbon dioxide. These gases were therefore referred to as "permanent gases."

Three of these gases (oxygen, nitrogen, and hydrogen) are pure substances, and they therefore received the most attention. (The noble gases helium, neon, argon, krypton, and xenon had not yet been discovered.)

Oxygen was first liquefied (a few droplets) in 1877 by Louis Paul Cailletet (1832–1913) [16] while measurable quantities were first successfully liquefied by the Polish scientists Zygmunt Florenty Wróblewski (1845–1888) and Karol Olszewski (1846–1915) in 1883.

These same scientists were the first to liquefy nitrogen in the same year (Table 1.1).

Hydrogen is the most abundant element in the observable universe, and it was first recognized to be a distinct element by the British scientist, Henry Cavendish (1731–1810) in 1766 [17,18].

Basic Properties of Hydrogen [10]

Symbol: H2
Atomic number: 1
Atomic mass: 1.0079 amu
Normal melting point: 14.01 K
Normal boiling point: 20.28 K
Normal latent heat: 443 kJ/kg

Phase Diagram of Hydrogen (Figure 1.8)

Table 1.1 Normal Boiling Point of Oxygen, Nitrogen, and Hydrogen

	Normal Boiling Point		
Cryogen	K	C	F
Oxygen	90.2	−183.0	−297.3
Nitrogen	77.4	−195.8	−320.4
Hydrogen	20.3	−252.9	−423.2

Source: Weast, R.C., *CRC Handbook of Chemistry and Physics*, 65th edition, Boca Raton, FL, CRC Press, 1984.

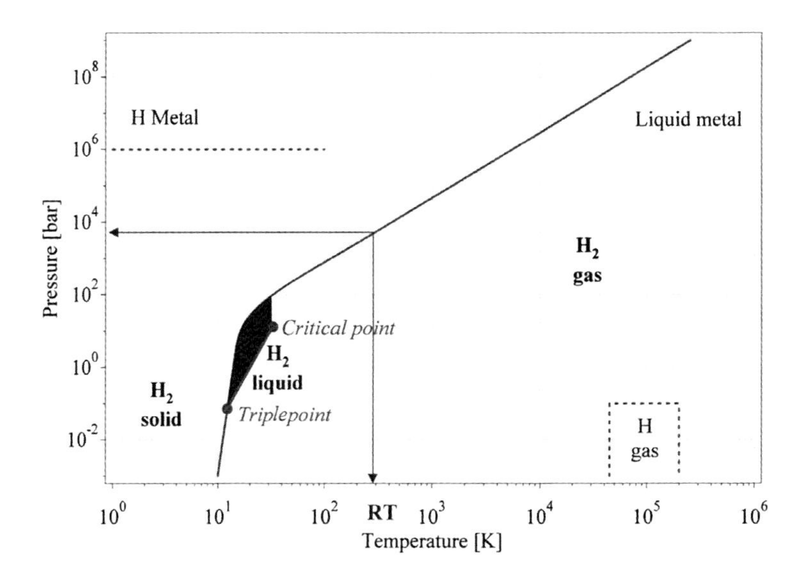

FIGURE 1.8 Hydrogen phase diagram at low temperatures; liquid only exists between the line defining the border of the solid state, and the line extending from the triple point at 21.2 K to the critical point at 33.19 K. (From Zuttel, A., *Mater. Today*, 24–33, 2003; Leung, W.B., et al., *Phys. Lett. A*, 56, 425, 1976.)

1.3.2 James Dewar and His Vacuum Flask

The effort to liquefy hydrogen was led by the Scottish chemist and physicist James Dewar (1842–1923) [21,22].

All cryogens, including liquid hydrogen, have low latent heat. A small quantity of heat flowing from a container into a liquid cryogen will cause the cryogen to evaporate in a short period of time. Cryogens were stored under very high pressure, in containers with significant reinforcement prior to Dewar's time. This approach was highly undesirable due to the high cost, as well as to safety concerns. Liquid hydrogen has a lower boiling temperature and smaller **mole** heat of evaporation (886 Joule) than other gases liquefied to that time, which made preservation of liquid hydrogen especially challenging (Figure 1.9).

Dewar understood that the fundamental solution was to minimize the heat exchange between the cryogen and its container. Following his idea that "vacuum" would work as a thermal barrier in between the container and the cold cryogens, Dewar reduced heat transfer into the cryogen by his invention of a **vacuum flask**.

This vessel consisted of a double-walled glass flask (or two-flask device), with the air removed from the space between the two walls, and then sealed at the neck.

There are three main channels of heat transfer: conduction, convection, and radiation. Vacuum prevents conduction and convection from transferring heat between the two flasks. The inner flask is therefore well thermally isolated from the outer flask, which is at room temperature. Any cryogen collected in the inner flask will remain cold, with a low evaporation rate, remaining in the liquid phase for a longer duration of "holding time."

FIGURE 1.9 (a and b) James Dewar (https://en.wikipedia.org/wiki/James_Dewar#/media/File:James_Dewar.jpg.)

The third channel of heat transfer (radiation) was later reduced by placing a silvered coating on the outer surface of the inner flask. Dewar's vacuum flask was exhibited for the first time on Christmas Day, 1892 (Figures 1.10 and 1.11).

The invention of the vacuum flask was an important historic milestone in cryogenic engineering. This breakthrough made it possible to preserve cryogenic liquids for

FIGURE 1.10 Sir James Dewar lecturing on liquid hydrogen at the Royal Institution in London in 1904. (https://en.wikipedia.org/wiki/Royal_Institution#/media/File:Henry_Jamyn_Brooks_-_A_Friday_Evening_Discourse_at_the_Royal_Institution;_Sir_James_Dewar_on_Liquid_Hydrogen,_1904.jpg.)

(a) (b)

FIGURE 1.11 (a and b) James Dewar's vacuum flask in the museum of the Royal Institution in UK. (https://en.wikipedia.org/wiki/James_Dewar#/media/File:Ri_2014_-_Thermos_flask_-_James_Dewar_(27).jpg.)

comparatively long periods of time, and it became a key tool used by Onnes for helium liquefaction in subsequent years. This invention allowed scientists to store a sufficient quantity of liquid gas for sufficient duration to study the properties of a variety of cryogens. The success of cryogenic engineering would soon lead to the birth of the "*Science of Low Temperatures.*"

1.3.3 Liquefaction of Hydrogen

Combining the large regenerative cooling refrigerating machine he had built at the Royal Institution in UK with the Joule–Thomson expansion technology, Dewar became the first person in history to successfully liquefy hydrogen on May 10, 1898. Dewar collected 20 cm^3 of liquid hydrogen in a double-walled insulated glass flask in the basement laboratory of the Royal Institution. The liquid was clear and colorless and had a relatively high refractive index. He found the boiling point of hydrogen to be 20 K. In the same year, Dewar succeeded in freezing hydrogen, thus reaching the lowest temperature achieved to that time, 14 K.

The first cryogenic thermometers were based on resistivity. They were subsequently replaced by the 4He vapor pressure thermometers, which provided more accurate readings of low temperatures. The invention of the Dewar flask and the liquefaction of hydrogen paved the way to the final goal: the liquefaction of the "noble gas" helium one decade later as the world entered the twentieth century!

Acknowledgments

The contributor acknowledges Xi Lin for reviewing this chapter on its scientific integrity.

Sincere acknowledgment goes to his colleagues at Janis Research Company:

Thanks Scott Azer for his detailed review and editing work; thanks Choung Diep for making drawings; thanks Jeonghoon Ha for his final proof reading; and last but not the least, thanks Ann Carroll for her constant help with software difficulties, reference searches, the copyright permission request, etc.

References

1. Timmerhause, Klaus D. and Reed, Richard P. (Eds.): *Cryogenic Engineering Fifty Year Progress*, New York: Springer Science+Business Media, LLC, 2007.
2. Chambers, R. and Thomson, T.: *A Biographical Dictionary of Eminent Scotsmen in Volumes*, Vol. 2, Cullen, UK: William, Encyclopedia Britannica, 1855.
3. Bathe, G. and Bathe, D. *Oliver Evans: A Chronicle of Early American Engineering*, Philadelphia, PA: Historical Society of Pennsylvania, 1935. Ferguson, E.: *Oliver Evans: Inventive Genius of the American Industrial Revolution*, Wilmington, DE: Eleutherian Mills-Hagley Foundation, 1980. Howe, H.: *Memoirs of the Most Eminent American Mechanics: Also, Lives of Distinguished European Mechanics*, Together with a Collection of Anecdotes, Descriptions, Etc., Etc. New York: W.F. Peckham, 1840.
4. Gorrie, J.: Encyclopedia Britannica, http://www.britannica.com/biography/John-Gorrie 2018.
5. Gorrie, J.: *ICE MACHINE*. No. 8,080. Patented May 6, 1851.
6. Watt, J.: *Dictionary of National Biography*, Vol. 60, London, UK: Smith, Elder & Co., 1885–1900.
7. Carnot, S.: *Reflections on the Motive Power of Fire*, Minoela, NY: Dover Publications, 1988.

8. Wisniak, J.: Nicolas Clement, *Educ. Quim*, 22(3), 254–266, 2011. Hargrove, J. L.: History of the calorie in nutrition, *The Journal of Nutrition*, 136(12), 2957–2961, 2006. Hargrove, J. L.: Does the history of food energy units suggest a solution to "Calorie confusion?" *Nutrition Journal* 6(1), 44, 2007. doi:10.1186/1475-2891-6-44.
9. Rogers, K. (ed.): *The 100 Most Influential Scientists of the All Time*, New York: Britannica Educational Publication, 2010.
10. Weast R. C.: *CRC Handbook of Chemistry and Physics*, 65th edition, Boca Raton, FL: CRC Press, 1984.
11. Griffiths, A. B.: *Biographies of Scientific Men*, London, UK: Robert Sutton, 1912, p. 185.
12. Thomson, W. (1854): *On the Dynamical Theory of Heat. Part V. Thermo-electric Currents*, Transactions of the Royal Society of Edinburgh 21 (part I): 123. reprinted in Sir William Thomson, LL.D. D.C.L., F.R.S. (1882): *Mathematical and Physical Papers* 1. London, UK: C.J. Clay, M.A. & Son, Cambridge University Press. p. 232.
13. Zemansky, M. Z. and Dittman, R. H.: *Hear and Thermodynamics*, New York: McGraw-Hill Education, 2011.
14. Qing, Y. H.: *Thermal Physics*, The Chinese Senior Education Publisher, 2011.
15. Lebrun, P.: *An Introduction to Cryogenics*, European Organization for Nuclear Research Laboratory for Particle Physics CERN, Accelerator Technology Department, 18 January (2007), CH – 1211 Geneva 23 Switzerland Departmental Report CERN/AT 2007-1.
16. Cailletet, L.: The liquefaction of oxygen, *Science*, 6(128): 51–52, 1885.
17. Byers, P.K.: *Henry Cavendish Biography, Encyclopedia of World Biography*, 2nd edition, Suzanne Michele Bourgoin, Gale Research, 1998.
18. Sloop, J.L.: *Liquid Hydrogen as a Propulsion Fuel, 1945–1959, Appendix A-1 Hydrogen through the Nineteenth Century*, NASA SP-4404, NASA History Series, Scientific and Technical Information Office, (1978).
19. Zuttel, A.: Hydrogen storage, *Material Today*, pp. 24–33, September 2003.
20. Leung, W.B., March, N.H., Motz, H.: Primitive Phase Diagram for Hydrogen, *Phys. Lett. A*, 56, 425, 1976.
21. Mottelay, P. F., Sir James Dewar, F.R.S.: Famous for his researches in low temperature phenomena, Scientific American, February 4, 1911, 104 Dewar.
22. Dewar, J.: *Collected papers of Sir James Dewar*, ed. Lady Dewar, Cambridge University Press, (1927), pp. 678–691.

2. Helium

Its Application, Supply, and Demand

Zuyu Zhao

2.1 Noble Gas of Helium

Breakthrough in helium liquefaction took placed in the early twentieth century, which is often referred to as the "***Century of Physics***."

Helium (*4He*) is one of the members of the noble gas family, also known as "inert gases" since they do not have chemical reactions with any other elements. Other noble gas family members include neon, argon, krypton, xenon, and radon. Helium has the lowest boiling temperature among the family, and it is also the last element to be liquefied.

Helium was first discovered in the atmosphere of the sun in 1868, as the French astronomer Pierre Jules Janssen and the British astronomer Joseph Norman Lockyer were studying the spectrum of the sunlight radiated by the sun in a solar eclipse. A new previously unknown line of bright yellow color was discovered, which could not be associated with any of the already known elements. A new element was discovered, and it was named "helium" after the Greek god Helios, who drew the sun across the sky every day behind his golden chariot.

Helium was first isolated by Scottish chemist Sir William Ramsay in 1895. In the same year, Swedish chemists Per Teodor Cleve and Abraham Langet discovered the gas independently. They collected enough of the gas and determined its atomic number: 2.

2.2 Basic Physical Properties of Helium

Symbol: He
Atomic number: 2
Atomic mass: 4.0026 amu
Normal boiling point: 4.216 K
Phase diagram:

We start from the 4He phase diagram (Figure 2.1) in zero magnetic field, which illustrates the locations of three different phases of helium, that is, solid, liquid, and gaseous, with parameters of pressures (P) and temperatures (T).

The following unique properties of 4He are illustrated:

- 4He does not solidify under normal (1.0 atm) pressure even at zero Kelvin.
- 4He solidifies under minimum 25 atm pressure at low temperatures.
- *Melting curve*: this curve illustrates the P-T relationship when solid phase and liquid phase coexist.

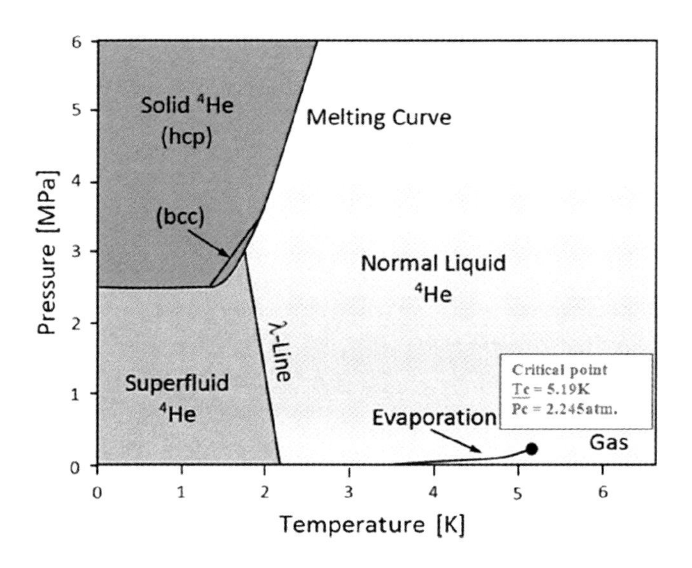

FIGURE 2.1 Phase diagram of helium-4.

- *Curve of evaporation*: this curve illustrates the saturated vapor pressure as a function of temperature of liquid 4He when liquid phase and gas phase coexist.
- Gaseous phase to liquid phase transition takes place at 4.2 K under normal (i.e., 1.0 atm.) pressure.
- Lower temperatures can be realized along the curve of evaporation by reducing the saturated vapor pressure of liquid helium-4.
- Liquid 4He becomes a *superfluid* below 2.17 K where the viscosity of the liquid helium becomes unmeasurable.
- Helium has no triple point.
- 4He has the critical temperature of 5.19 K. Helium does not have liquid phase above the critical temperature even under high pressures.

The curve of evaporation is of great importance since our discussion will be focused on the liquid and gaseous phases.

4He will be referred to as ***Helium*** for our discussion in this chapter.

2.3 Heike Kamerlingh Onnes and Helium Liquefaction [1–5]

The world was ready for the long-waited victory of helium liquefaction when it entered the twentieth century—the last battle of gas liquefaction. This battle was not only challenging but also important because the success of helium liquefaction would provide scientists with a tool to pursue research at much lower temperatures.

It was Heike Kamerlingh Onnes (Figure 2.2) at Leiden University in the Netherlands who fought this battle and won the victory. Onnes was well aware of the challenges and difficulties of this battle. With his great vision he had taken different approaches that

FIGURE 2.2 Heike Kamerlingh Onnes. (Courtesy of National Museum Boerhaave, Leiden, the Netherlands, https://en.wikipedia.org/wiki/Heike_Kamerlingh_Onnes#/media/File:Kamerlingh_Onnes_signed.jpg.)

distinguished himself from Dewar, the winner of the race of hydrogen liquefaction in 1898. His approaches that had played a crucial role to his success of helium liquefaction are subsequently discussed.

2.3.1 Building a Well-Planned Cryogenic Laboratory and Training the Team with Different Talents

Onnes had visions that the success of this level of research needed a first-rate team with different expertise. He was focused on building up a cryogenic laboratory almost right after he was appointed as the professor in experimental physics in 1882. This lab was not built just for one particular project, such as "helium liquefaction," but for all challenging low-temperature researches to come. With his unusual vision and perseverance, Onnes built the most advanced and sophisticated cryogenic lab of his time in the world. The lab included an advanced glass shop to produce Dewar glass with sophisticated designs, since it was indispensable to cope with the challenges he would encounter soon.

Onnes has also founded a ***School of Instrument Makers*** where he had trained and orchestrated a research team, including technicians, instrument makers (also called "blue boys" when they were working in Onnes's lab wearing the blue uniforms), engineers, graduate students, glassblowers, and managers—people with various kinds of talents, skills, and techniques. Onnes's lab was a "cold factory," and his "industrial approach" was the first decisive step for the success. This was the attitude lacking in his competitors (Figure 2.3).

(a)

FIGURE 2.3 (a) Leiden Cryogenic Laboratory, 1895. *(Continued)*

(b)

(c)

FIGURE 2.3 (Continued) (b) The machine shop with trainees, 1896. (c) The glassblower shop, 1902. (Courtesy of Museum Boerhaave, Leiden, the Netherlands.)

2.3.2 Making Scientific Research Plans on Solid Theoretical Foundations

Although Onnes was an experimentalist, he had always paid close attention to theory and made his research plan on deliberated theoretical analysis. As van der Waals's student, Onnes gave van der Waals's theory a high credit that it "***had guided me to the end of my work on the liquefaction of gases***." van der Waals's theorem was used to calculate the key parameters in his research of gas liquefaction. The results gave Onnes the light in the tunnel that it was plausible to liquefy helium with the Joule–Thomson expansion starting from the freezing temperature of hydrogen. He estimated the inversion temperature of helium and got 23 K at approximately 37 atmospheric pressures, a pressure he felt confident to be able to reach. Onnes has also estimated the critical temperature of helium and got it around 5.3 K, which was very close to the accurate number of 5.19 K!

Onnes had also kept himself up with the contemporary theories, including the "old quantum theory" as well as the newly born quantum mechanics at the beginning of the twentieth century.

2.3.3 Tireless Renovation of Technology

Onnes was a tireless explorer for renovation and improvement, since he knew that advanced facilities were crucial to the success. For example, he modified the Cailletet (who first liquefied oxygen with the cascade process of adiabatic expansion of compressed gases) compressor that was used to compress and circulate helium at 100 atm. He has also expanded the cascade approach to four cycles with methyl chloride, ethylene, oxygen, and air by 1894, which was able to produce approximately 14 liters of liquid air per hour.

Onnes did not simply duplicate Dewar's approach of hydrogen liquefaction after he lost the race. Instead, he developed systems with larger productivity of liquid hydrogen in a continuous operation. His hydrogen liquefier was fully operational at the University of Leiden in 1906, producing 4 liters of liquid hydrogen per hour (before long, this number went up to 13 liters per hour). His efforts were soon rewarded during the helium liquefaction experiment.

2.3.4 Collection and Purification of Enough Amount of Helium Gas

Liquefaction of helium needed a sufficient amount of helium gas.

His brother, Mr. O. Kamerlingh Onnes, was the director of the Department of Foreign Trade Relations in Amsterdam at that time, and he was able to send Onnes dozens of sacks of thorium-bearing monazite sand from the US, a mineral that contained helium. Onnes succeeded in extracting a few cubic centimeters of "raw" helium gas from each gram of sand. The "raw" helium gas needed to be purified since the impurities like argon, neon, or hydrogen would freeze and cause blockage in the cooling process of Onnes's experiments. The purification was an extremely difficult and formidable task. Onnes took the challenges and had collected over 300 liters of pure helium gas by the time he gave the push for helium liquefaction.

2.3.5 Sophisticated Experimental Setup

Onnes's basic idea was to take the cascade approach and reach the temperature needed for helium liquefaction. It was a step-by-step cool-down approach with multiple cycles.

Five cycles with various gases, including methyl chloride (cooled by water), ethylene, oxygen, air, and hydrogen, were included in Onnes's cascades. Each cycle was a refrigerator with a closed loop consisting of a compressor, an expansion tank, and pumps. All cycles were operated in the continuous mode. The evaporated gas pumped out from each coolant bath was compressed, expanded, and re-liquefied. The bath temperature of the previous cycle was the starting point for the next one. Each bath served as a refrigeration reservoir to liquefy the gas with a lower condensing temperature. Hydrogen was the last cycle. The compressed hydrogen was liquefied from the Joule–Thomson expansion (refer to Section 6.2). The liquid hydrogen was first stored in a silver-coated vacuum glass dewar, then was transferred to a second bath where the liquid hydrogen was pumped and cooled down to approximately 15 K as illustrated in Figure 2.4.

Helium had its own cycles and iterative JT cooling process, which was indispensable since helium was not able to be liquefied through simple free expansion.

By the time of the experiment, Onnes was prepared with 75 liters of liquid air, 20 liters of liquid hydrogen, and 360 liters of pure helium gas (200 liters were used for circulation and liquefaction, and 160 liters were stored in a reserve container as spare).

The helium liquefaction experiment was performed on the July 10, 1908. It took approximately 13 hours before Onnes collected approximately 60 mL liquid helium with its surface standing ***"sharply against the vessel like the edge of a knife against the glass wall"*** [2].

The temperature of the helium was found at 4.25 K measured by a helium gas thermometer. Onnes pumped away the vapor above the liquid and tried to freeze the helium. The liquid helium temperature went down below 1.8 K based on Onnes's best knowledge

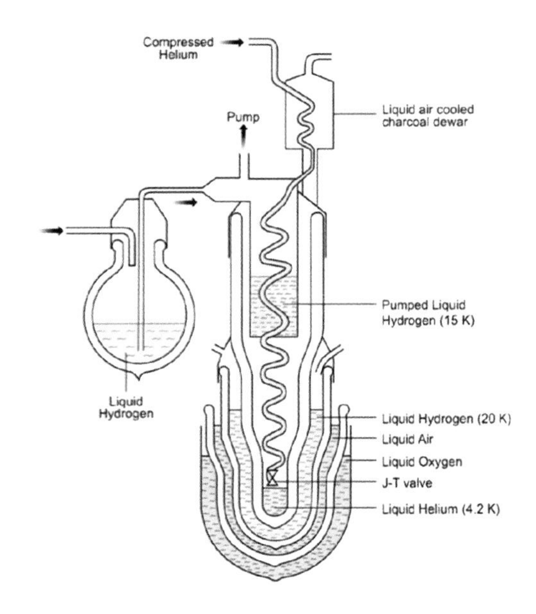

FIGURE 2.4 A schematic diagram of the helium liquefier used by Kamerlingh Onnes. (From Reif-Acherman, S., *Rev. Bras. Ensino Fis.*, 33, 1–17, 2011; Onnes, H.K., *KNAW Proceedings*, 11, 168–185, 1908–1909.)

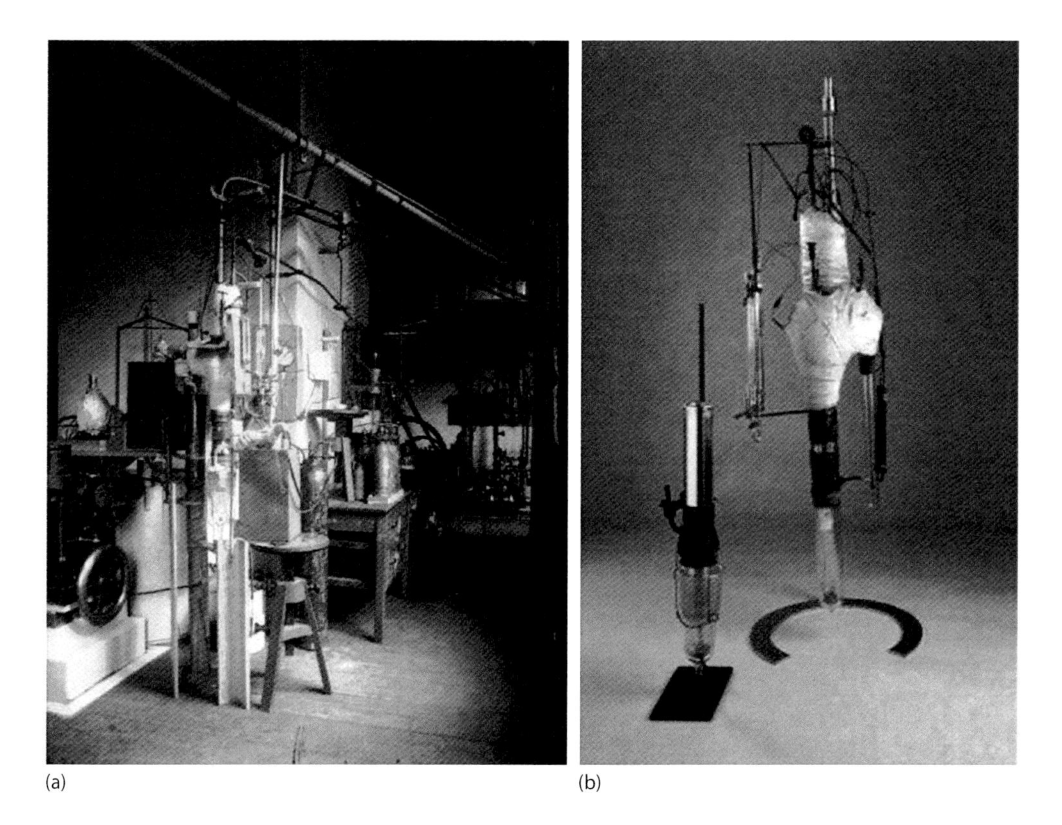

(a) (b)

FIGURE 2.5 (a) The 1908 Leiden helium liquefier. (Courtesy of National Museum Boerhaave, Leiden, Netherlands). (b) Onnes's historical helium liquefier (last stage) in Museum Boerhaave. (Courtesy of Museum Boerhaave, Leiden, the Netherlands).

at that time but did not freeze. He also concluded the critical pressure of helium being between 2 and 3 atmospheres (**the accurate number is 2.25 atm**). The experiment was terminated 16 hours after it started (Figure 2.5).

The success of helium liquefaction was doubtless one of the most important milestones in cryogenic engineering, and it ended the race of gas liquefactions. Its impact and significance were far beyond the race itself. This well-planned project with great success had led the unexpected discovery of superconductivity on mercury at Leiden on April 8, 1911 (Onnes's research associate, Gilles Holst, should share the credit with him). By the way, mercury was chosen for the measurement because of its purity. It could be repeatedly distilled to make it as pure as possible.

Following the discovery of superconductivity, multiple quantum phenomena were observed, such as the superfluidity of helium, the persistent current, critical current density, and the relationship between magnetic fields and superconductivity.

These two almost side-by-side breakthroughs, that is, helium liquefaction and the discovery of superconductivity, manifested the beginning of a new era of the low-temperature physics in the twentieth century. The milestone of the modern cryogenic engineering and low-temperature physics has been erected (Figure 2.6).

Figure 2. A terse entry for 8 April 1911 in Heike Kamerlingh Onnes's notebook 56 records the first observation of superconductivity. The highlighted Dutch sentence *Kwik nagenoeg nul* means "Mercury['s resistance] practically zero [at 3 K]." The very next sentence, *Herhaald met goud,* means "repeated with gold." (Courtesy of the Boerhaave Museum.)

(a)

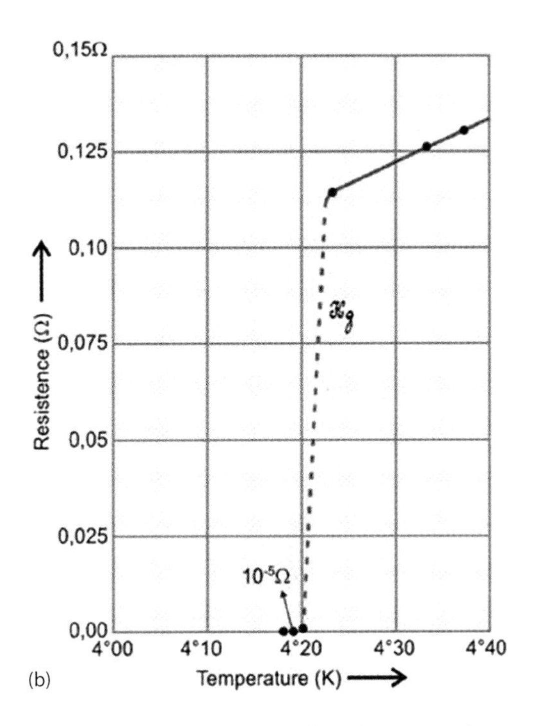

(b)

FIGURE 2.6 (a) Onnes's notebook #56 recording the first discovery of superconductivity. (From van Delft, D. and Peter Kes, P. *Phys. Today*, 63, 38–43, 2010.) (b) Discovery of superconductivity of Hg on April 8, 1911. (From Tinkham, M., *Introduction to Superconductivity*, McGraw-Hill, New York, 1995; Courtesy of Museum Boerhaave, Leiden, the Netherlands.)

2.4 Issue with Helium

Due to its uniquely low boiling point, liquid helium had become an indispensable tool, and actually the professional lifeblood, in low-temperature physics research and the cryogenic industry during the entire twentieth century.

However, helium is a scarce and non-renewable natural resource. Once helium is released into the atmosphere it is gone forever. There is no chemical way of manufacturing helium.

After one century of using liquid helium as the major approach to maintain at 4.2 K or lower, the growing unbalance between the supply and demand has raised serious concerns on resource shortage, higher price, etc.

At the same time, the concerns have stimulated a faster pace of cryocooler development as solutions.

This chapter is to present a brief outline of the status of helium, including its applications, demands, supplies, concerns and solutions.

Most of the information used in this section came from the following sources as listed in [8]:

Campbell, J.R. Garvey, M., *A Worldwide & US Perspective on Helium Demand & Supply, gasworld Global Helium Summit 2.0* (2016).

Kramer, D., *Erratic helium prices create research havoc* by from Physics Today, Volume 70, Issue 1 (2017).

Responding to The U.S. Research Community's Liquid Helium Crisis, a report overseen by POPA (Panel of Public Affairs), APS publication date: October 2016.

US Geological Survey, Mineral Commodity Summaries, January 2017, US Department of Interior, US Geological Survey.

Economics, Helium, and the U.S. Federal Helium Reserve: Summary and Outlook by Steven T. Anderson for the near future outlook of helium.

Some information will be outdated when this book is published. Interested readers can refer to the future *Global Helium Summit* for updated information.

2.4.1 Applications

Helium has some special applications due to its unique properties. A few examples are mentioned below.

- *Cryogenics*: Helium has a uniquely low boiling point and is the only element that remains in the liquid phase at absolute zero degrees under normal pressure. In this case, helium is the only and irreplaceable material (coolant) available to maintain temperatures near 4 K or lower. Liquid helium exclusively serves as the coolant for the magnetic resonance imaging (MRI) system (Figure 2.7a), the International Linear Collider (ILC), the conventional superconducting magnets in the National High Magnetic Field Laboratory in Florida, etc.
- *Aerostatics*: Helium is lighter than air. It is often used as the lifting gas for meteorological balloons, airships, blimps, etc. Helium is the preferred candidate for lifting gas because it is not flammable. As an example, helium-filled airships developed by Lockheed Martin to carry and deliver approximately 20 tons of cargo supplies are illustrated in Figure 2.7b.
- *Big balloons*: The amount of helium to inflate the large balloons at holiday festivities is significant.

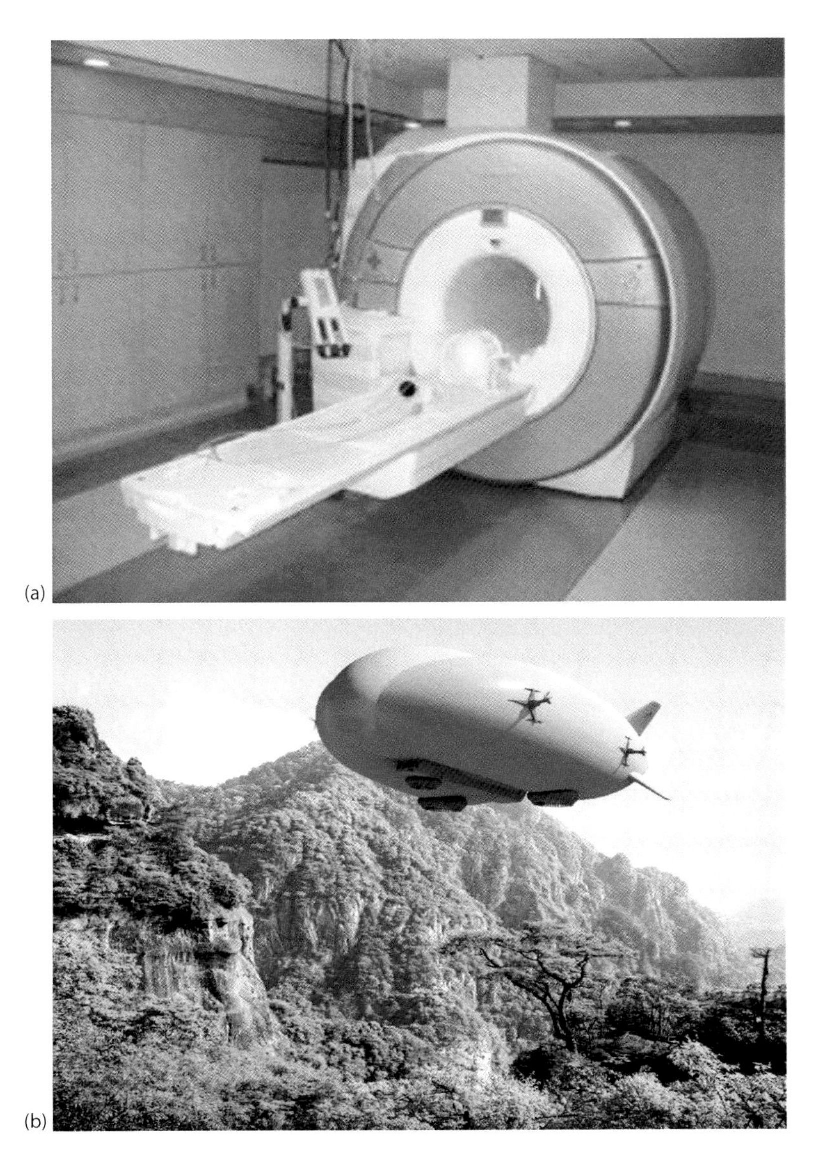

FIGURE 2.7 Applications of helium: (a) A helium-cooled MRI machine. (b) Flight carrier from Lockheed Martin.

A well-known example is the Macy's Thanksgiving Day Parade in New York City. Typically, 300,000 cubic feet of 4He gas are used to fill the balloons for the parade, which consumes nearly 1% of the annual helium gas production in USA!

Big balloons also include weather balloons, research balloons, Department of Defense survey balloons, etc.

- *Welding*: Due to its high thermal conductivity and its inertness, helium gas has been used as a protective gas in welding to prevent the metal being welded from oxidation in the molten state.
- *Leak detection*: Due to its low viscosity, large diffusion coefficient, and small molecular size, helium gas is the exclusive candidate used for leak detection.

- *Semiconductor manufacturing*: Helium is an essential component in the manufacturing process for semiconductors.
  During the processing of semiconductor chips fabrications, helium is used as a protective gas for flushing vessels. Furthermore, helium is used as protective gas and cooling gas when growing silicon and germanium crystals.
- *Fiber optics manufacturing*: Helium is used as a coolant during the manufacture of fiber optic cable due to its very high thermal conductivity. Liquid helium's low temperature and inertness make it ideal to cool silica strands rapidly in a cooling tube as glass fibers are drawn from a glass billet.
  It is important to keep in mind that helium is a non-renewable resource, and no substitute exists for helium for many of the above listed applications. All of these applications will be affected, even stopped, if the supply of helium ceases to exist.

2.4.2 Demand

The worldwide demand (including the United States) for helium was 3.89 billion cubic feet (107.9 million cubic meter) in 1995 and 5.9 billion cubic feet (167.6 million cubic meter) in 2016, respectively [9]. The worldwide demand for helium has increased by about 55% during that time.

The health-care industry has become the largest consumer of helium, and this industry will continue to grow.

US demand of helium in 2016 is 1.99 billion cubic feet (54 million cubic meter). Although the United States remains the largest helium-consuming country, the helium demand in the United States and Western European appears to be stabilized, and even started tapering since 2010. Consumption of helium in other parts of the world, such as China and the Republic of Korea, has increased significantly.

The conference of **Helium Summit** provides the most important forum to review the status of helium in the world of the year, including demand, supply, expansion activities of helium production, new discoveries of helium resources, etc (Figure 2.8).

For example, the **Helium Summit 2018** has sent the following messages:

1. The worldwide helium supply in 2018 was approximately 6.2 billion cubic feet (BCF).
2. The total global helium gas demand in 2018 is 6.2 BCF as listed in Table 2.1, that is, 2% increase per year from 2014.
3. Mainly driven by China, Asia represents the largest market with 33.9% of the worldwide demand, which has surpassed the United States represented with 32.2%.

2.4.3 Resource and Supply [10–15]

1. The National Federal Helium Reserve in the United States
   The National Federal Helium Reserve, also known as the Federal Helium Reserve, is a strategic reserve of the United States holding for helium gas, including all operations of the Cliffside Field helium storage reservoir in Potter County, Texas, and the government's crude helium pipeline system. This facility is operated by the US Department of the Interior's Bureau of Land Management (BLM). The BLM manages the Federal Helium Program, which is the first major helium gas supplier for both US and worldwide.

FIGURE 2.8 Helium Summit 2018.

Table 2.1 Global Helium Gas Demand in 2018

Region	Amount (BCF)	Total Worldwide Demand (%)
United States	2.0	32.2
America except USA	0.7	11.3
Asia (except India)	2.1	33.9
Europe	1.2	19.4
Africa/Middle East/India	0.2	3.2

2. By-product of natural gas
3. New discovery and facilities

   A new finding of helium resource located in the Rift Valley in Tanzania was reported in 2016. Its capacity of helium is estimated as 54 BCF (1.5 billion cubic meters). Although this new discovery may change the game play, helium price reduction is not expected to drop soon. Investment, construction, and time will be required for the industry to produce commercially refined helium from Tanzania.

   Exploration activities also take place in the United States, Canada, etc.

   New facilities include, but are not limited to, Qatar 3 and 4 projects in Qatar, Gasprom's Amur project, as well as another project by Irkutsk Oil Corporation in Russia, are underway.

   These projects are scheduled to be completed by 2023.

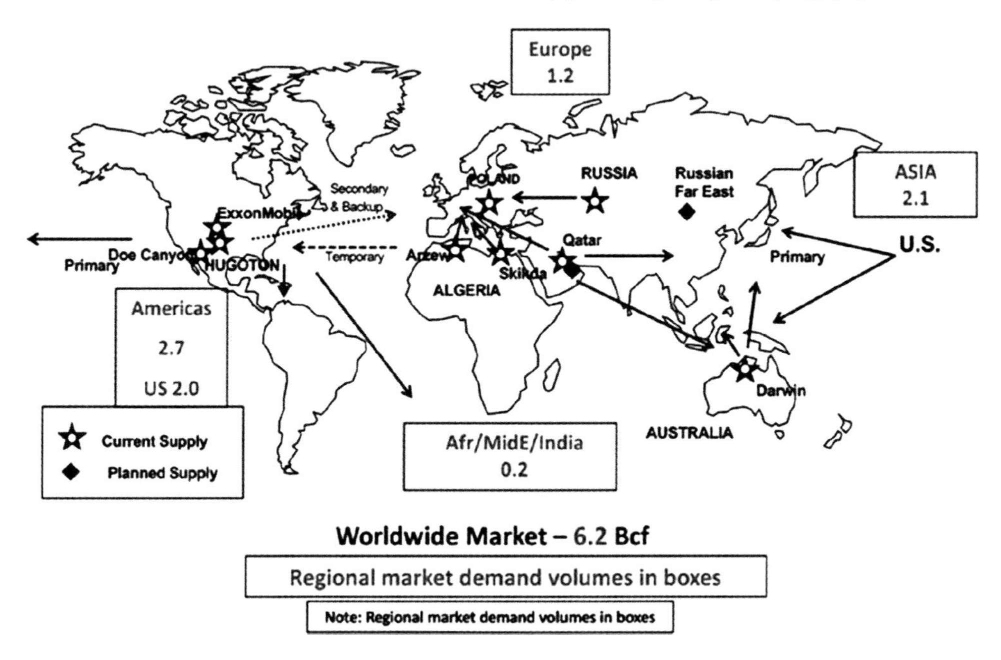

FIGURE 2.9 Worldwide helium demand and supplies by region (BCF) in 2018. (From Worldwide helium demand and supplies by region (BCF) in 2018, *gasworld US*, 56, November 2018.)

The production of helium is expected to increase dramatically after these projects are completed [17]. As a summary, (Figure 2.9) illustrates the worldwide helium demand and supplies by region (BCF) in 2018.

4. Outlook and speculation on year 2025 [17,18].

The helium disruption caused by different reasons in 2006–2007, 2011–2013, 2017, and early 2018 had a serious impact on the whole community, both users and suppliers. The helium shortage remains a deep concern due to the growing imbalance between the supply and demand.

The helium situation is under close monitor by multiple organizations. Analysis and speculations on the helium resource, supply, demand, and possible shortage are published from time to time so that the community may take preventive actions to minimize the impact when the shortage happens. For example, a speculation on helium supply and demand made in 2016 is illustrated in Figure 2.10. Thanks to the expansion activities and the new discovered resources, the helium demand may remain tight but likely not exceed the supply before 2025, unless some major shutdown happens from the suppliers' sides. The year 2018 data supports this speculation.

At the same time, the world may encounter issues for the helium supply during 2020–2022 due to the uncertainties caused by the sales of the BLM crude helium to the private industry and the privatization of the BLM's helium assets, which are scheduled to take place between late 2020 and 2021.

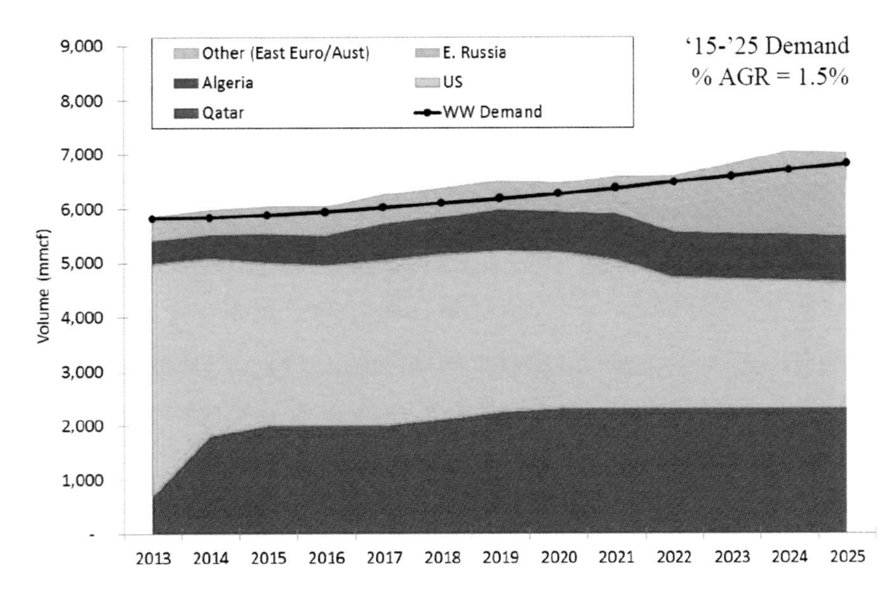

FIGURE 2.10 Speculation on helium supply and demand made in 2016.

We may conclude from preceding discussions that the concern of helium shortage due to the growing imbalance between supply and demand is unlikely to go away soon. The helium shortage will still haunt the community, in particular when any major helium production facility shuts down for maintenance.

The helium shortage situation is expected to gradually ease along with more expansion activities to boost the helium production, more discovery of helium resources, more infrastructures of helium recovery and liquefaction facilities, and above all further development and usages of cryocoolers.

The bottom line is that the 1.4 trillion cubic feet of global helium reserves and the worldwide production expansion should be able to provide enough helium for the market for decades.

Some optimistic speculation even expects a shift from helium shortage to oversupply after 2021.

2.4.4 Impact [19]

The very first impact from the growing imbalance between supply and demand is the price spike. Since 2009, the market price for crude helium from the reserve has risen more than 60%. The price spike of liquid helium for the laboratory around 2012 and 2013 was truly shocking (Private Communications with Janis Research Company, December 31, 2018.).

The first victims of the price spike are many small academic laboratories and groups in universities that need helium for their research. Most of these groups operate with constrained budgets from the federal science agencies. Their research programs need to be cut down in order to balance the budget for the rising helium cost.

Unreliable supply and delivery are additional concerns that make the operation of low-temperature research groups more difficult.

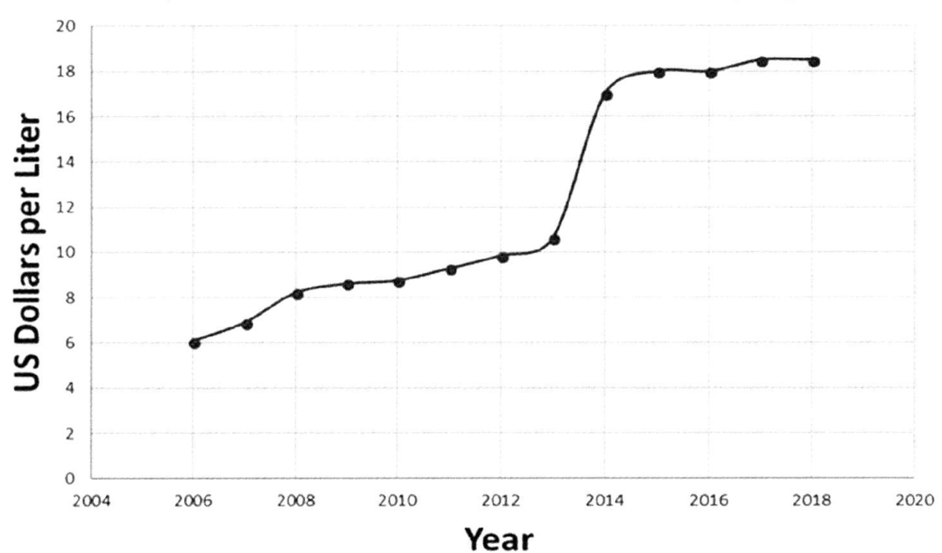

FIGURE 2.11 LHE-4 price for Janis Research Company during the past decade. (Courtesy of Janis Research Company, Private Communications.)

The private sector and industries are also victims of the price increase. Janis Research Company, a leading manufacturer of cryogenic equipment, is a good example. A significant portion of Janis products require liquid helium for the performance test before the products are shipped to the customers. Figure 2.11 illustrates the prices Janis Research Company has been paying for liquid helium. In 2013, Janis not only suffered the helium shortage but also faced a 70% price spike from $10 per liter liquid helium to $17 per liter. As a matter of fact, the price Janis paid for liquid helium had gone up by 300% during the past 12 years.

2.4.5 Solutions

The crisis must be resolved! Various approaches have been developed to minimize the impact of the helium shortage.

2.4.5.1 Helium Recovery System

This approach is to connect a helium recovery system to the experimental facility. The vented helium gas is collected by the recovery system (it is often contaminated with air) instead of venting into the atmosphere. Figure 2.12 depicts the helium recovery process at the National High Magnetic Field Laboratory (NHMFL) in Florida, United States.

The NHMFL continues making improvements to increase the helium recovery rate. According to their report up to 2015 [20], their recovery system has reduced the amount

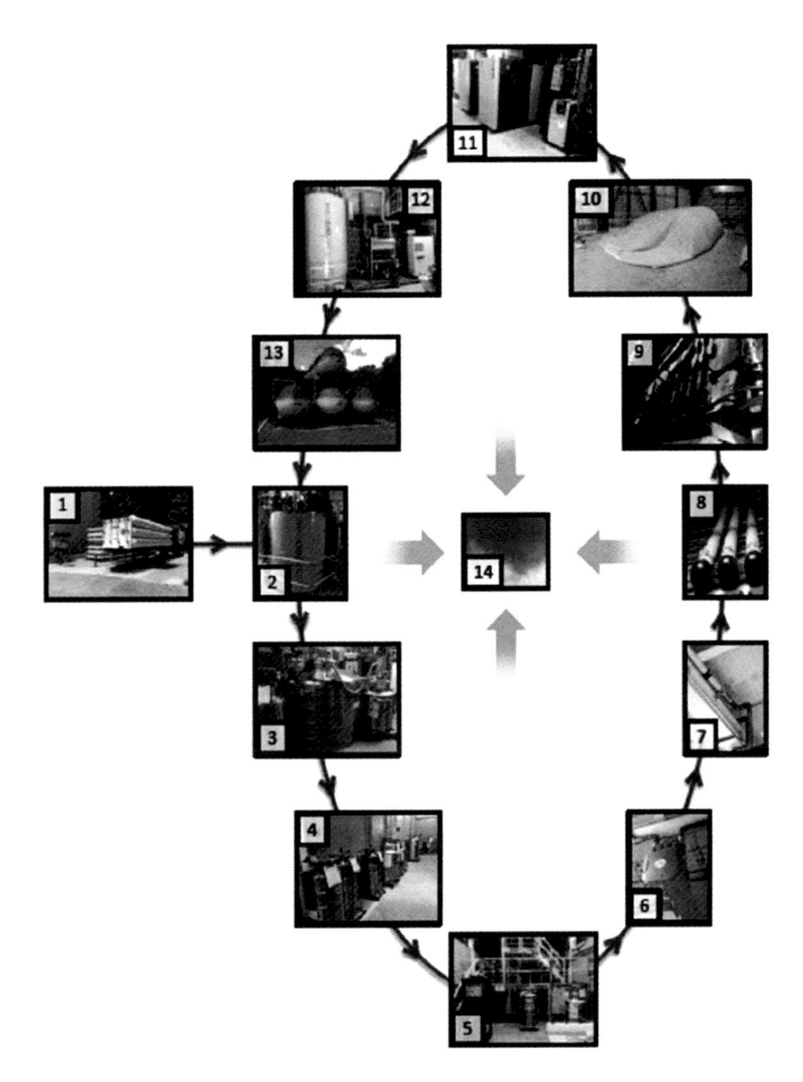

FIGURE 2.12 Helium cycling used for research at the National High Magnetic Field Laboratory (NHMFL) in USA. 1. High pressure storage trailers make up for helium losses, 2. LR280 liquefier, 3. 3,000 L dewar, 4. Helium retrieval station, 5. Magnet cell, 6. Gas meter, 7. Copper to HDPE transition, 8. HDPE piping, 9. HDPE to stainless transition, 10. Gas bag, 11. Compressors, 12. Purifier, 13. Medium pressure storage tanks, 14. Losses to atmosphere can occur at any stage of the life cycle. (From Wang, C., A helium re-liquefier for recovery and liquefying helium vapor from cryostat, in *Advances in Cryogenic Engineering*, Vol. 55A, AIP Publisher, Melville, NY, pp. 687–694, 2010; US patent, 8,375,742.)

of purchased helium by 60% while helium usage has increased by roughly 40%, as illustrated in Figure 2.13.

The helium recovery system operated at the NHMFL is actually a "shop" that requires full-time operators and can produce around 50 liters of liquid helium per hour.

The recovery rate cannot be 100% since some helium gas always gets lost during the process. New pure helium gas always needs to be added into the system when needed.

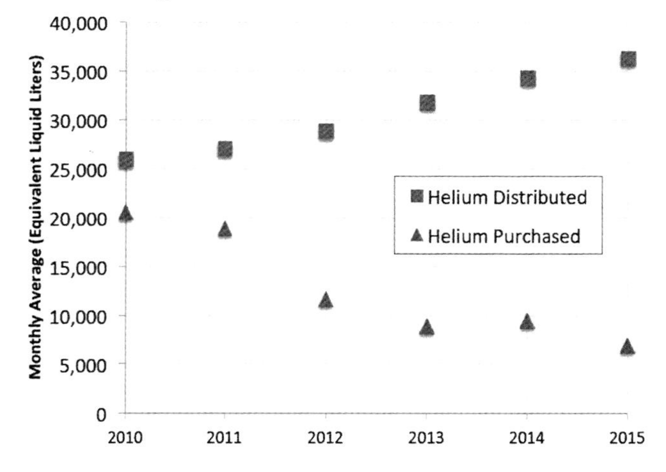

FIGURE 2.13 Helium recovery at the National High Magnetic Field Laboratory.

2.4.5.2 Laboratory-Scale Helium Recovery and Liquefaction

Commercial, laboratory-scale, compact helium liquefiers using 4 K pulse-tube cryocoolers are offered by multiple venders. As an example, we take the systems offered by Cryomech Inc. for our discussion [21].

Figure 2.14 shows a photo of the helium liquefier. The liquefier system integrates the pulse tube cryocooler, a dewar, helium extraction line, and control system of automatic operation. The entire system is portable and can be moved around the laboratory. Figure 2.15 shows a schematic of the helium liquefaction system. The system mainly consists of a liquefier [a pulse-tube cryocooler (7), (12) and (13)], a liquid helium dewar (4), an extraction line (6), a liquid helium level sensor (10), and a controller (11). The pulse-tube cold head (7) resides in the neck of the dewar where it liquefies the helium gas. Helium gas enters the neck of the dewar and is first precooled by the heat exchangers on the pulse-tube cold head. Helium gas is then condensed on the "heat exchanger" on the 4 K stage of the cold head. The condensed liquid helium drops into the dewar belly where it is stored for later use. The liquid extraction line and liquid level probe are inserted into the same neck of the dewar and reach the bottom of the dewar.

With a single 4 K pulse-tube cryocooler, the liquefier can provide liquefaction rates of 15, 22, 28 L/day. With two or three 4 K pulse-tube cryocoolers in a liquefier, it can have liquefaction rates of 40, 55, 80 L/day. Figure 2.16 shows a liquefier with three cryocoolers.

The compact liquefier along with a helium recovery system can be used for recycling helium for users' single or multiple cryostats. The boiled-off helium from the cryostats is collected in an atmospheric bag. A helium recovery compressor compresses the atmospheric helium and stores it in pressurized storage cylinders. When the recovered helium is ready to be re-liquefied, the gaseous helium first flows through a helium purifier to increase the purity (>99.999%). Then it is ready to be supplied to the liquefier where it is liquefied and stored in the liquefaction dewar. The helium recovery rate is normally >99%. Hundreds of the laboratories around world are benefitting from the laboratory-scale helium recovery system and liquefiers.

FIGURE 2.14 Photo of a compact helium liquefier. (Courtesy of Cryomech, Inc.)

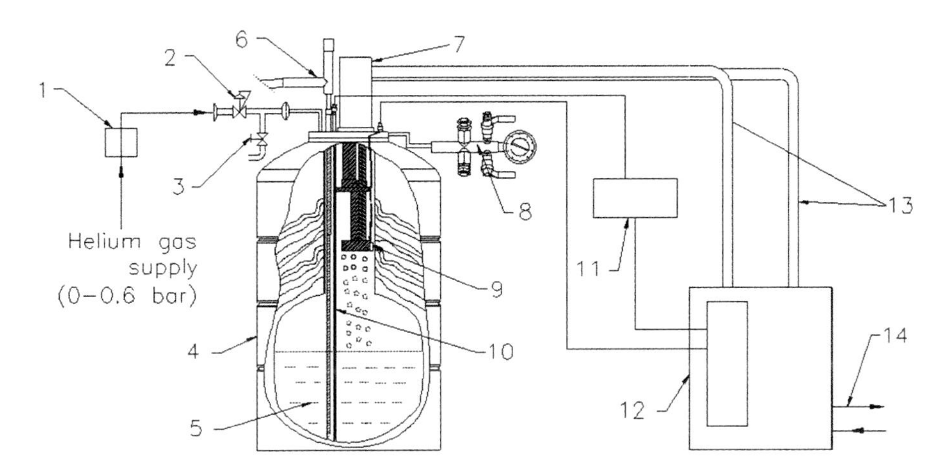

FIGURE 2.15 Schematic of helium liquefaction system.

FIGURE 2.16 An 80 L/day liquefier with three 4 K pulse-tube cryocoolers. (Courtesy of Cryomech, Inc.)

2.4.5.3 Zero Boil-Off Systems

Integrate 4 K cryocoolers in the liquid helium cryostat: Owing to the commercialization of 4 K cryocoolers, many forms of low-temperature instruments such as MRI and laboratory cryostats can have zero helium boil-off by using one or more 4 K cryocoolers to recondense the boiled-off helium (refer to Figure 2.17a). A 4 K cryocooler with a 4 K stage condenser is integrated in a cryostat. The cryocooler cold head is surrounded by helium gas. The condenser attached on the second stage recondenses evaporated helium into liquid and returns the liquid helium to the cryostat. Other designs utilizing 4 K cryocoolers exist for zero boil-off cryostats.

An alternative design of a two-neck zero-helium-loss cryostat is shown in Figure 2.17b. The cryostat has one neck for installing a 4 K cryocooler and one neck for an instrumentation probe. The design has been successfully employed in several commercial products utilizing 4 K pulse-tube cryocoolers.

Cost reduction of liquid helium for operations of larger open-loop liquid helium cryostats, such as dilution refrigerators, MRI, NMR, and SQUIDs, is highly desirable. Helium reliquefiers were developed at Cryomech, Inc. [22] to provide a solution for

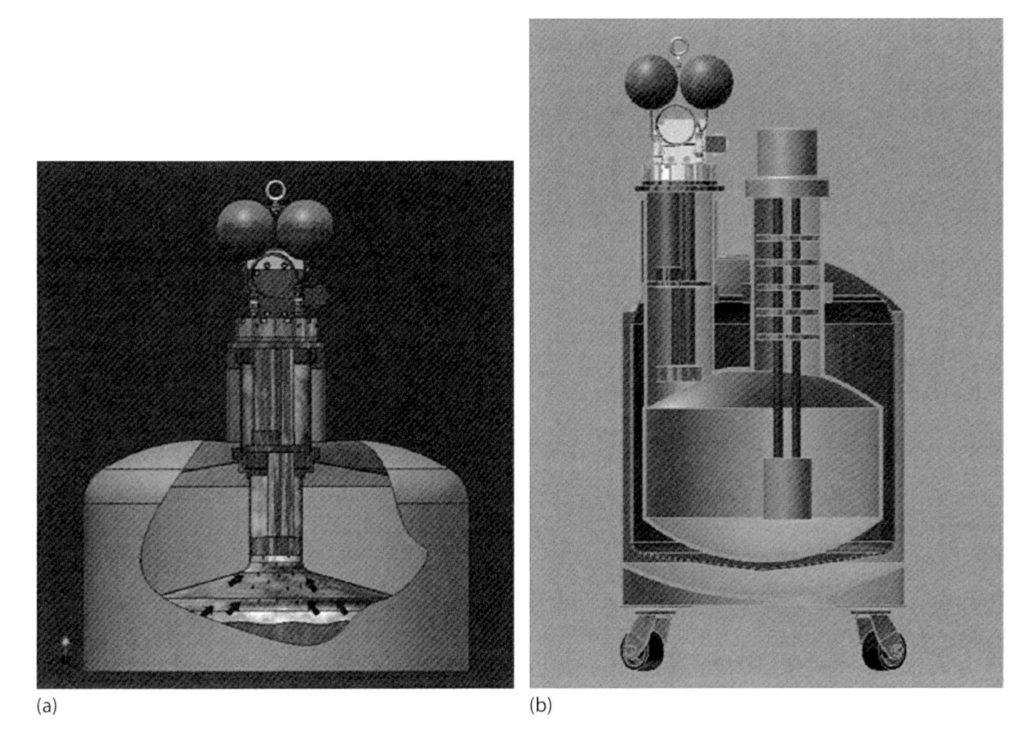

(a) (b)

FIGURE 2.17 (a) Zero boil-off cryostats with a 4 K cryocooler to recondense helium boil. (b) Two-neck zero-helium-loss cryostat. (Courtesy from the website of Cryomech, Inc.)

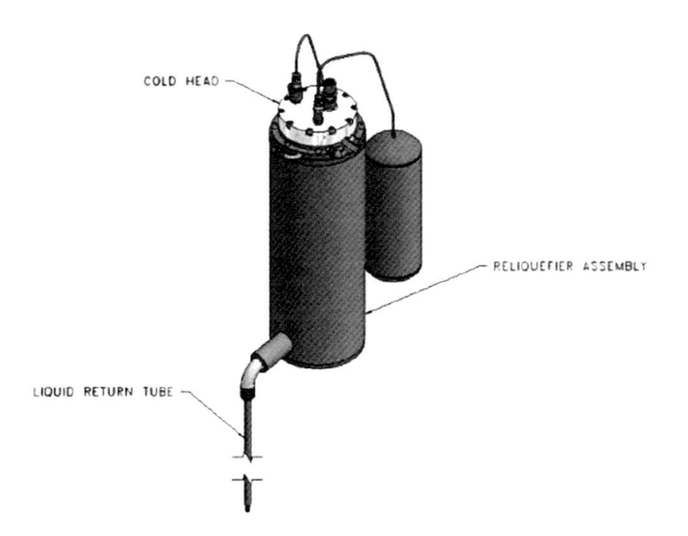

FIGURE 2.18 A helium reliquefier with 4 K pulse-tube cryocooler.

recovering and liquefying helium vapor from open-loop helium cryostats. Cryomech systems have been designed to retrofit into the cryostat and form a closed helium loop for the system as illustrated in Figure 2.18.

The reliquefier head includes a 4 K pulse-tube cryocooler and reliquefier assembly with a liquid helium return tube. The pulse-tube cold head resides in the vacuum-insulated

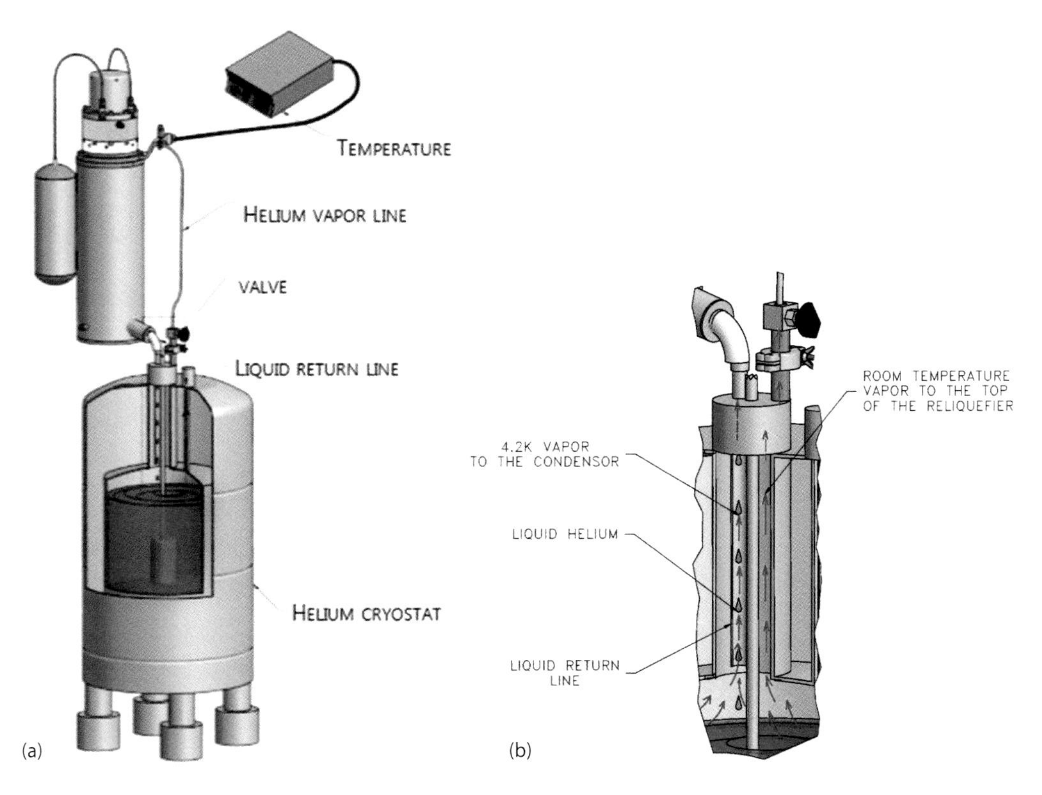

FIGURE 2.19 (a) and (b) A reliquefier used in an open liquid helium cryostat. (Courtesy of Cryomech, Inc.)

neck of the reliquefier assembly. Figure 2.19 shows a reliquefier installed in an open-loop liquid helium cryostat. A vacuum-insulated liquid return tube on the main assembly is inserted into the neck (or fill tube) of the cryostat. A flexible stainless-steel hose with a control valve connects the vent port on top of the cryostat to the gas inlet of the reliquefier. The reliquefier essentially converts the open-loop cryostat into a closed-loop system with no helium leaving the system. Reliquefiers have different reliquefaction capacities from 10 to 60 L/day.

The reliquefiers have been used in many liquid helium systems, such a MEG, MPMS, PPMS, MRI, NMR and dilution refrigerators.

2.4.5.4 Development of "Cryogen-Free" (or Called "Dry") Systems

The fundamental solution is to completely abandon liquid helium and use cryocoolers instead (i.e., closed-cycle refrigerator) to (pre-) cool the system to 4.2 K or lower temperature.

Everything boils down to development of reliable, efficient, and high-power closed-cycle refrigerators that can reach below 4 K. The development of sub-4 K cryocoolers is one of the most important breakthroughs of cryogenic engineering in the twentieth century.

Details about the cryocooler will be discussed in the next chapter.

Acknowledgments

The contributors sincerely acknowledge Chao Wang for the final editing work of this chapter.

The contributors also acknowledge his colleagues at Janis Research Company for their help.

Sincere thanks to Dan Logan for his final read-proof. Many thanks to Choung Diep for making drawings. Many thanks go to Ann Carroll for her constant help with software difficulties, reference searches, and copyrights requests.

References

1. van Delft, D., Little cup of helium, big science, *Physics Today*, 61(3), 41, 2008.
2. van Delft, D., Heike Kamerlingh Onnes and the Road to Liquid Helium. IEEE/CSC & ESAS European Superconductivity News Forum (ESNF), No. 16, April 2011.
3. Onnes, H.K., Investigations into the properties of substances at low temperatures, which have led, amongst other things, to the preparation of liquid helium, *Nobel Lecture*, 4, 306–336, 1913.
4. Reif-Acherman, S., Liquefaction of gases and discovery of superconductivity: Two very closely scientific achievements in low temperature physics, *Revista Brasileira de Ensino de Física*, 33(2), 1–17, 2011.
5. Onnes, H.K., The liquefaction of helium, *KNAW Proceedings*, 11, 168–185, 1908–1909.
6. van Delft, D. and Peter Kes, P., The discovery of superconductivity, *Physics Today*, 63(9), 38–43, 2010.
7. Tinkham, M., *Introduction to Superconductivity*, New York: McGraw-Hill, 1995.
8. Campbell, J.R. Garvey, M., *A Worldwide & US Perspective on Helium Demand & Supply, gasworld Global Helium Summit* 2.0 (2016) and other sources as listed on page 34.
9. Garvey, M.D., Global helium market, *Gasworld*, 1 November 2017.
10. National Research Council, Overview, conclusions, and recommendations, In: *Selling the Nation's Helium Reserve*, Washington, DC: The National Academies Press, 2010. http://www.nap.edu/read/12844/chapter/1.
11. Federal Helium Program, *Department of the Interior's Bureau of Land Management (BLM)*, https://www.blm.gov/programs/energy-and-minerals/helium.
12. Zhang, S., United State extents life of helium reserve, *Nature News*, September 26, 2013. http://www.nature.com/news/united-states-extends-life-of-helium-reserve-1.13819.
13. USGS, 2012. *Mineral Yearbook* (Helium). Reston, VA: U.S. Department of the Interior U.S. Geological Survey: Mineral Commodity Summaries 2017.
14. Phetteplace, T., *Examining the Geological Potential for Helium Production in the United States*, Senior Thesis, The Ohio State University, 2014.
15. Piccirillo, C., Helium shortage: Situation update one year later, *Decoded Science*, 2014. http://www.decodedscience.org/helium-shortage-situation-update-one-year-later/42314.
16. Garvey, M., Worldwide helium demand and supplies by region (BCF) in 2018, *gasworld US*, 56(11), 2018.
17. Mohr, S. and Ward, J., Helium production and possible projection, *Minerals*, 4, 130–144, 2014.
18. Kaplan, K.H., Helium shortage hampers research and industry, *Physics Today*, 60(6), 31, 2007.
19. Cramer, D., Erratic helium prices create research havoc, *Physics Today*, 70(1), 26, 2017.
20. Barrios, M.N., Helium recovery at the national high magnetic field laboratory, *Material Science and Engineering*, 101(1), 012103, 2015.
21. Wang, C., Small-scale helium liquefaction systems, *Journal of Physics: Conference Series*, 150(1), 012053, 2009.
22. Wang, C., A helium re-liquefier for recovery and liquefying helium vapor from cryostat. In: *Advances in Cryogenic Engineering*, Vol. 55A, AIP Publisher, Melville, NY, 2010, pp. 687–694. US patent, 8,375,742.

3. 4 K Regenerative Cryocoolers

Chao Wang

3.1 Introduction of Cryocoolers

3.1.1 Classification of Commercial Cryocoolers

A cryocooler is a refrigeration system to achieve cryogenic temperatures below 123 K. All mechanical cryocoolers generate cooling by basically expanding a gas from a high pressure to a low pressure. This process is also called "Simon expansion."

The fundamental distinction between cryocooler types is the nature of the refrigerant flow within the cryocooler: either alternating flow (AC systems) or continuous flow (DC systems). This distinction is also denoted as regenerative cryocoolers versus recuperative cryocoolers based on the type of heat recovery heat exchanger that is applicable: regenerators for an alternating flow (AC system) or recuperators for a continuous flow (DC system).

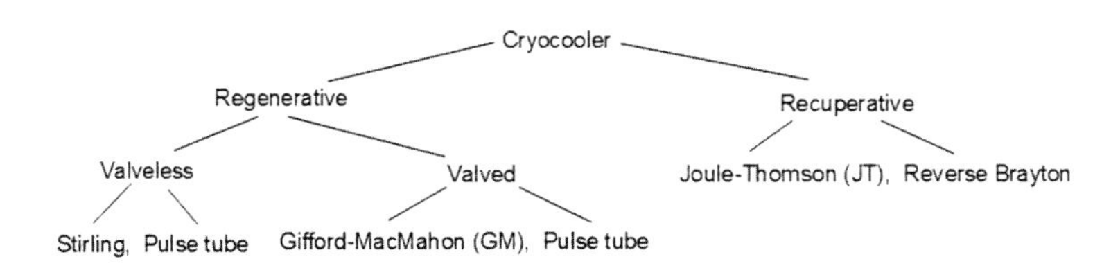

FIGURE 3.1 Classification of commercial cryocoolers.

Figure 3.1 lists categories of commercially available cryocoolers. The regenerative cryocooler employs a regenerative heat exchanger, called a "regenerator," between the compression and expansion spaces. The regenerator, made of, for example, fine-mesh screens or densely packed particles with good specific heat properties, performs as a thermodynamic sponge, alternatively releasing and absorbing heat in a cycle.

In a recuperative cryocooler, a recuperative heat exchanger, also called "counter flow heater exchanger," exchanges energy between two opposing streams of flowing gas or liquid. Of the common cooler types, Stirling, pulse tube, and Gifford McMahon (GM) use regenerative (AC flow) cycles, while Joule–Thomson and reverse turbo-Brayton cryocoolers use recuperative (DC) flows. Cryogen-free systems introduced in this book use 4 K regenerative cryocoolers. The recuperative cryocooler will not be introduced in this chapter. More detail introductions of these types of cryocoolers can be found in references [1,2].

The regenerative cryocoolers, valveless and valved systems, are illustrated in Figure 3.2. The pulse-tube cryocooler is a new generation refrigeration system without cryogenic moving parts. Figure 3.2 also shows similarity between the Stirling cryocooler and Stirling-type (valveless) pulse-tube cryocooler as well as GM cryocooler and GM-type (valved) pulse-tube cryocooler. The pulse-tube cryocooler eliminates mechanical displacers in the Stirling and GM cryocoolers and uses a pneumatic pulse-tube (PT) circuit (regarded as gas displacer) to achieve the thermodynamic cycle.

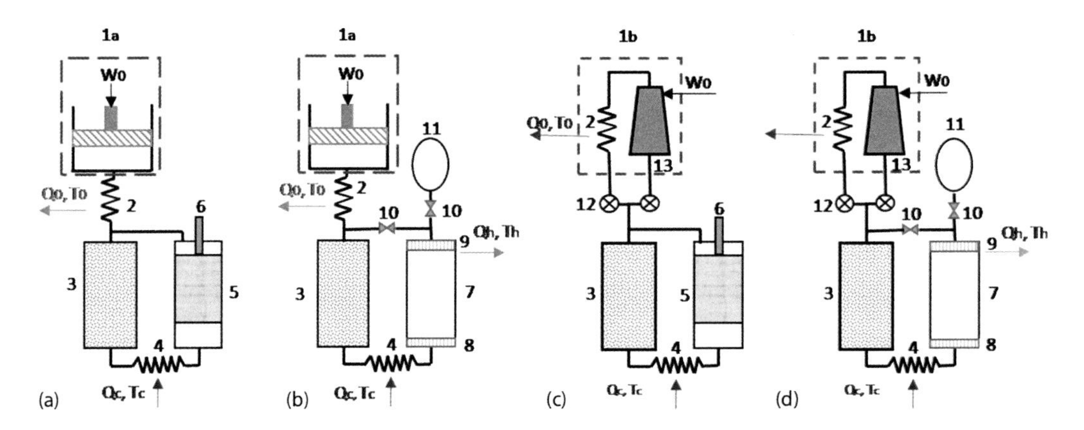

FIGURE 3.2 Regenerative cryocoolers: (a) Stirling cryocooler. (b) Stirling-type PT cryocooler. (c) GM cryocooler. (d) GM-type PT cryocooler. (1a) Valveless compressor. (1b) Valved compressor package. (2) Warm-end heat exchanger. (3) Regenerator. (4) Cold-end heat exchanger. (5) Displacer. (6) Displacer drive (mechanical or pneumatical). (7) Pulse tube displacer (also called "pulse tube"). (8) Cold-end flow straightener. (9) Warm-end flow straightener. (10) Orifices and/or inertance tubes. (11) Gas reservoir. (12) Switching valves (rotary valve). (13) DC flow compressor module.

The Stirling cryocooler and Stirling-type PT cryocooler are valveless systems (without rotary valve). Both use an oscillating-flow compressor to generate the AC flow needed by the cold head. In a valveless system, a compressor (1a) generally must be quite close to the cold head since the compressor piston generates oscillation gas flow directly to the cold head. The connecting line from the compressor to the cold head introduces some empty volumes (so-called dead volume), which reduces the amplitude of oscillating pressure and the cryocooler efficiency. In a Stirling cryocooler, the piston receives work, W_0, from an external source. The cold head includes a warm-end heat exchanger (2) for removing the compression heat Q_0 (generated by the compressor) to ambient at temperature T_0, a regenerator (3) to release and absorb heat in a cycle, and a cold-end heat exchanger (4) to provide a cooling station for instrumentation. The motion of a displacer (5) is controlled by a mechanical or pneumatic displacer drive (6) to have a correct phase angle between dynamic pressure and motion of the displacer for generating expansion cooling Q_c at the cooling temperature T_c.

The PT cryocooler replaces the mechanical displacer (5) of the classic Stirling cycle with a pneumatic (no moving part) displacer (7). The motion of gas displacer is controlled by orifices (10) and a gas reservoir (11) to achieve the desired mass flow/gas pressure-phase relationship needed for refrigeration. The benefit of the PT cryocooler is lower displacer vibration and elimination of complexity and mechanical wear associated with the moving displacer for high reliability and long mean time between maintenance (MTBM). These days, most of the industry is moving to the use of PT cryocoolers. In Stirling-type PTs, the orifice into the reservoir is often replaced by an inertance tube. These are not discussed here, as they are not used in commercial GM-type cryocoolers.

Because of the direct coupling between the compressor drive frequency and the cold-head drive frequency, most Stirling-based cryocoolers operate at between 30 and 100 Hz using helium in the 10 to 35 bar pressure range as the refrigerant gas. This relatively high AC frequency is an advantage for cooling in the temperature range above 80 K but serves as a disadvantage for obtaining high efficiency at very low operating temperatures of below 20 K, where the reduced specific heat of regenerator materials and small thermal penetration depth of regenerator material drastically limits heat storage between cycle phases.

In valved regenerative cryocoolers, such as the GM and GM-type PT cryocoolers, they use a constant flow (also called DC flow) compressor package (1b). The compression heat Q_0 generated by the compressor module (13) is rejected at T_0 by a heat exchanger (2) inside of compressor package. More details of the DC compressor package will be introduced in Section of 3.3. The compressor is connected to the cold head through supply and return stainless steel (SS) flexible lines. Switching valves (12) (or a rotary valve) chop the DC gas flow into a periodic oscillating gas flow (also called AC flow) within the cold head itself. The cold head normally includes switching valves, a regenerator (3), cold-end heat exchanger (4), a displacer (5), and a displacer drive (6). The operation of the GM cold head is similar to that of the Stirling cold head. The compressed gas is expanded through the expender to produce a low temperature fluid. The low temperature fluid introduces a cooling Q_c at temperature T_c. Again, the GM-type PT cold head employs the gas displacer (7) and the phase shifter (controller) of orifices (10) and the gas reservoir (11) to replace the solid displacer (5) and driving part (6) used in the GM cold head.

GM or GM-type PT cryocoolers normally operate with pulsation frequencies of 1–2.4 Hz. Because the compressor is a DC-flow device, it can be located remotely from the cold head/actual cryogenic application from few meters to few hundred meters, connected only by supply and return SS flexible lines.

The two-stage cryocooler with these low operating frequencies has better efficiencies at low temperatures below 20 K. Owing to the discovery of rare earth regenerator materials with higher specific heat at temperatures below 10 K (see introduction in Section 3.4.3), the valved regenerative cryocoolers, with low operating frequencies around 1 Hz, are successfully developed and commercialized and being used for many 4 K applications. These cryocoolers include 4 K GM cryocoolers and 4 K PT cryocoolers. This chapter will focus on the introduction of GM and GM-type PT cryocoolers. The GM-type PT cryocooler will be called pulse tube cryocooler below.

3.1.2 Development History of GM and Pulse-Tube Cryocoolers

The GM cryocooler cycle was first proposed by William Gifford and Howard McMahon of A. D. Little, Inc. in the late 1950s [3,4]. The GM cycle is based on an AC oscillating flow, typically using helium in the 10–30 bar range as the refrigerant gas with a working frequency of 1–2.4 Hz. The GM cryocooler uses a low-cost high-availability DC flow compressor (typically modified from a commercial air conditioning compressor) to provide the primary gas-compression function. The alternating flow needed by the GM cold head is provided by a rotary valve mounted on the cold head assembly. This valve chops the DC flow into an AC flow by alternately connecting the cold head's refrigeration portion (defined as "expander" in this book) to the high- and low-pressure sides of the compressor at the required oscillatory frequency of 1–2.4 Hz. This low frequency results in low maintenance and high reliability. It is particularly useful for obtaining improved efficiency at very low operating temperatures where the reduced specific heat of regenerator materials limits heat storage between cycle phases. The required phase relationship between refrigerant pressure and mass flow is achieved by synchronizing the rotary valve with the pneumatic driven or motor-driven motion of the displacer.

The GM cryocooler was conceived at A.D. Little, Inc. and rapidly became a commercial product. The low frequency GM cryocooler requires minimal maintenance, which makes it the system choice where size, weight, and efficiency are not of prime importance.

The first PT refrigerator was accidently discovered at Syracuse University by Gifford and Longsworth [5] in the mid-1960s, as they were developing the GM cryocooler. This basic PT cryocooler is illustrated in Figure 3.3a. They noticed that the closed end of a pipe became very hot when there was a pressure oscillation inside, whereas the open end toward the compressor was cool. This basic PT cryocooler was based on surface heat pumping for refrigeration. It was buried for nearly two decades because of its poor cooling performance.

The modern PT cryocooler was invented by Mikulin et al. by introducing an orifice and reservoir in the basic pulse tube in 1984 [6], see Figure 3.3b. The orifice and reservoir in the PT cryocooler performs as a phase shifter of the mass flow and pressure for obtaining the maximum PV-work (expansion work). Therefore, the orifice PT cryocooler could operate as traditional cryocoolers, such as the GM, Stirling, and Solvay cryocooler. This invention ignited worldwide development of the PT cryocoolers.

Following after, Radebaugh et al. [7] relocated the warm heat exchanger before the orifice and allowed the warm heat exchanger to act as a flow straightener, see Figure 3.3c.

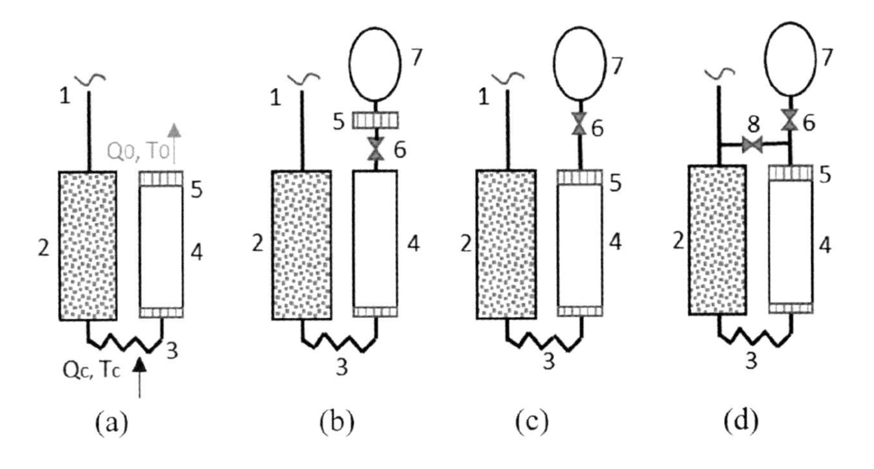

FIGURE 3.3 Development history of pulse tube cryocooler: (a) Basic pulse tube. (From Gifford, W.E. and Longsworth, R.C., *J. Eng. Ind.*, 86, 264–268, 1964.) (b and c) Orifice pulse tube. (From Mikulin, E.I., et al., Low-temperature expansion pulse tubes, *Advances in Cryogenic Engineering*, 29, Plenum Press, New York, 629–637, 1984; Radebaugh, R., et al., A comparison of three types of pulse tube refrigerators: New methods for reaching 60 K, *Advances in Cryogenic Engineering*, 31, Plenum Press, New York, 779–789, 1986.) (d) Double-orifice pulse tube. (From Zhu, S.W., Wu, P.Y. and Chen, Z.Q., *Cryogenics*, 30, 514–520, 1990.) (1) Pressure wave generator. (2) Regenerator. (3) Cold-heat exchanger. (4) Pulse tube. (5) Warm-heat exchanger. (6) Orifice. (7) Reservoir. (8) Second orifice.

The performance of PT cryocooler was further improved with the invention of new phase shifters called double-inlet type by Zhu et al. [8], see Figure 3.3d.

At present, few different types of phase shifters located at the warm end can be used for a PT cryocooler to efficiently control the phase between the pressure and mass flow in the pulse tube for obtaining maximum PV work (gross cooling capacity). Two popular phase shifters used in commercial 4 K PT cryocoolers are given in Figure 3.4. Figure 3.4a is orifice-type (with two orifices) in which the warm end of the pulse tube is connected to a gas reservoir through an orifice. The regenerator inlet is connected to the pulse tube warm end through a second orifice (so-called double-inlet [8]) for improving the phase angle control. The orifice-type phase shifter can be used for both Stirling-type and GM-type PT cryocoolers. 4 K PT cryocoolers manufactured by Cryomech, Inc. are orifice-type PT cryocoolers [9].

Figure 3.4b is a four-valve type, introduced by Matsubara et al. [10], which is only used for GM-type PT cryocoolers. Two valves, connected to the warm end of the pulse tube and the high/low pressures from the compressor, perform as phase shifters. Considering two more valves at the regenerator inlet for cold-head pressure oscillation, there are a total of four switching valves for this cold head. The timings for valve open/close are properly chosen to control the phase of pressure and mass flow rate in the pulse tube. The four-valve-type phase shifter is employed in the 4 K PT cryocoolers made by Sumitomo Heavy Industries (SHI) [11].

The cooling cycle for the four-valve-type PT-type cryocooler is essentially identical to a GM cryocooler cycle, except that the phasing of the gas flow in the cold head is controlled by the PT gas displacer instead of by the motion of the mechanical displacer.

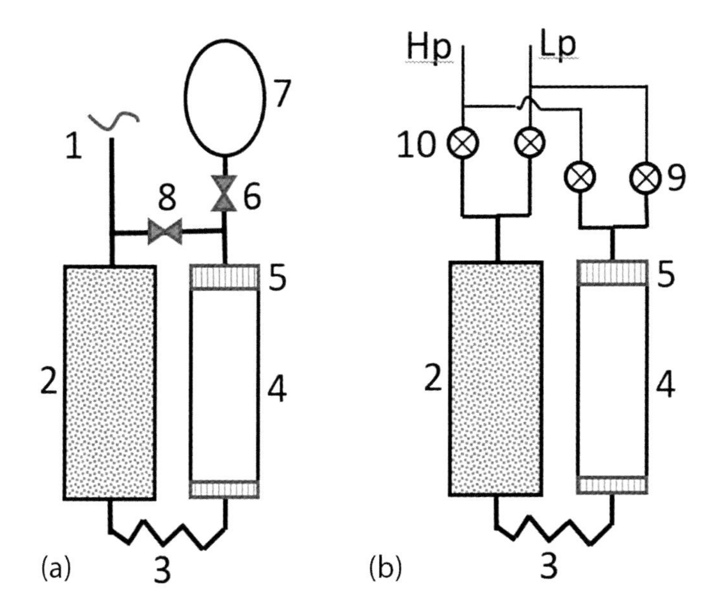

FIGURE 3.4 Two popular phase shifters for PT cryocoolers: (a) Orifice type. (b) Four-valve type. (1) Pressure wave generator. (2) Regenerator. (3) Cold-heat exchanger. (4) Pulse tube. (5) Warm-heat exchanger. (6) Orifice. (7) Reservoir. (8) Second orifice. (9) Two valves for phase shifter. (10) Two valves for cold head.

The cooling cycle of the orifice-type PT cryocooler with the orifice and reservoir is essentially identical to a Solvay cryocooler cycle [1].

A numerical modeling of orifice PT cryocooler by Wu and Zhu [12] first demonstrated the internal working process and revealed a concept of "gas displacer" in the pulse tube where a portion of gas (gas displacer) never leaves the pulse tube. The concept of gas displacer is very useful to understand the operation of PT cryocooler and compare with traditional cryocoolers in this chapter.

3.2 Thermodynamic Cycles of GM and PT Cryocoolers

3.2.1 GM Cryocooler

Figure 3.5 schematically illustrates the thermodynamic cycle of the GM cryocooler. For the GM cryocooler, the regenerator and the displacer are generally combined into one displacer/regenerator unit as noted, not like the schematic in Figure 3.2. The combined design makes the GM cryocooler compact and easy for manufacture.

The cold head of the GM cryocooler consists of a cylinder, closed at both ends, and displacer containing a regenerator in its center. The displacer is connected to a displacer drive (mechanical drive or pneumatic drive) so that it can be driven back and forth in the cylinder. Two volumes, volume 1 and volume 2, are connected through a regenerator and to a gas supply. The gas supply system consists of a gas compressor, inlet/outlet switching valves, high and low pressure connecting lines. A heat exchanger is located downstream of the compressor to cool the gas to ambient temperature after compression.

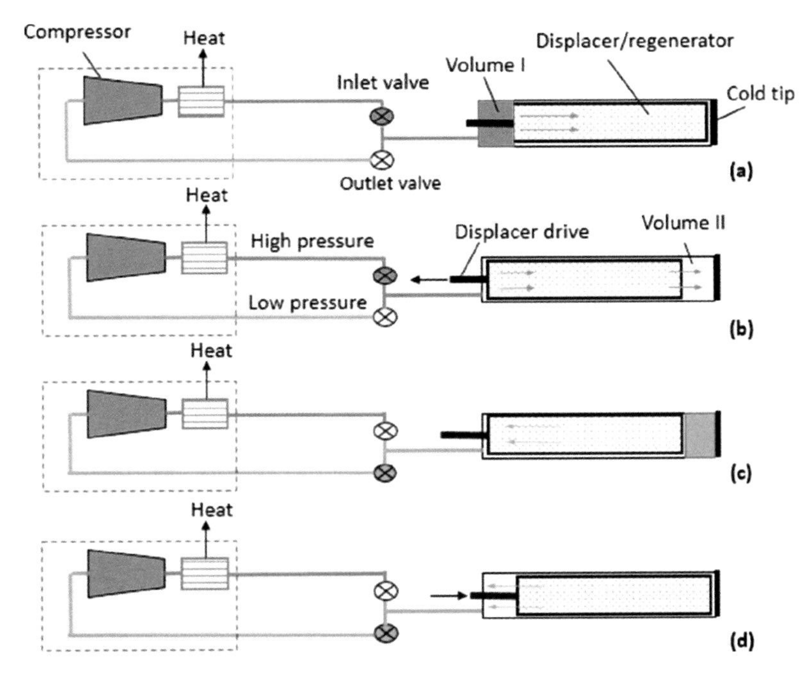

FIGURE 3.5 (a–d) Thermodynamic cycle of GM cryocooler.

The pressure in volumes 1 and 2 are almost the same except for small pressure drops across the regenerator when gas is flowing through it. The displacer is a loose fit in the cylinder except at the top and where it is equipped with a dynamic (sliding) seal to prevent gas leakage past the displacer from one space to the other.

The ideal GM cooling cycle can be divided into four steps:

1. *Pressure Buildup*: The cycle starts with the inlet valve connecting the expander (outlet valve closed) to the high-pressure room temperature gas from the compressor with the displacer at the bottom of the cylinder. High-pressure refrigerant fills the expander's gas pocket—volume 1 at the room temperature end of the cold head and regenerator.
2. *Intake Stroke*: When the inlet valve opens, the displacer moves to the left to reposition the expander gas pocket from volume 1 to the volume 2 at the cold end of the cold head. It is ready for the upcoming expansion phase. During this part of the cycle, the gas passes through the regenerator, entering the regenerator at ambient temperature $T_{ambient}$, releasing heat to it and leaving it at temperature T_{low}. Thus, the heat storage feature of the regenerator retains the temperature gradient between the warm and cold ends of the cold head and smooths out the cyclic temperature variation of the gas.
3. *Pressure Release and Expansion*: With the displacer at the top of the cylinder, the inlet valve closes, and the outlet valve connects to the low-pressure suction from the compressor, thus expanding and cooling the gas in volume 2—adjacent to the cryocooler's cold tip where the useful cooling power is produced.
4. *Exhaust Stroke*: In the final portion of the cycle with the outlet valve open, the displacer moves to the right to reposition the expander's gas pocket from volume 2 to

volume 1. It is ready for the upcoming high-pressure gas-filling phase. During this part of the cycle, the expansion gas passes through the regenerator, adsorbs heat from it, and leaves at ambient temperature. Again, the heat storage feature of the regenerator smooths out the cyclic temperature of the gas as it flows the regenerator.

The idea P-V work diagrams for volume 1 and volume 2 are given in Figure 3.6a and b. The refrigeration cycle of the GM cryocooler includes four thermal dynamic processes: (1) pressurizing at constant volume, (2) intake at constant pressure, (3) pressure release and expansion at constant volume, and (4) exhaust at constant pressure. At the beginning of a cycle, volume 1 is maximum and volume 2 is minimum. The pressure is low at P_l, represented on Figure 3.5 by state (a). The inlet valve opens and fluid enters volume 1 and the regenerator. The fluid contained therein is compressed, and its temperature is, consequently, increased. It mixes with incoming fluid to a final temperature of mixed fluid shown by state (b). Now the displacer moves from the bottom to the top of the cylinder and fluid is displaced from volume 1–2 through the regenerator. In passing through the regenerator, the fluid cools to temperature (c) and more fluid enters through the inlet valve to maintain the pressure constant.

At state (c), the inlet valve closes and outlet valve opens. The pressure falls as fluid leaves the cold head through the outlet valve. Fluid in volume 2 expands as pressure decreases and the temperature of the gas in volume 2 decreases to state (d). Finally, the displacer moves from top to bottom of the cylinder causing fluid to return back through regenerator to volume 1 exhausting at constant pressure. In passing through the regenerator, the temperature of gas increases to a value slightly less than the matrix, following admission of high-pressure gas. The gas leaving the exhaust is, therefore, cooler than the intake feed gas. Heat removed in this way is equal to the total refrigeration achieved. The useful product is the cooling effect generated during expansion.

The work done by the gas on the displacer is shown by the PV work diagram for volume 2. The work done by the displacer on the gas is shown by the PV work diagram for volume 1. Based on the first law of thermodynamics, the ideal refrigeration can be expressed as:

$$\Delta U = Q_c - W \tag{3.1}$$

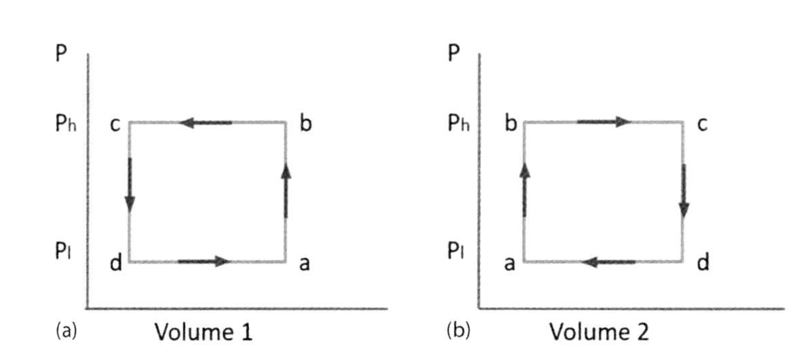

FIGURE 3.6 (a) and (b) PV work diagram for the GM cryocooler.

where ΔU is internal energy variation, Q_c is a gross refrigeration capacity, and W is the work done by the expansion volume. In the ideal cycle, the internal energy U is constant,

$$Q_c = \oint dW = (P_h - P_l)V \tag{3.2}$$

where P_h is discharge pressure, P_l is the suction pressure, and V is the expansion volume (volume 2). A real operation GM cryocooler has different losses associated with pressure drop, heat transfer efficiency, dead volumes, seals, regenerator efficiency, etc., that result in a much lower net refrigeration capacity than the gross refrigeration capacity.

3.2.2 Pulse Tube Cryocooler

Few types of phase shifters can be used for a PT cryocooler, which has been described in Section 3.1.2. An orifice-type PT cryocooler will be used to describe the thermodynamic process in the following. Figure 3.7 shows schematics of a refrigeration cycle for the PT cryocooler. The PT cold head consists of a regenerator, a cold-heat exchanger, a pulse tube, a warm-heat exchanger, and phase shifter.

In a PT cryocooler, the displacer is replaced by a pulse tube in which a slug of helium gas essentially acts as the displacer and separates the cooling and heating aspects of the cycle. It insulates the thermal processes at the cold end and warm end. In this way, gas inside of

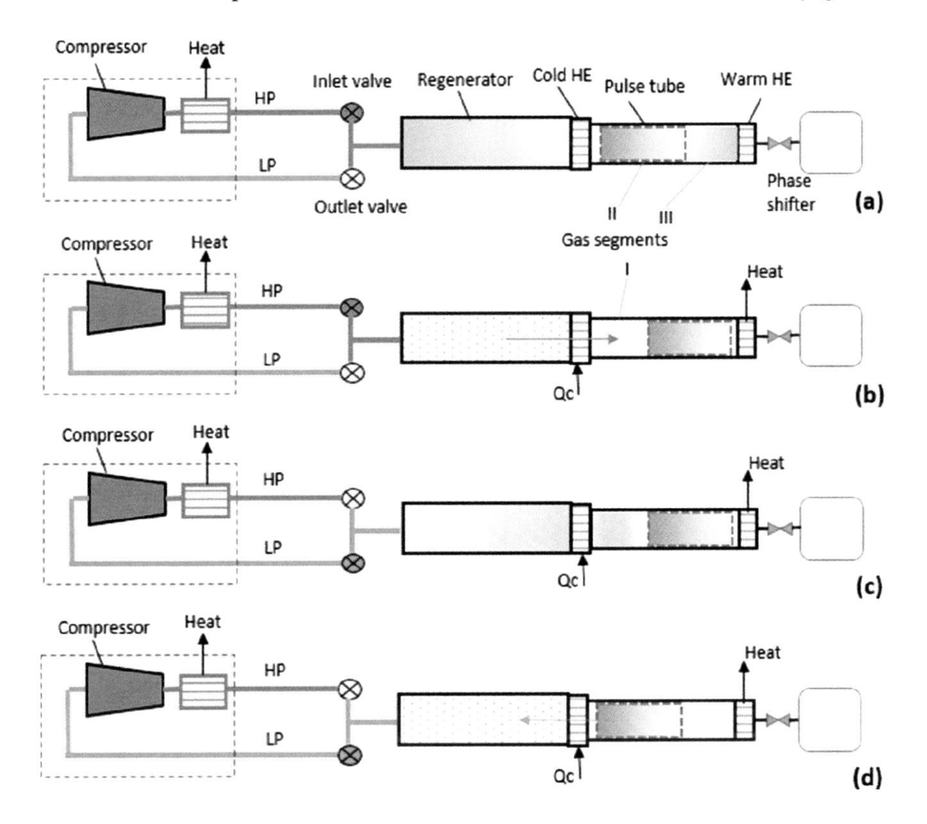

FIGURE 3.7 (a–d) Schematics of pulse tube cryocooler refrigeration cycle. HP, high pressure; LP, low pressure; HE, heat exchanger; Q_c, cooling capacity.

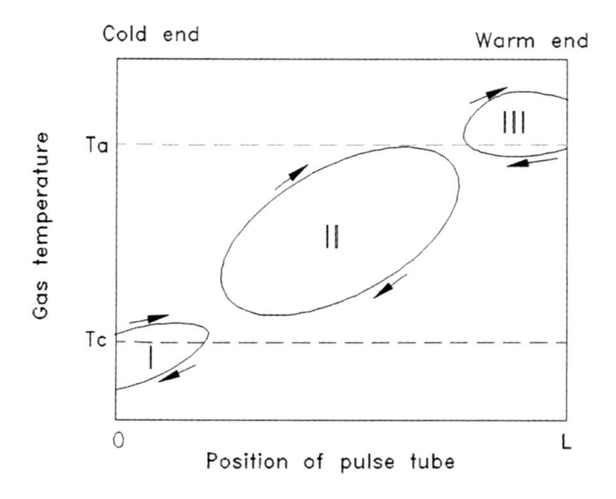

FIGURE 3.8 Motion of three gas segments inside of the pulse tube during a cycle: (I) gas segment flows in and out of the cold end. (II) Gas segment never leaves the pulse tube. (III) gas segment flows in and out of the warm end; T_a, ambient temperature; T_c, cooling temperature.

pulse tube can be roughly divided into three segments. Figure 3.8 shows the movement of three gas segments inside of the pulse tube in a cycle. The gas segment I leaves the cold-heat exchanger and enters the tube when the pressure is high. It returns to the cold-heat exchanger with lower temperature ($<T_c$) when the pressure is low during the expansion process. Hence, it produces cooling on the cold-heat exchanger. The gas segment II is a gas displacer that forms a temperature gradient to isolates the two ends. It operates ideally with adiabatic compression and expansion in the pulse tube. The gas segment III enters the pulse tube through the warm-heat exchanger when the back pressure is higher, leaves the pulse tube, and passes through the warm-heat exchanger with higher temperature ($>T_a$) when the gas pressure in the tube is higher than the back pressure during the compression process. Heat from the gas segment III is removed by the warm-heat exchanger.

The four steps can be used to describe the thermodynamics of the PT cryocooler shown in Figure 3.7. The regenerator functions the same as that in GM refrigeration cycle. The processes below are focused on the pulse tube. The reservoir is large enough that its pressure is near constant.

1. The cycle starts with the inlet valve open (outlet valve close), and the cold head connects with high pressure gas from the compressor. The gas is heated by compression. The pulse tube's gas displacer (gas segment II) is located near the cold end.
2. At the compression phase ends, the gas displacer moves to the right to compress the gas (helium) in the pulse tube. Because this heated, compressed gas is at a higher pressure than the pressure in the reservoir, the gas of segment III flows through the orifice into the reservoir and exchanges heat with the ambient through the warm-heat exchanger. The flow stops when the pressure in the pulse tube is reduced to the pressure in the reservoir. The expansion gas segment I moves inside of the pulse tube and is ready for the upcoming expansion phase.
3. Next, the inlet valve is closed, and the outlet valve is connected to the low-pressure suction from the compressor, thus expanding and cooling the gas segment I. The cold gas flows through the cold-heat exchanger, where the useful cooling power is produced.

4. In the final portion of the cycle, the gas displacer moves to the left. The gas flows from the reservoir into the pulse tube through the orifice until the reservoir gas pressure equals to the pressure in the pulse tube, which forms gas segment III in the pulse tube. The system is ready for the upcoming compression phase.

The cooling cycle of the orifice-type PT cryocooler with the orifice and reservoir is essentially identical to a Solvay cryocooler cycle [1], except that the phasing of the gas flow in the cold head is controlled by the gas displacer instead of the mechanical displacer. The gas displacer is compressible, and the volume of it varies with pressure in a cycle.

The cooling cycle for the four-valve-type PT-type cryocooler is essentially identical to a GM cryocooler cycle.

The enthalpy flow model for understanding refrigeration of the PT cryocooler is derived using the First Laws of thermodynamics for an open system. The expressions of the oscillating flow can be simplified if averages over one cycle are made. Even though the time-averaged mass flow rate is zero, other time-averaged quantities, such as enthalpy flow and entropy flow, will have non-zero values in general. Assuming the process in the pulse tube is adiabatic and reversible, the time average enthalpy flow is the same in the entire tube.

The First Law balance for the cold section in Figure 3.9 shows the time averaged heat flow $\dot{Q}_c$ is given by

$$\dot{Q}_c = \left\langle \dot{H}_p \right\rangle - \left\langle \dot{H}_r \right\rangle - \dot{Q}_l \tag{3.3}$$

where $\left\langle \dot{H}_p \right\rangle$ is the time-averaged enthalpy flow in the pulse tube, $\left\langle \dot{H}_r \right\rangle$ is the average enthalpy flow in the regenerator, and $\dot{Q}_l$ are the losses of the system except regenerator loss of $\left\langle \dot{H}_r \right\rangle$. For a perfect system with a perfect regenerator, $\dot{Q}_l = 0$ and $\left\langle \dot{H}_r \right\rangle = 0$, the maximum, or gross, refrigeration power is

$$\dot{Q}_c = \left\langle \dot{H}_p \right\rangle \tag{3.4}$$

The average enthalpy flow, assuming an ideal gas, is given by

$$\left\langle \dot{H}_p \right\rangle = \left(\frac{c_p}{\tau} \right) \int_0^{\tau} \dot{m} T \, dt \tag{3.5}$$

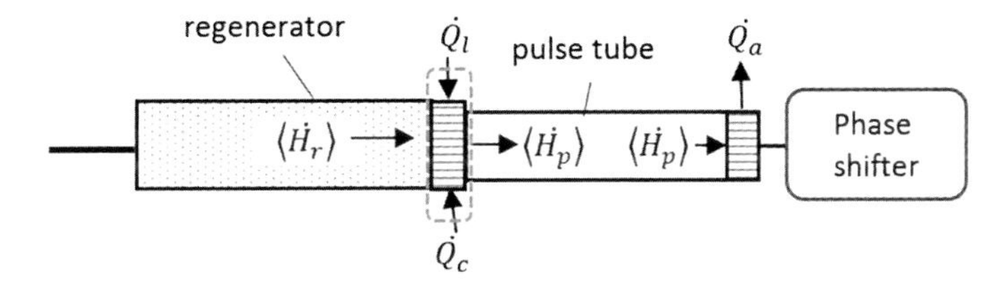

FIGURE 3.9 Time-averaged heat and enthalpy flow in a PT cryocooler.

where τ is the period of cycle, c_p is the specific heat, $\dot{m}$ is the mass flow rate, T is the gas temperature, and t is time. The average enthalpy flow also can be given by

$$\left\langle \dot{H}_p \right\rangle = \left(c_p * \frac{A_p}{R\tau} \right) \int_0^{\tau} p\,u\,dt \tag{3.6}$$

where A_p is the cross area of the pulse tube, R is the gas constant, p is the gas pressure, and u is the gas velocity. The dynamic gas pressure and velocity contribute to $\left\langle \dot{H}_p \right\rangle$ in Equation 3.6. The phase shifter at the warm end of the PT cold head is used to control the phase between the gas pressure and velocity for maximum cooling capacity. It is crucial to build an efficient phase shifter in a PT cryocooler.

3.3 Engineering Aspects of GM-Type Cryocoolers

A GM-type cryocooler consists of three components: (1) a DC compressor package, (2) a set of SS flexible lines for connecting the compressor to the cold head, (3) and a cold head for generating cooling for application. High-pressure helium gas is transferred to the cold head through the supply line. After expansion in the cold head, low-pressure helium is returned to the compressor through the return line.

The three components are equipped with Aeroquip self-sealing connectors, which allow them to be connected and disconnected at the Aeroquip connectors without losing helium or introducing contaminations to the system.

3.3.1 Compressor Package

A key engineering attribute of the GM cryocooler is the use of commercially available, mature, relatively inexpensive, oil-lubricated compressor modules used in air conditioning for the compression part of the cycle. This allows GM-type cryocoolers to directly benefit from the years of reliability and cost reduction of these mass-produced compressors. In addition, the DC-flow nature of these compressors allows them to be remotely located from the application cold load by few meters to few hundred meters, a big advantage in many applications. The one downside is that the modules have to be in a large compressor package.

The ability to use these oil-lubricated compressor modules results from the development of an efficient oil separator and an adsorber that reliably keep oil out of the cold head. The commercial compressor packages normally draw input powers from a few hundred watts to ~14 kW with single-phase or three-phase electrical supplies.

Figure 3.10 show a flow diagram of a water-cooled compressor package and its major components. It consists of three flow loops of helium gas, oil, and cooling water. The compressor module uses oil for lubrication and cooling.

1. *Helium gas loop*: A suction line (LP) brings low-pressure gas from the cold head to the compressor module (5). The high-temperature compressed gas containing carryover oil from the compressor module passes through aftercooler I (6) that transfers heat to the cooling water. The gas and oil leaving the aftercooler enter an oil separator (7).

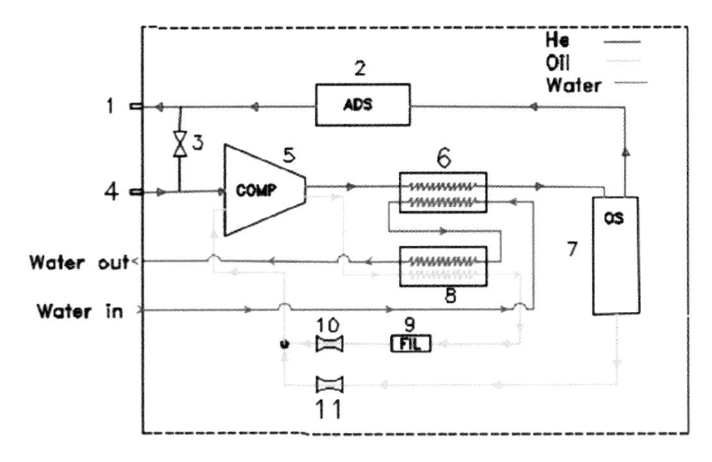

FIGURE 3.10 Flow diagram of a DC flow compressor package: (1) Aeroquip connector for HP gas. (2) Adsorber. (3) Bypass valve. (4) Aeroquip connector for LP gas. (5) Compressor module. (6) Aftercooler I (gas). (7) Oil separator. (8) Aftercooler II (oil). (9) Filter. (10–11) orifices.

The oil is separated from the compressed gas in the oil separator. The separated oil leaves the separator through an orifice (11), enters the oil return manifold, and returns to the compressor module. The orifice (11) meters the oil flow rate to match the amount of oil separated. The separated helium gas with a very small amount of oil mist enters an adsorber (2) to remove all of the oil residue. Thus, oil-free compressed gas can connect with the cold head through the high-pressure Aeroquip connector (1).

2. *Oil circulation loop*: There are normally two oil sumps in low- and high-pressure chambers in the compressor module. During the operation, the gas–oil mixture is compressed from the low-pressure side of the compressor module and discharged into the module housing where most of the oil separates from the gas and collects in the high-pressure sump. Oil from the high-pressure sump flows through the aftercooler II (oil) (8), then through a filter (9) and a metering orifice (10) in the oil return manifold. Thereby, oil is recirculated to the low-pressure sump. The filter blocks the solid particles in the oil to prevent the orifice blockage and compressor module damage. This oil removes most of the heat from the compressor module, lubricates and seals the compression parts, and cools the helium gas as it is being compressed.
3. *Cooling water loop*: The compressor package requires a large heat dissipation because of high-power input. The chilled water first flows through aftercooler I to remove heat from the helium gas, then through the aftercooler II to cool the circulation oil. The heat dissipation to the cooling water is almost the same as the power input to the compressor package.

Some monitoring components and safety protection components are also installed in the compressor package to prevent damage of the compressor module. The components include a bypass valve (3), pressure sensors, temperature sensors, and controllers.

The compressor package can be air-cooled instead of water-cooled. In the air-cooled compressor package, the aftercoolers I and II are finned heat exchangers. A fan is used to force air flowing through the aftercoolers to cool the gas and oil. The air-cooled compressor eliminates the chilled water supply. However, it dissipates heat to the environment and has higher acoustic noise due to the fan operation. Selection of air-cooled or water-cooled compressors should include consideration of the installation environment.

3.3.2 GM Cold Head

In commercial GM cryocoolers, the regenerator and the displacer are combined into one unit (called the displacer) as noted in Figures 3.11 and 3.12. The displacer body is made of non-metallic materials, such as linen phenolic or linen epoxy, with regenerator materials densely packed in the center.

A rotary valve instead of two switching valves is used in a GM cryocooler to chop the DC gas flow into an AC oscillating flow by alternately connecting the cold-head expander to the high- and low-pressure sides of the compressor. The rotary valve is coupled to the drive mechanism so that its operation is synchronized with the position of the displacer. More details of the rotary valve will be described in the following section.

A copper cap is thermally installed at the bottom of cold-head cylinder as a cooling station. The internal cold-heat exchanger for cold gas is a "gap" between the displacer and cold-head cylinder. The "gap" heat exchanger transfers heat from the cap when the cold gas flows through it in a cycle. With these practical designs, a GM cryocooler is compact, user friendly, and easy for manufacturing.

A single-stage GM cryocooler normally provides cooling capacity at temperatures between ~11 and 180 K. And a two-stage GM cryocooler usually provides the first-stage cooling capacity at temperatures of 25 K–100 K and the second-stage cooling capacity at a temperature of $\geq$ ~2.1 K. The lowest attainable temperature of the GM-type cryocooler is limited at where the thermal expansivity is = 0, which occurs at the density maximum at a temperature slightly above the helium Lambda line. The regenerators will be discussed in Section 3.4.

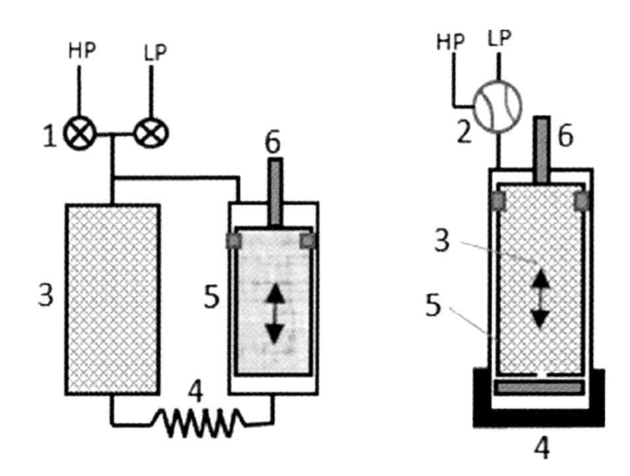

FIGURE 3.11 Practical, compact design of a single-stage GM cryocooler: (1) Switching valves. (2) Rotary valve. (3) Regenerator. (4) Heat exchanger. (5) Displacer. (6) Displacer drive.

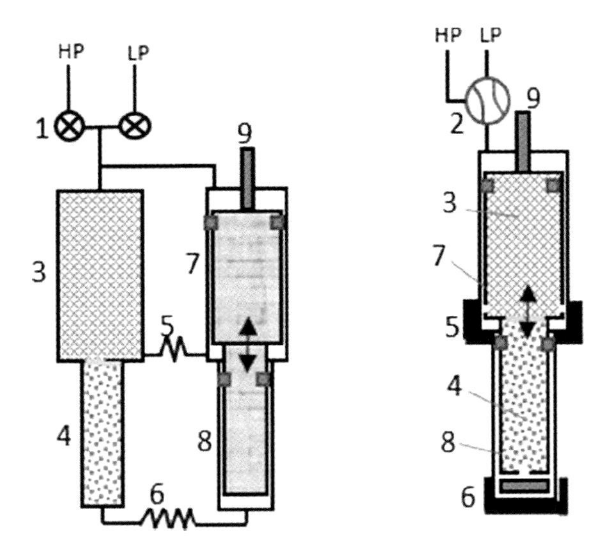

FIGURE 3.12 Practical, compact design of a two-stage GM cryocooler: (1) Switching valves. (2) Rotary valve. (3) First-stage regenerator. (4) Second-stage regenerator. (5) First-stage heat exchanger. (6) Second-stage heat exchanger. (7) First-stage displacer. (8) Second-stage displacer. (9) Displacer drive.

3.3.3 Displacer Drive

The displacer drive in Figures 3.11 and 3.12 can be a pneumatic drive or mechanical drive. Figure 3.13 shows the cross section of a mechanically driven GM cryocooler, also called a Scotch–Yoke-driven cryocooler. A low-speed motor (1) (normally 50 or 60 rpm) connects a crank (3) through the motor shaft (2). The crank connects the rotating part in a slot of the sliding yoke (4). A displacer (7) is directly coupled to the sliding yoke. The Scotch–Yoke, a reciprocating motion mechanism, converts the motor rotational motion into linear motion of the displacer.

The rotary valve system is furnished to control the helium gas intake and exhaust timing. It is also coupled to the drive motor through a crank, so that intake and exhaust

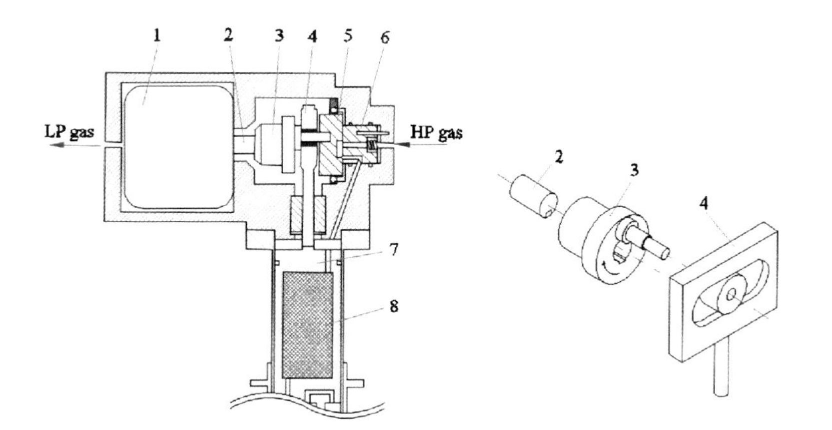

FIGURE 3.13 Cross section of a Scotch–Yoke-driven GM cryocooler: (1) Cold-head motor. (2) Motor shaft. (3) Crank. (4) Scotch–Yoke. (5) Rotary valve. (6) Valve plate. (7) Displacer. (8) Regenerator.

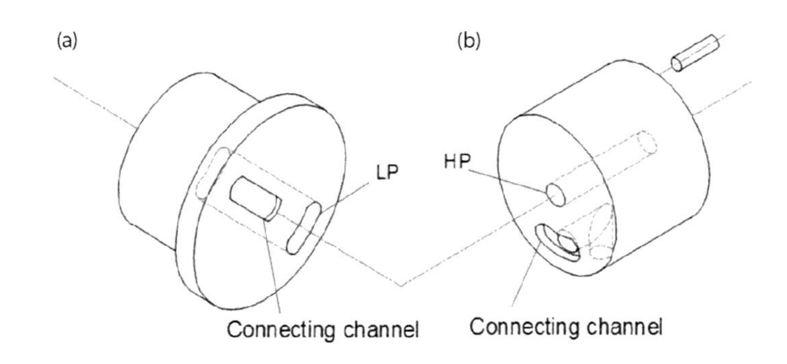

FIGURE 3.14 Rotary valve and valve plate for mechanically driven GM cryocooler: (a) Rotary valve. (b) Valve plate.

operations are synchronized with the position of the displacer. The required phase relationship between refrigerant pressure and mass flow is achieved by this synchronization.

A rotary valve and valve plate for a Scotch–Yoke-driven GM cryocooler is given in Figure 3.14. It is a single-rotation valve. This means that for every one rotation of the valve, the displacer undergoes one full cycle. It is different to the double-rotation valve, which is used for a pneumatically driven GM cryocooler (discussed later in this chapter). Operation of the single-rotation valve is illustrated in Figure 3.15. The valve connected the motor shaft spins on the valve plate. In the high-pressure period, the rotary valve connects the displacer to the high-pressure supply gas. In the low-pressure period, the rotary valve connects the displacer to the low pressure of return gas. The Scotch–Yoke-driven displacer is used in GM cryocoolers from SHI cryogenics and CTI cryogenics.

A pneumatically driven GM cryocooler is different from a mechanically driven GM cryocooler in that it uses an internal pressure differential to move the displacer. The driving force (pressure differential) acting on the displacer is obtained from in and out gas through a simple rotary valve. A cross section of a GM cryocooler with pneumatically driven displacer is shown in Figure 3.16.

The displacer (9) has a coaxial extension called a "stem" (7) to fit the stem cylinder. This creates two chambers I (6) and II (8) of variable volumes, depending on the position of the displacer. Chamber I is the volume between the top of the stem and the stem cylinder head. Chamber II is the volume between the top of the displacer and the cold-head cylinder. A gas seal is provided on the stem to prevent a gas leak between chamber I

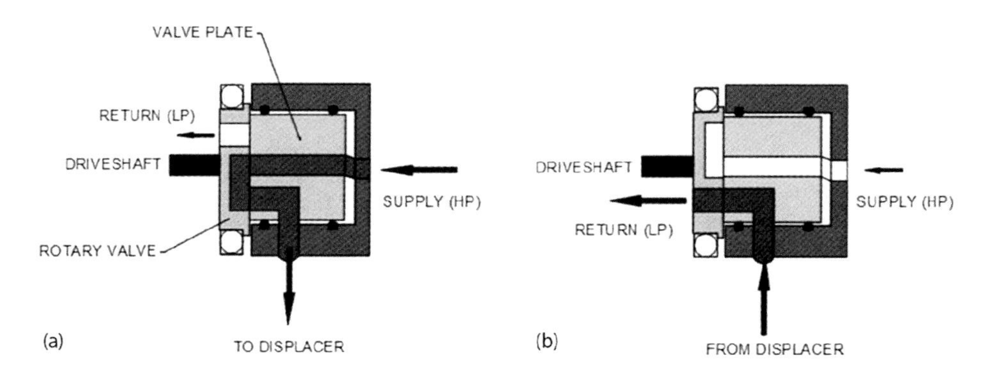

FIGURE 3.15 Operation of rotary valve for a mechanically driven GM cryocooler: (a) Discharge process. (b) Exhaust process.

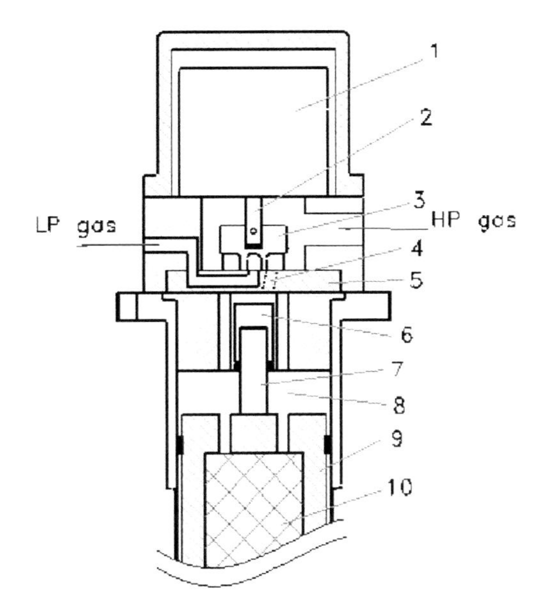

FIGURE 3.16 Cross section of a pneumatically driven GM cryocooler. (1) Cold-head motor. (2) Motor shaft. (3) Rotary valve. (4) Driving hole. (5) Valve plate. (6) Chamber I. (7) Stem. (8) Chamber II. (9) Displacer. (10) Regenerator.

and chamber II. There is a driving hole (4) connecting chamber I and the rotary valve assembly (3, 4) with proper timing. Gas is admitted and exhausted from chamber I and chamber II (regenerator conduit) though the rotary valve assembly.

The rotary valve and valve plate for the pneumatic-drive GM cryocooler is shown in Figure 3.17. It is a double-rotation valve. The displacer undergoes two full cycles for every one rotation of the rotary valve. The rotary valve spins on the valve plate, connecting the cold-head expander alternately to the high- and low-pressure sides of a compressor to generate the pressure variations. For the pneumatic-driven GM displacer, there is a drive hole on the valve plate (see Figure 3.17b). The position of the rotary valve and valve plate is synchronized with the motion of the displacer by timing the rotary valve connection to the drive hole and regenerator inlet.

Figure 3.18 shows an example of the rotary valve timing for gas cycling in a GM cryocooler. One rotation of the rotary valve is 360°, given on the X-axis. The grey line is for the HP or LP gas connecting to chamber 1. The black line is for the HP and LP gas connecting to the chamber 2 (regenerator). The proper valve timing is required to perform the four-steps of the GM cycle described in Section 3.2.1. The steps of "Pressure Buildup" and "Intake Stroke" take 160° of 360° for a cycle. At 50°, chamber 1 connects to the low pressure. The pressure differential drives displacer from the bottom to the top (Intake Stroke). The steps of "Pressure Release/Expansion" and "Exhaust Stroke" start from 160° to 360° At 253°, chamber 1 connects to the high pressure, and the displacer starts to move from the top to the bottom (Exhaust Stroke). Due to the difference of pressure drops in the intake and exhaust processes, the design of the rotary valve timing is slightly different from the theoretical number under the ideal conditions.

GM cryocoolers made by Cryomech, Inc., APD-SHI, Oerlikon Leybold Vacuum, etc. are the pneumatic-driven type.

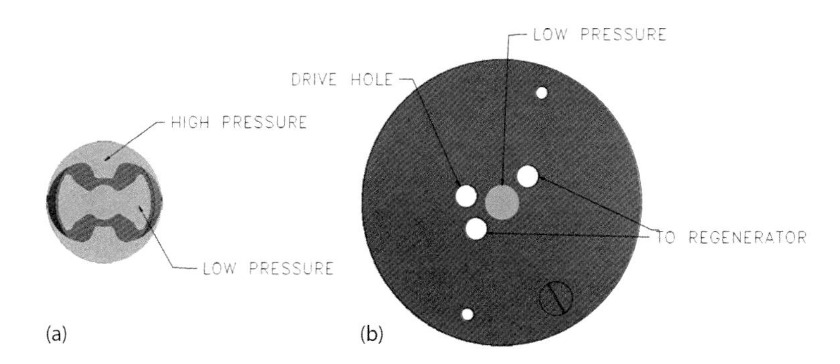

FIGURE 3.17 Rotary valve and valve plate for pneumatic-driven GM cryocooler: (a) Rotary valve. (b) Valve plate.

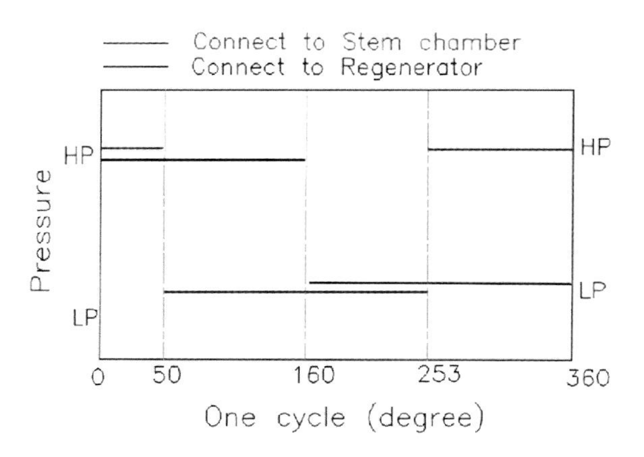

FIGURE 3.18 Rotary valve timing for a pneumatic-drive GM cryocooler.

3.3.4 Pulse Tube Cold Head

Most commercial PT cryocoolers use an orifice type. In this section, the orifice-type PT cryocooler will be introduced. The PT cryocooler employs the identical compressor and rotary valve (double-rotation valve) as used in a GM cryocooler.

The pulse tube should be carefully designed so that the helium gas doesn't mix or set up convective heat transfer loops, reducing the efficiency of the cycle. The pulse tube must be long enough so that the gas flow is reversed before the gas has traversed the length of the tube. The PT cold head could be designed in three configurations: U-type, coaxial type, and inline type.

Schematics of the U-type and coaxial type are given in Figure 3.19. The inline type for the cold head (see Figure 3.7) is the most energy efficient, but it has the disadvantage of having the cold region located between the two warm ends. So far, the inline type is only used in Stirling-type PT cryocoolers for space applications. Both U-type and coaxial-type PT cold heads are employed in GM-type PT cryocoolers, in which the cold head must be mounted with the cold end down to prevent natural convection losses caused by gravity in the pulse tube; see detailed discussion in Section 4.6. The warm-end heat exchanger (3), orifices (2), and a reservoir (1) are integrated together to dissipate heat to ambient. The rotary valve assembly including the motor 8 is almost the same as that in the GM cryocooler in Figure 3.17 without the drive hole.

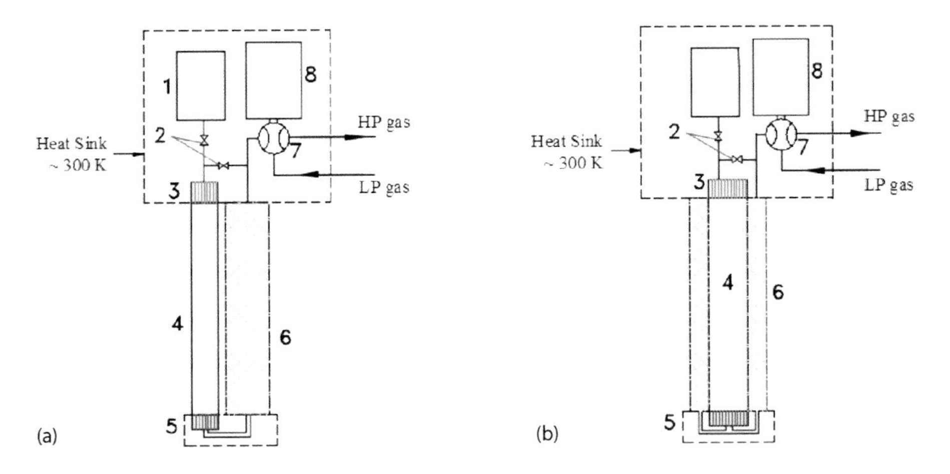

FIGURE 3.19 Two practical designs of a PT cryocooler: (a) U-type design. (b) Coaxial design. (1) Reservoir. (2) Orifices. (3) Warm-heat exchanger. (4) Pulse tube. (5) Cold-heat exchanger. (6) Regenerator. (7) Rotary valve. (8) Motor.

In the coaxial-type configuration, the pulse tube (4) is located inside of the "donut-shaped" regenerator (6). The coaxial type is compact, but loss associated with the heat transfer between the pulse tube and regenerator must be carefully managed in the design. The U-type is easier to build, but it has a bigger size than the coaxial-type. So far, commercial two-stage PT cryocoolers only employ the U-type configuration because of the design complexity and the heat-transfer losses associated with the coaxial type.

Figure 3.20 shows schematics of two-stage PT cryocoolers. The first and second stages have independent phase shifters of orifices and reservoirs. The first-stage pulse tube (12) and second-stage pulse tube (13) are arranged in parallel. The second-stage pulse tube extends from 4 K to room temperature. This design allows the second-stage phase shifter to mount at the warm end for easy access. Both the first- and second-stage warm-heat exchangers (10,11) can be integrated at the same warm end to dissipate heat. The standard version of the two-stage PT cryocooler integrates the rotary valve/motor assembly at the warm end of the pulse tube cold head; see Figure 3.20a.

A unique feature of the PT cryocooler is the ability to separate the rotary valve/motor assembly from the PT expander. The configuration is called "Remote Motor" in Figure 3.20b. The rotary valve/motor assembly is connected to the PT expander through a SS flexible line (15) of ~1 meter. An electrical isolator (14), made of non-metallic material, is installed between the rotary valve/motor assembly and the SS flexible line to isolate the electromagnetic interference (EMI) introduced by the cold-head motor (1) and the compressor. The vibrations, generated by the cold-head motor (1) and rotary valve (2), as well as the large flexible lines connecting to the compressor, are significantly reduced through the small flexible line (15). The EMI and vibrations from cryocoolers are two major concerns for most sensitive applications. The cooling performance of this remote motor version is approximately 5%–10% less than that of the standard version. The remote motor PT cryocoolers have been used for cooling many sensitive devices, such as SQUID magnetometers, NMRs, MRIs, precooling dilution refrigerators, and ADRs.

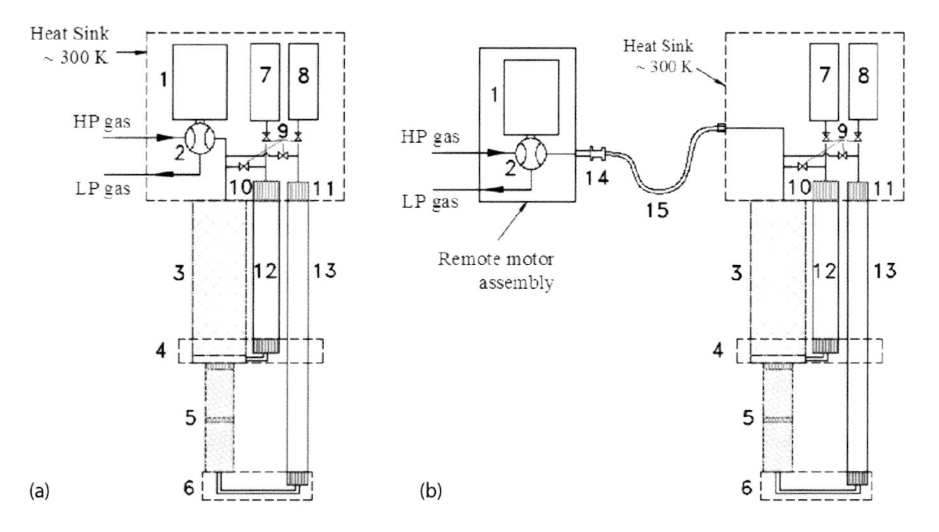

FIGURE 3.20 Schematic of a two-stage PT cryocooler: (a) Standard. (b) Remote motor. (1) Motor. (2) Rotary valve. (3) First-stage regenerator. (4) First-stage heat exchanger. (5) Second-stage regenerator. (6) Second-stage heat exchanger. (7, 8) First- and second-stage reservoirs. (9) Orifices. (10–11) First- and second-stage warm-heat exchangers. (12–13) First- and second-stage pulse tubes. (14) Electrical isolator. (15) Remote motor line.

3.4 Development of 4 K Cryocoolers

3.4.1 4 K Regenerators

The regenerator is a very important component for the cryocooler efficiency and lowest attainable temperature. An effective regenerator should have following features:

1. Much higher heat capacity of the regenerator matrix than that of the gas;
2. Large surface area of the matrix for heat transfer between the gas and the matrix;
3. Low axial thermal conductivity along the regenerator matrix;
4. Low gas flow resistance in the regenerator; and
5. Low void volume of the matrix.

The first stage regenerator in a two-stage cryocooler normally uses a regenerative material made of phosphor bronze screen or SS screen of 150–200 mesh. Before the rare earth regenerative materials were used in the cryocoolers, the second-stage regenerator was filled with lead spheres. The lead spheres were, later on, replaced by tin or bismuth spheres when the Restriction of Hazardous Substances Directive (RoHS 1) took effect on July 1, 2006 in the European Union.

The regenerator ineffectiveness is closely related to the capacity ratio C_r/C_g, which is given by

$$\frac{C_r}{C_g} = \frac{m_r c_{pr}}{\dot{m}_g c_{pg} \theta_H} \tag{3.7}$$

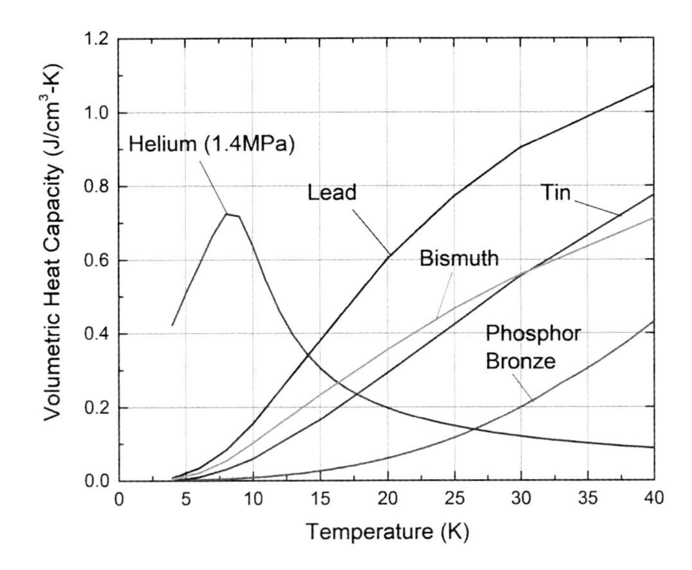

FIGURE 3.21 Specific heat of helium gas and regenerative materials.

where C_r is the heat capacity of the regenerator matrix, C_g is the heat capacity of the gas flowing per pulse period, m_r is the mass of matrix, C_{pr} is the specific heat of the matrix, $\dot{m}_g$ is the mass flow rate of the gas, C_{pg} is the specific heat of the gas, and θ_H is the period of heating. C_r should be few times higher than C_g for an effective regenerator [13].

Figure 3.21 shows specific heat of helium gas and a few regenerative materials. The pressure of helium gas is at 1.4 MPa, which is close to an average pressure in a cryocooler. Due to very low specific heat of lead and the real gas effects of helium gas below 10 K, the ratio of C_r/C_g is very small (<<1). The poor regenerator effectiveness with the lead material results in the lowest attainable temperature of ~7 K for two-stage regenerative cryocoolers.

3.4.2 Early 4 K Cryocooler Development

Development and commercialization of small superconducting devices around 1980 required closed-cycle cryocoolers that permit such a device to be maintained at liquid helium temperature continuously. Hybrid GM/JT cryocoolers for 4 K operation were developed at ~10 organizations at that time.

Figure 3.22 shows a flow schematic of the GM/JT hybrid cycle, used by Air Products and Chemicals, Inc. (APCI) [14]. A two-stage GM cryocooler consists of the compressor and cold head, which operates with a helium high pressure of ~2.2 MPa and low pressure of 0.8 MPa. The first stage and second stage of the cold head are used to precool helium in the J-T loop, which consists of three counterflow heat exchangers, three heat exchangers at the temperatures of ~70 K, ~16 K, and ~4 K, respectively, as well as a JT valve and a JT compressor. The JT compressor brings the return helium pressure at ~ 0.1 MPa up to the return pressure of the GM cold head at ~0.8 MPa. One APCI model, CS-308, had cooling capacities of 2.5 W at 4.5 K plus 2 W < 20 K and 10 W < 77 K.

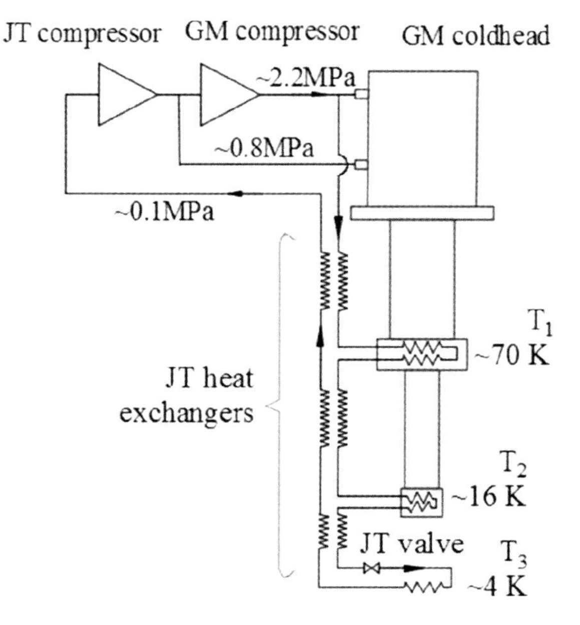

FIGURE 3.22 Schematic of a GM/JT hybrid 4 K cryocooler.

It was envisioned as being suitable for cooling the superconducting Josephson Junction device computer being developed by IBM [15]. Reliability of this hybrid 4 K cryocooler was a big concern in the test program because of the JT valve applied. The use of a 4 K superconducting computer did not materialize. Attempts to use the GM/JT hybrid 4 K cryocooler for recondensing helium in MRIs proved impractical because of the poor reliability and high cost [16]. The GM/JT hybrid 4 K cryocooler essentially doubles the complexity of the cryocooler. The system is bulky and complicated, which results in less reliability, high manufacture cost, and difficulty for system integration. The hybrid systems were not commercially successful, and their productions were discontinued.

Magnetic resonance imaging (MRI) had come into widespread use in the 1980s. Most of the MRIs at that time incorporated superconducting magnets, which required a cooling capacity between 0.5 and 2 W at 4.2 K plus additional cooling at higher temperatures to cool the radiation shield and current leads. The requirements of MRI had spurred more development of 4 K cryocoolers.

3.4.3 Development of Rare Earth Regenerative Materials

In general, the specific heat of a normal solid material is contributed by lattice vibration and conduction electrons. The volumetric specific heat can be simply expressed as:

$$C_v = 234\left(\frac{T}{\theta_D}\right)^3 \tag{3.8}$$

where C_v is the volumetric specific heat, T is the material temperature, and θ_D is the Debye temperature. Most metals (including lead) have a lowest θ_D of ~100 K. The specific heat, C_v, is proportional to T^3; thus, the heat capacities at 4–10 K are very low.

It has been necessary to find a material whose specific heat could be contributed by other mechanisms. An entropy effect associated with magnetic phase transition could produce an anomaly in the specific heat around the phase transition temperature. Based on thermodynamics, the specific heat is

$$C_v = T\left(\frac{\partial S_J}{\partial T}\right) \tag{3.9}$$

where S_J is the magnetic entropy. It is well-known that large heat capacities occur when the material has the order-to-disorder magnetic phase transition. At the transition temperature of T_c, the S_J changes suddenly from the order phase ($S_J = 0$) to the disorder phase (large value of S_J). Considering Equation 3.9, the magnetic specific heat will have large peak value near the transition temperature T_c.

Daniels et al. first noticed a rare earth material, europium sulfide (EuS), which shows a high-peak specific heat due to the magnetic phase transition at ~16 K and has a specific heat higher than lead at the temperatures between 4 and 16 K [17]. They tested EuS as the regenerative material in a three-stage Stirling cryocooler in 1971. The lowest temperature of the Stirling cryocooler was reduced from 9.0 to 7.8 K by replacing the lead spheres with EuS material.

In 1975, Buschow et al. demonstrated the pseudobinary compounds $Gd_x\ Er_{1-x}\ Rh$ to have high volumetric heat capacities as compare to lead in the temperature range of 4–10 K due to the magnetic ordering transition of rare earth materials below 20 K [18]. However, these materials are too costly to use in commercial products because of the high cost of rhodium. Figure 3.23 shows the specific heat of early rare earth materials of EuS, GdRh, and $Gd_{0.5}Er_{0.5}Rh$.

The intermetallic compounds of rare earth materials, $Er_{1-x}Dy_xNi_2$ and $Er(Ni_{1-x}Co_x)_2$, were introduced in 1988 by Prof. Hashimoto's group at Tokyo Institute of technology [19]. They have large specific heats below 20 K because of magnetic transition. Soon after, the rare earth materials of Er_3Ni, ErNi, $ErNi_{0.9}Co_{0.1}$, and $HoCu_2$ were discovered to have large heat capacities and

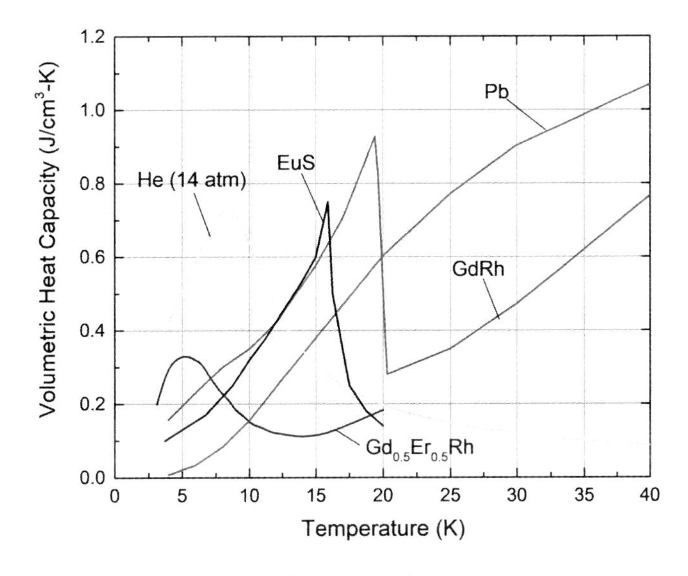

FIGURE 3.23 Specific heat of EuS, GdRh, and $Gd_{0.5}Er_{0.5}Rh$.

were applied to the second-stage regenerator in GM cryocoolers [20–22]. These works inspired worldwide developments of 4 K cryocoolers using rare earth regenerative materials.

A group of ceramic rare earth materials was proposed for an adiabatic demagnetization refrigerator in 1984 by J. A. Barclay [23]. The ceramic rare earth materials of $GaAlO_3$ (GAP) and Gd_2O_2S (GOS) were investigated and introduced for a 4 K cryocooler by Numazawa et al. in 2000 and 2003 [24,25]. The GAP and GOS show the magnetic transition at the temperatures of 3.8 and 5.2 K, respectively, and very high specific heat near 4 K. Small spheres of polycrystal GAP and GOS were developed and fabricated by using a chemical process. Both GM and PT cryocoolers used GOS in the second-stage regenerator and demonstrated higher cooling capacity and efficiency at 4 K. The GOS also has good magnetic properties with less magnetic distortion to instruments and less magnetic field dependence of heat capacity in high magnetic field (see introductions in Sections 4.4 and 4.5).

There are some basic requirements for rare earth regenerative materials used in commercial 4 K cryocoolers:

1. High specific heat from 4 to 10 K. The specific heat in this temperature range is very important for the regenerator efficiency;
2. Enough mechanical strength. The rare earth materials should be strong enough to withstand the acceleration of displacer and gas pulsation to ensure reliable, long-term operation of cryocooler; and
3. Antiferromagnetic materials. Antiferromanetic materials have less degradation of specific heat in high magnetic field (up to few Tesla) than ferromagnetic materials.

Based on these requirements, $HoCu_2$, Er_3Ni, and GOS have been chosen to be used in commercial products. Their specific heats are given in Figure 3.24.

Figure 3.25 shows three packing configurations of the second-stage regenerator in commercial 4 K cryocoolers. The regenerator in Figure 3.25a was used in SHI 4 K GM cryocoolers and might be still used in some SHI 4 K models. The regenerator in Figure 3.25b

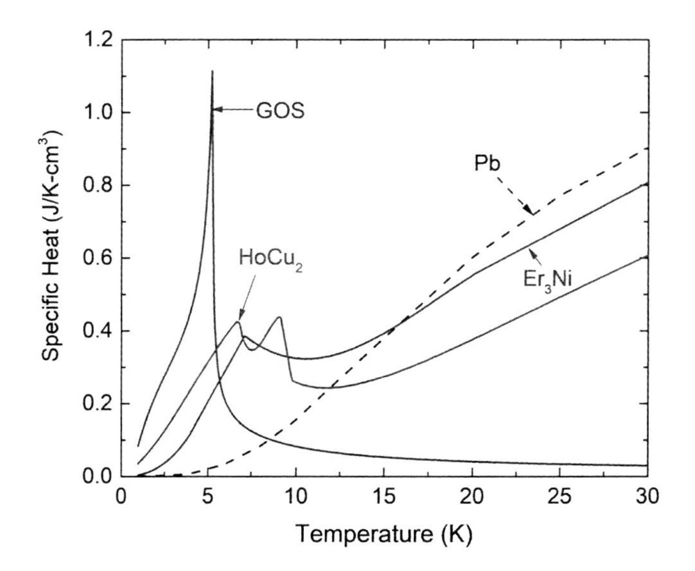

FIGURE 3.24 Specific heat of rare earth materials used in commercial 4 K cryocoolers.

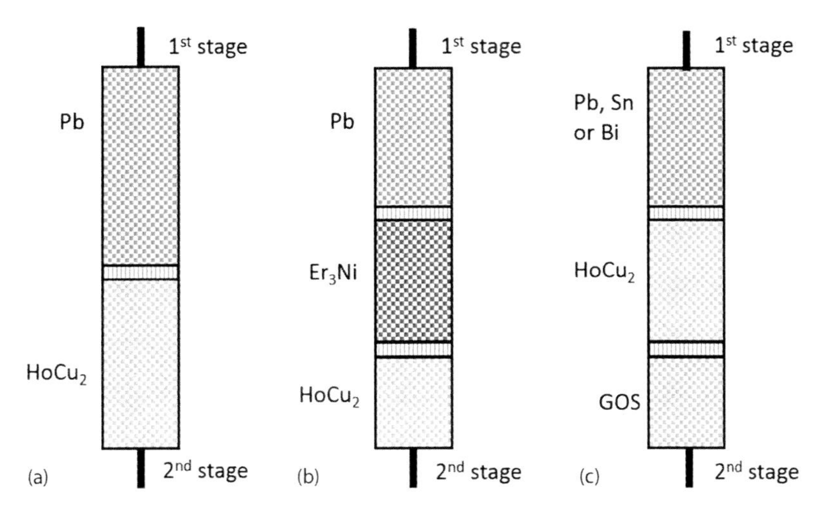

FIGURE 3.25 Regenerative materials packed in the second-stage regenerator: (a) packed with Pb and $HoCu_2$; (b) packed with Pb, Er_3Ni and $HoCu_2$; and (c) packed with Pb (or Sn or Bi), $HoCu_2$ and GOS.

was used in Cryomech 4 K PT cryocoolers. The regenerator with GOS, $HoCu_2$, and Pb (Sn or Bi) in Figure 3.25c is used in the latest 4 K cryocoolers. The packings of regenerative materials are based on the temperature distribution of the second-stage regenerator and their specific heat capacities at different temperatures. GOS spheres are located at a lower temperature section, $HoCu_2$ spheres are packed in the middle section, and Pb (or Sn or Bi) spheres are at an upper section close to the first stage.

3.4.4 Development and Commercialization of 4 K GM Cryocoolers

The intermetallic rare earth materials are very attractive to use in a 4 K cryocooler because of their specific heats and acceptable prices. Many companies and research laboratories had research and development activities on 4 K GM cryocoolers in the late 1980s and in 1990s, which include Toshiba Corporation, Sumitomo Heavy Industries, Mitsubishi Electric Corporation, US David Taylor research center, Balzer Inc., Leybold GmBH, and so on. Two-stage GM cryocoolers were improved to have cooling capacities of >1 W at 4.2 K by using regenerative materials of Er_3Ni, $ErNi_{0.9}Co_{0.1}$, and $HoCu_2$, which were powerful enough for a lot of applications, including medical MRIs.

However, reliability of 4 K GM cryocoolers as a commercial product became a big challenge due to the cryogenics seal issue for the second-stage displacer. Ductility and mechanical strength of the intermetallic rare earth regenerative material were also concerns. A traditional 10 K cryocooler is not sensitive to a performance degradation by a few tenths of Kelvin. But for the 4 K cryocoolers, a small temperature degradation of a few tenths of Kelvin will significantly impact performance at 4.2 K because of the small temperature differential between the no-load temperature (~3.0 K) and 4.2 K.

An early commercial 4 K GM cryocooler of 1 W at 4.2 K was made by Balzers Inc. by using Nd spheres in the second-stage regenerator. The Nd spheres are easy to produce and have as strong of mechanical strength as lead spheres. But it has unstable chemical properties when exposed to an air and moisture environment. The 4 K GM cryocooler had faced issues of long-term reliable operation because of the second-stage cryogenic seal and unstable chemical properties of Nd. This 4 K GM cryocooler was discontinued after losing the battle to other commercial 4 K GM cryocoolers.

Various efforts had been made by GM cryocooler manufacturers to improve the second-stage displacer seals. In 1994, an innovative sealing technology of adding a spiral groove on the second-stage displacer was developed by Sumitomo Heavy Industries [26]. Figure 3.26 shows a traditional displacer seal and the new spiral groove on the second-stage displacer. Figure 3.27 shows the photo of the new displacer and cold head from Sumitomo Heavy Industries. The new displacer was made of metal instead of non-metallic material of linen phenolic. It was found that using the

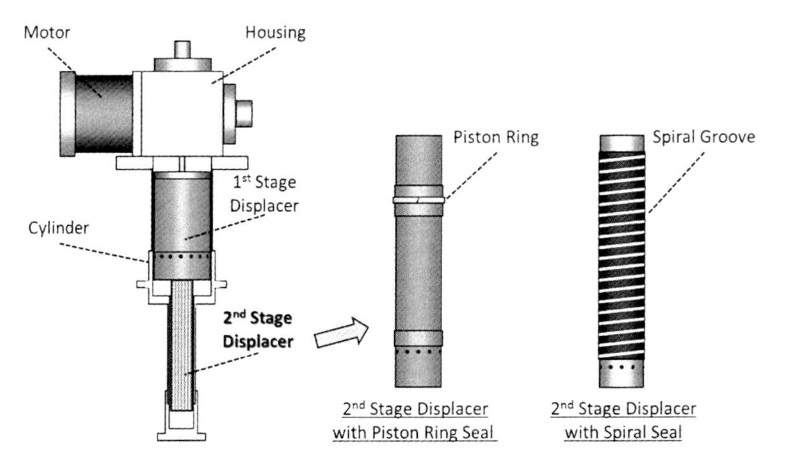

FIGURE 3.26 Replacing the second-stage displacer seal with a spiral groove in SHI 4 K GM cryocooler. (Reprinted with permission from Ikeya, Y., Expanding market of 4 K GM cryocooler, *Presented at 24th IIR International Congress of Refrigeration*, 2015.)

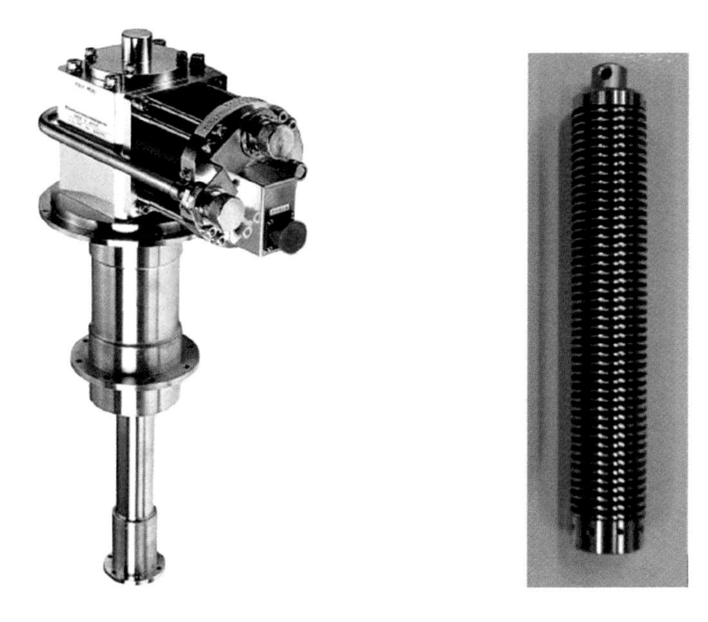

FIGURE 3.27 Photos of a SHI 4 K GM cryocooler with the novel second-stage displacer. (Courtesy of SHI Cryogenics with permission from Ikeya, Y., Expanding market of 4 K GM cryocooler, *Presented at 24th IIR International Congress of Refrigeration*, 2015.)

spiral groove to replace the displacer seal resulted in drastically improving of the 4 K temperature stability and cooling performance [27].

The second-stage displacer has no seal between the first-stage expansion chamber and the second-stage expansion chamber. The spiral groove design allows a very small portion of gas flow through the long path of the spiral groove. The large surface area of the spiral groove provides enough heat transfer between gas and the displacer surface. Therefore, the "leakage" gas through the spiral groove has the similar temperatures as that of the majority gas flow through the regenerator at the both ends of the regenerator. So that the thermal loss of the gas flow in the spiral groove is negligible when compared to the performance with the displacer seal. The spiral groove eliminates the failure mode of the displacer seal and ensures long-term operation of the 4 K GM cryocooler.

Figure 3.28 shows good temperature stability of the SHI 4 K GM cryocooler with the new second-stage displacer over 15,000 hours. This innovative design made the SHI 4 K GM cryocoolers commercially successful. The first large application of SHI 4 K GM cryocoolers is for the GE zero boil-off MRI [16]. Developments of 4 K GM cryocoolers started to slow down at the late of the 1990s.

Table 3.1 lists all models and performances of 4 K GM cryocoolers manufactured by SHI Cryogenics. Currently, only a few manufactures produce 4 K GM cryocoolers. Ulvac Cryogenic Inc. manufactures five models of 4 K GM cryocoolers from 0.1 to 1.5 W at 4.2 K. Advanced Research Systems, Inc. manufactures four models from 0.1 to 1.5 W at 4.2 K. CSIC Pride Cryogenic Technology Co. manufactures four models from 0.25 to 1.8 W at 4.2 K.

3.4.5 Development and Commercialization of 4 K Pulse Tube Cryocoolers

There are a few configurations for staging PT cryocoolers. Choosing the right staging configuration is a crucial to developing a 4 K PT cryocooler. An important staging configuration of the PT cryocooler was first suggested by Chan and Tward [28] by considering that a two-stage PT cryocooler could work as a two-stage GM cryocooler; see Figure 3.12. An old Cryomech model GB04, two-stage GM cryocooler had employed this configuration. In this configuration for the pulse tube, the phase shifter of the second-stage pulse tube can be located at room temperature, in which, the heat generated by the second-stage pulse tube can be dissipated to the ambient. The design is convenient for optimizing the phase shifter at room temperature. Figure 3.29 shows milestones of early development of 4 K PT cryocoolers with different stages and configurations.

A temperature below 4 K was first obtained in 1994 by Matsubara and Gao with a three-stage PT cooler [29]. The schematic of it is given in Figure 3.29a. The regenerators

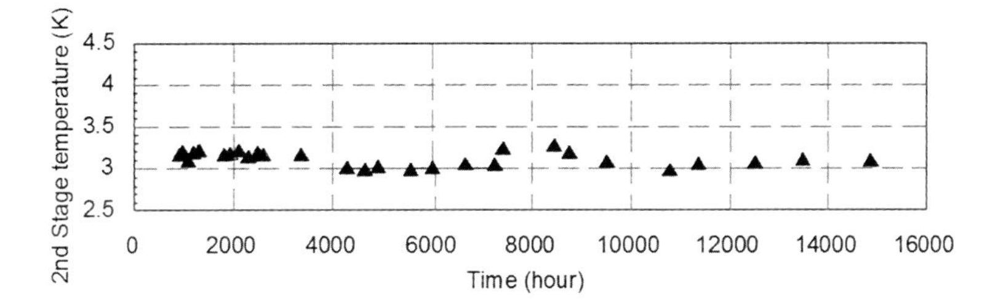

FIGURE 3.28 A long-term test result of a SHI 4 K GM cryocooler. (Date from reference Ikeya, Y., Expanding market of 4 K GM cryocooler, *Presented at 24th IIR International Congress of Refrigeration*, 2015.)

Table 3.1 Model and Performance of 4 K GM Cryocoolers Manufactured by SHI Cryogenics

		RDK-101D	RDK-305D	RDK-205D	RDK-408D2	RDE-412D4	RDK-415D
1st stage	50 Hz	3.0 W@60 K	15 W@40 K	3.0 W@50 K	40 W@43 K	53 W@43 K	35 W@50 K
Q_c	60 Hz	5.0 W@60 K	20 W@40 K	4.0 W@50 K	50 W@43 K	60 W@43 K	45 W@50 K
2nd stage Q_c		0.1 W@4.2 K	0.4@4.2 K	0.5 W@4.2 K	1.0 W@4.2 K	1.25 W@4.2 K	1.5 W@4.2 K
No-load T_c		<3.0 K	<3.5 K	<3.5 K	<3.5 K	<3.5 K	<3.5 K
Power input		~1.3 kW	~4.2 kW	~4.2 kW	~7.5 kW	~7.5 kW	~7.5 kW

Note: Q_c, cooling capacity; T_c, cooling temperature; second-stage Q_c is the same with 50 and 60 Hz power. The power input is an average for 50 and 60 Hz.

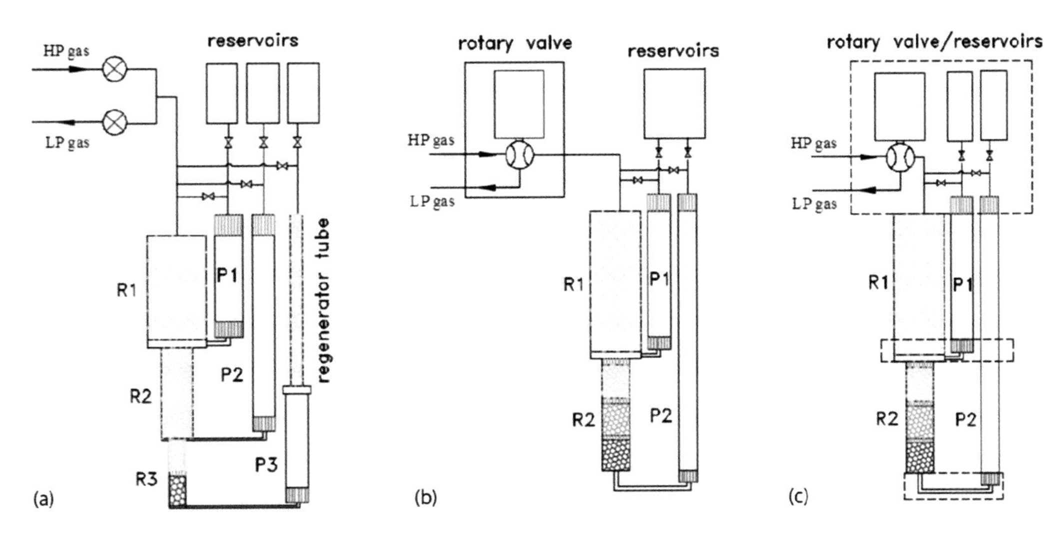

FIGURE 3.29 Schematics of three milestone 4 K PT cryocoolers: (a) The first three-stage PT cryocooler below 4 K. (b) The first two-stage PT cryocooler below 4 K. (c) The first commercial 4 K pulse tube cryocooler.

of the three stages were packed with various regenerative materials: phosphor bronze screen for the first stage, lead spheres for the second stage, and lead and Er_3Ni for the third stage. A special regenerator portion filled with lead spheres was added at a warmer section of the third-stage pulse tube to help "heat pumping." The three-stage PT cryocooler reached a no-load temperature of 3.6 K and 30 mW at 4.2 K.

In 1997, a two-stage 4 K PT cryocooler with the configuration in Figure 3.29b was first developed by Wang, Thummes, and Heiden at the University of Giessen [30]. The first-stage and second-stage pulse tubes shared one reservoir volume. The first-stage regenerator was packed with a SS screen, and the second-stage regenerator was packed with three layers of material, $ErNi_{0.9}Co_{0.1}$, ErNi, and lead spheres, from the 4 K station to the first station. The two-stage PT cryocooler could reach a no-load temperature of 2.23 K and provide 370 mW at 4.2 K with a power input of 6.3 kW.

In 1999, the world's first two-stage 4 K PT cryocooler was commercialized at Cryomech, Inc., model PT405, which can provide 0.5 W at 4.2 K with a power input of 4.7 kW [31]. The schematic of the 4 K PT cryocooler is given in Figure 3.29c. For the regenerators in this PT cryocooler, the first stage has 200 mesh phosphor bronze screens, and the second stage has three layers of regenerative material, lead, Er_3Ni, and $HoCu_2$. The first- and second-stage pulse tubes are connected with two separated reservoirs through orifices. The PT cryocooler had the design of a commercial product to provide user friendly first- and second-stage cooling stations. Two reservoirs, four orifices, and a rotary valve, controlling the helium flow in the cold head, are integrated inside the warm-end assembly. After that, the compact warm-end design and cooling station design became the standard for 4 K PT cryocoolers. Cooling performances of three milestone developments of 4 K PT cryocoolers are given in Figure 3.30. Figure 3.31 is a photo of the first commercial 4 K PT cryocooler.

Later on, a series of 4 K PT cryocoolers from 0.25 to 2.0 W at 4.2 K (models PT403, PT405, PT407, PT410, PT415, and PT420), have been developed and commercialized at Cryomech, Inc. These 4 K PT cryocoolers have efficiencies that are close to that of the 4 K GM cryocoolers.

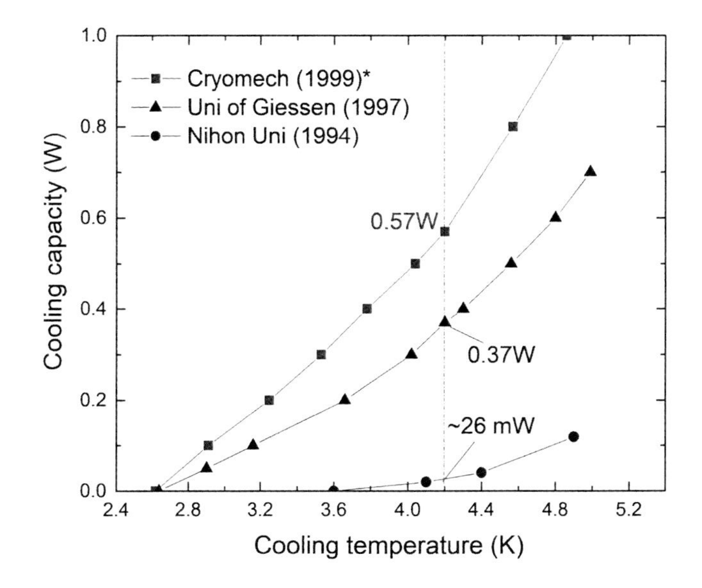

FIGURE 3.30 Performance improvement in three milestones of 4 K PT cryocooler development. *Commercial product, model PT405 from Cryomech.

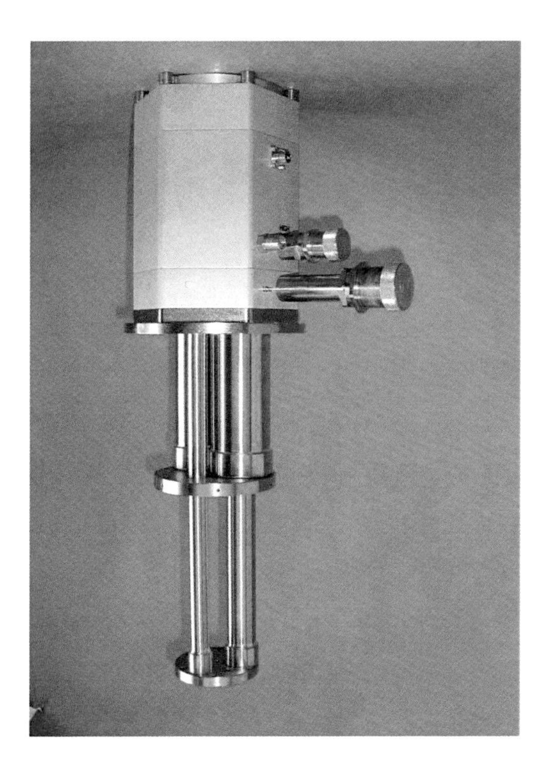

FIGURE 3.31 Photo of the first commercial 4 K PT cryocooler, Cryomech model PT405. (Courtesy of Cryomech, Inc.)

4 K PT cryocoolers are so attractive for cooling low-temperature superconductors and some sensitive sensors that quite a few cryogenic industries have been involved in the development in late 1990 and early 2000, such as Leybold Vacuum Ltd., Sumitomo Heavy Industry, APD cryogenic, Oxford Magnet Technology, Daikin Industries Ltd., and Aisin Seiki Co., as well as several research organizations. So far, Cryomech, Inc. and Sumitomo Heavy Industry have successfully developed and commercialized a few models of 4 K PT cryocoolers. Tables 3.2 and 3.3 list models and performances of these 4 K PT cryocoolers.

Typical cooling load maps of a 1 W at 4 K PT cryocooler, Cryomech model PT410, are given in Figure 3.32. Figure 3.32(a) is a load map near 4 K, and (b) is a cooling load map from bottom temperature to room temperature. The load maps are useful to design a cryogenic system and estimate cool-down time of the whole cryogenic system.

Table 3.2 Model and Performance of 4 K Pulse Tube Cryocoolers Manufactured by Cryomech, Inc.

	PT403	PT405	PT407	PT410	PT415	PT420
1st stage Q_c	7 W@65 K	25 W@65 K	25 W@55 K	40 W@45 K	45 W@45 K	50 W@45 K
2nd stage Q_c	0.25 W@4.2 K	0.5 W@4.2 K	0.7 W@4.2 K	1.0 W@4.2 K	1.5 W@4.2 K	2.0 W@4.2 K
No-load T_c	<2.8 K	<2.8 K	<2.8 K	<2.8 K	<2.8 K	<2.8 K
Power input	~3.0 kW	~4.8 kW	~7.0 kW	~8.0 kW	~10.0 kW	~12.0 kW

Source: Product catalog from Cryomech Inc., http://www.cryomech.com/.
Note: Q_c, cooling capacity; T_c, cooling temperature. The power input is an average for 50 and 60 Hz.

Table 3.3 Model and Performance of the 4 K Pulse Tube Cryocoolers Manufactured by SHI Cryogenics

	RP-062BS	RP-082B2	RP-182B2S
1st stage Q_c	30 W@65 K	40 W@45 K	36 W@48 K
2nd stage Q_c	0.5 W@4.2 K	1.0 W@4.2 K	1.5 W@4.2 K
No-load T_c	<3.0 K	<3.0 K	<2.8 K
Power input	~7.4 kW	~7.6 kW	~14.0 kW

Source: Product catalog from SHI cryogenics, http://www.shicryogenics.com/.
Note: Q_c, cooling capacity; T_c, cooling temperature. The power input is an average for 50 and 60 Hz.

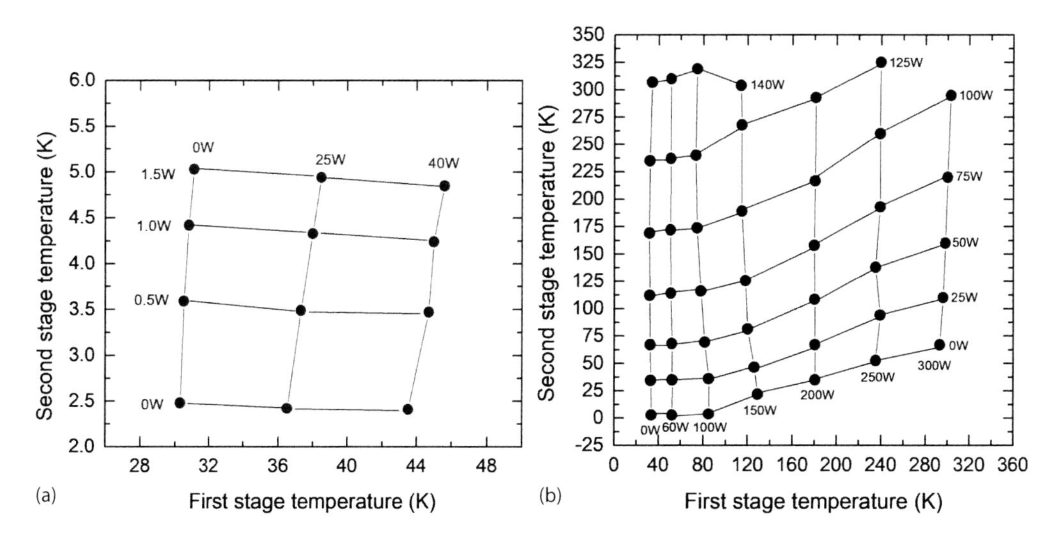

FIGURE 3.32 Cooling load maps of a 1 W@4 K PT cryocooler, Cryomech Model PT410: (a) Load map near 4 K. (b) Load map from bottom temperature to room temperature.

Acknowledgments

The author would like to thank Mr. Jeff Raab, retired from Northrop Grumman, and Dr. Peter Kittel, retired from NASA Ames Research Center, for reviewing and improving the manuscript.

References

1. Walker, G., *Cryocoolers, Part 1 and 2*, International Cryogenics Monographs Series, Plenum Press, New York (1983).
2. Walker, G. and Bingham, E. R., *Low-capacity Cryogenic Refrigeration*. No. 9, Oxford University Press, Oxford (1994).
3. McMahon, H. O. and W. E. Gifford., "A new low temperature gas expansion cycle-Part I," *Advances in Cryogenic Engineering*, vol. 5, Plenum Press, New York (1960), pp. 354–367.
4. Gifford, W. E. and McMahon, H. O., "A new low-temperature gas expansion cycle—Part II," *Advances in Cryogenic Engineering*, vol. 5, Plenum Press, New York (1960), pp. 368–372.
5. Gifford, W. E. and Longsworth, R. C., "Pulse tube refrigeration," *Journal of Engineering for Industry*, 86.3 (1964), pp. 264–268.
6. Mikulin, E. I., Tarasov, A. A. and Shkrebyonock, M. P., "Low-temperature expansion pulse tubes," *Advances in Cryogenic Engineering*, vol. 29, Plenum Press, New York (1984), pp. 629–637.
7. Radebaugh, R., Zimmerman, J., Smith, D. R., and Louie, B., "A comparison of three types of pulse tube refrigerators: New methods for reaching 60 K," *Advances in Cryogenic Engineering*, vol. 31, Plenum Press, New York (1986), pp. 779–789.
8. Zhu, S. W., Wu, P. Y. and Chen, Z. Q., "Double inlet pulse tube refrigerators: An important improvement," *Cryogenics*, vol. 30 (1990), pp. 514–520.
9. Product catalog from Cryomech, Inc. Website: http://www.cryomech.com/.
10. Matsubara, Y, Gao, J. L., and Tanida, K., "An experimental and analytical investigation of 4 K pulse tube refrigerator," *7th International Cryocooler Conference Proceedings*, Air Force Phillips Laboratory Report PL-CP--93-1001, Kirtland Air Force Base, NM, April 1993, pp. 166–186.
11. Product catalog from SHI cryogenics group. Website: http://www.shicryogenics.com/.

12. Wu, P. Y. and Zhu, S. W., "Mechanism and numerical analysis of orifice pulse tube refrigeration with a valveless compressor," *Proceedings of International Conference of Cryogenics and Refrigeration*, International Academic Publishers (1989), pp. 85–90.
13. Gifford, W.E and Acharya, A., "Low-temperature regenerator test apparatus," *Advances in Cryogenic Engineering*, vol. 15, Plenum Press, New York (1976), pp. 436–442.
14. Longsworth, R. C., "Serviceable refrigerator system for small superconducting devices," *Refrigeration for Cryogenic Sensors and Electronic Systems*, vol. 607, US Department of Commerce, National Bureau of Standards (1981), pp. 82–92.
15. Flint, E. B., Jenkins, L. C. and Guernsey, R. W., "Performance of a 1 Watt 4 K cryosystem suitable for a superconducting computer," *Refrigeration for Cryogenic Sensors and Electronic Systems*, vol. 607, US Department of Commerce, National Bureau of Standards, Gaithersburg, MD (1981), pp. 93–102.
16. Ackermann, R., Herd, G., and Chen, W., "Advanced cryocooler cooling for MRI systems," *Cryocoolers 10*, Kluwer Academic/Plenum Publishers, New York (1999), pp. 857–867.
17. Daniels, A. and du Pre, F. K., "Triple-expansion Stirling-cycle refrigerator," *Advances in Cryogenic Engineering*, vol. 16, Plenum Press, New York (1971), pp. 178–184.
18. Buschow, K. H. J., J. F. Olijhoek and Miedema, A. R., "Extremely large heat capacities between 4 and 10 K," *Cryogenics*, vol. 15 (1975), pp. 261–264.
19. Li, R., Ogawa, M. and Hashimoto, T., "Magnetic intermetallic compounds for cryogenic regenerator," *Cryogenics*, vol. 30 (1990), pp. 521–526.
20. Sahashi, M., et al., "New magnetic material R_3T system with extremely large heat capacities used in heat regenerators," *Advances in Cryogenic Engineering*, vol. 35, Plenum Press, New York (1990), pp. 1175–1182.
21. Nagao, M., Inaguchi, T. and Yoshimura, H., et al., "Helium liquefaction by a Gifford-McMahon cycle cryocooler," *Advances in Cryogenic Engineering*, vol. 35, Plenum Press, New York (1990), pp. 1251–1260.
22. Kuriyama, T., Hakamada, R., Nakagome, H., Tokai, Y., Sahashi, M., Li, R., Yoshida, O., Matsumoto, K. and Hashimoto, T., "High efficient two-stage GM refrigerator with magnetic material in the liquid helium temperature region," *Advances In Cryogenic Engineering*, vol. 35, Plenum Press, New York (1990), pp. 1261–1269.
23. Barclay, J. A., "Wheel-type Magnetic refrigerator," US patent No. 4,408,463, Washington, DC: U.S. Patent and Trademark Office (1993).
24. Numazawa, T., et al., "New regenerator material for sub-4 K cryocooler," *Cryocoolers 11*, Kluwer Academic/Plenum Publishers, New York (2001), pp. 465–474.
25. Numazawa, T., et al., "A new ceramic magnetic regenerator material for 4 K cryocoolers," *Cryocoolers 12*, Kluwer Academic/Plenum Publishers, New York (2003), pp. 473–482.
26. Hiroshi, A. and Suzuki, M., "Refrigerator having regenerator." U.S. Patent No. 5,481,879, Washington, DC: U.S. Patent and Trademark Office (1996).
27. Ikeya, Y., "Expanding market of 4 K GM cryocooler," *Presented at 24th IIR International Congress of Refrigeration* (2015).
28. Chan, C.K and Tward, E., "Multi-stage pulse tube cooler," U.S. Patent No. 5,107,683, Washington, DC: U.S. Patent and Trademark Office (1992).
29. Matsubara, Y. and Gao, J. L., "Novel configuration of three-stage pulse tube refrigerator for temperature below 4 K," *Cryogenics*, vol. 34 (1994), pp. 259–262.
30. Wang, C., Thummes G. and Heiden C., "A two-stage pulse tube cooler operating below 4 K," *Cryogenics*, vol. 37 (1997), pp. 159–164.
31. Wang, C. and Gifford, P. E., "0.5W Class Two-Stage 4 K Pulse Tube Cryorefrigerator," *Advances in Cryogenic Engineering*, vol. 45B, Kluwer Academic/Plenum Publishers, New York (2000), pp. 1–7.

4. Features and Characteristics of 4 K Cryocoolers

Chao Wang

4.1 Introduction

4 K pulse tube cryocoolers and 4 K GM cryocoolers have different behaviors in operation. It is very important to understand these operating features and characteristics when choosing a 4 K cryocooler for an application. Generally speaking, the 4 K GM cryocoolers have slightly higher efficiencies, the cold head can be mounted in any orientation, the 4 K pulse tube cryocoolers have less vibration, higher reliability, and longer mean time between maintenance (MTBM), and the 4 K pulse tube cold head must be installed with the cold end down. A small tilting angle from the vertical position of $\leq 30°$ is accepted for an installation.

Table 4.1 gives some general operating information of the 4 K GM and pulse tube cryocoolers. The cryocoolers being compared are the SHI 1W@4 K GM cryocooler, model RDK408, and the Cryomech 1W@4 K pulse tube cryocooler, model PT410. More details of

Table 4.1 Comparison of 4 K GM and 4 K Pulse Tube Cryocoolers

		GM: SHI Model RDK408	PT: Cryomech Model PT410
Vibration	Second-stage displacement	± ~13 μm	± ~10 μm
	Mounting force[a]	~178 N	~4.5 N
Magnetic distortion from rare earth material in the second-stage regenerator		~ ± 200 nT	~ ± 20 nT
Cold-head MTBM[b]		~10,000–25,000 hr	~35,000–85,000 hr
Cold-head orientation		Any	Vertical position with cold end down[c]

[a] Vibration force on the room temperature flange.
[b] MTBM: mean time between maintenance.
[c] Small offset from vertical is allowed.

mechanical vibration, temperature oscillation, magnetic distortion, performance variation in high-magnetic field, cold-head orientation, and intercept cooling at the temperatures between 4 and 40 K are analyzed theoretically and investigated experimentally in the following sections below.

4 K GM cryocoolers normally have MTBM from 10,000 to 25,000 hours. 4 K pulse tube cryocoolers have much longer MTBM than the manufacture's expectation (20,000 hours in the brochure). 4 K pulse tube cold heads manufactured by Cryomech, Inc. are approved to have MTBM in the range of 35,000–85,000 hours.

4.2 Mechanical Vibrations

Vibration of 4 K GM cryocoolers is mainly the result of the displacer motion. The pressure oscillation inside of the cold head generates periodic elastic deformation of the thin-wall tubes. The oscillating pressure and cold-head displacer driving mechanism generate lower level vibrations when compared to that generated by the motion of displacer.

Vibrations of 4 K pulse tube (PT) cryocooler are mainly generated by the pressure oscillations in the thin wall stainless steel (SS) cold-head tubes. Cryocoolers employ thin wall SS tubes from the room temperature flange to the first-stage cooling station and from the first-stage cooling station to the second-stage cooling station to minimize thermal conduction losses.

The displacement of the PT cooling stages can be estimated by a normal stress equation. Figure 4.1 shows the structure of a PT cold head for displacement calculation. Tensile or compressive stress of the tube is usually denoted "normal stress" and can be expressed as

$$\sigma = F / A_w \tag{4.1}$$

where $F = \Delta P\, \pi d^2/4$, which is a normal force acting perpendicular to the cross-sectional area $A_w = \pi S(2d + S)/4$ of the tube wall. Normal strain of the tube (elongation or contraction) can be expressed as

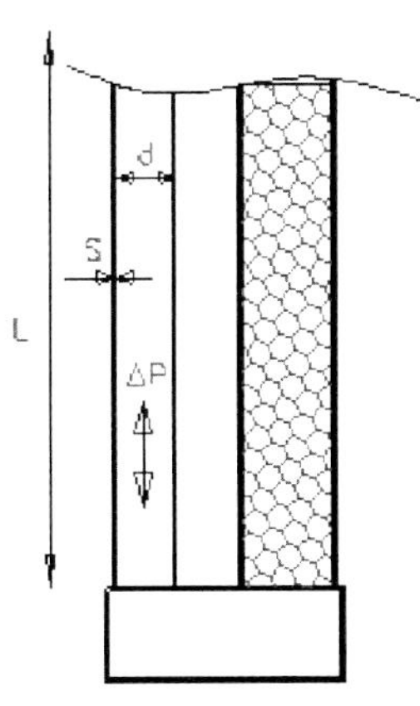

FIGURE 4.1 Structure of the PT cold head: d, ID of the tube; L, overall length of the tube; S, thickness of the tube; ΔP, pressure differential inside of the tube.

$$\varepsilon = \Delta L / L = \sigma / E \tag{4.2}$$

where ΔL is change of length, and E is Young's modulus. In a pulse tube cold head,

$$\Delta L = FL / \left(EA_w \right) \tag{4.3}$$

Assuming $S << d$, ΔL can be written as

$$\Delta L = \frac{d \Delta P L}{2SE} \tag{4.4}$$

The displacement of the PT cold head is proportional to the tube ID, overall length, pressure differential, and inversely proportional to the wall thickness. The vibration calculation using Equation 4.4 is in a good agreement with experimental measurements.

The cold-head motor/rotary valve and flexible lines connecting the cold head to the compressor also contribute to a lower level and higher frequency vibration. For a 4 K GM cryocooler, the vibrations generated by displacer motion or periodic oscillating pressure at frequencies of 1–1.4 Hz are over an order of magnitude higher than that generated by the cold-head motor and flexible lines.

The vibration on the room temperature flange and the second cooling station are of high concern of users since the room temperature flange of the cold head is mounted on a cryostat, and the second-stage cooling station normally has mechanical contact with the device to be cooled. Figure 4.2 shows a test rig for measuring the vibrations of 4 K cryocoolers. The displacements of the cooling stations are measured by a length gauge or an optical sensor. A load cell is mounted below the room-temperature flange of the cold head to measure the vibration force exerted on the cryostat. This force is also called the "mounting force." An accelerometer is installed on the top of the cryostat to measure the cryostat vibration generated by the cryocooler. The X-axis in Figure 4.2 is the flow direction between the regenerator and pulse tube, the Y-axis is the direction perpendicular to X-axis, and the Z-axis is the vertical direction.

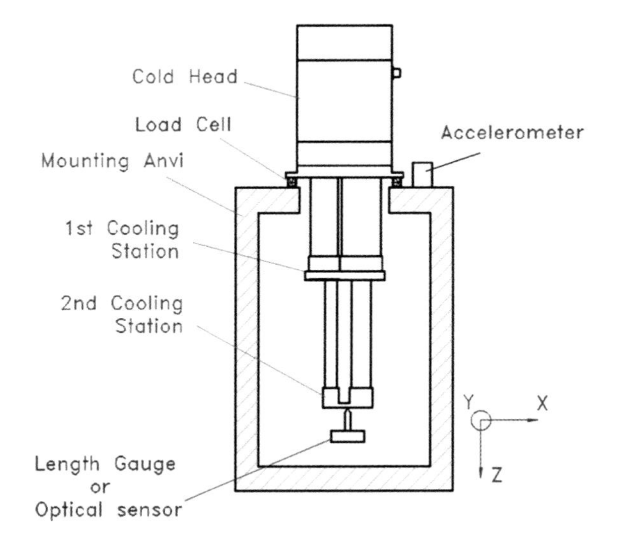

FIGURE 4.2 Vibration measurement setup for cold-head testing.

Displacement of the second cooling station of the GM cryocooler (SHI model RDK408) is measured by an optical sensor at steady operation while the second stage is at a cryogenic temperature [2]. Figure 4.3 shows the measured vertical displacement of the second stage with an amplitude of ±~13 μm. Horizontal vibrations of the GM cryocooler have not been measured. It is believed that the vibration of the GM cryocooler at the horizontal position should be smaller because of the concentric design of the cold head.

The second-stage displacements of the 4 K PT cryocooler (Cryomech model PT410) have been measured in three directions using a length gauge [3]. The displacement measurements are taken for 300 s, starting from room temperature because of restrictions

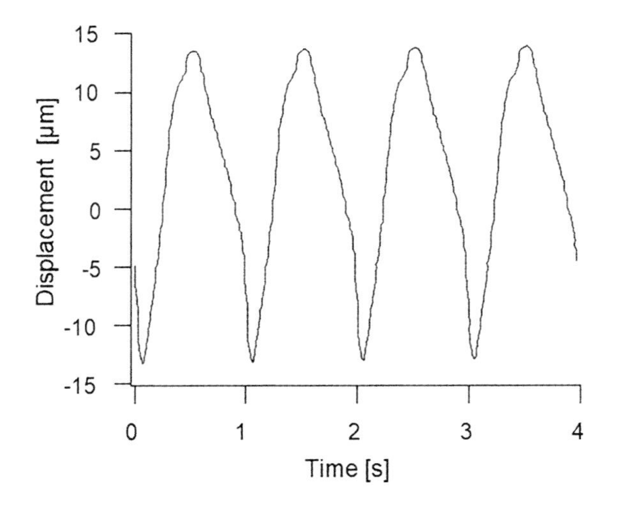

FIGURE 4.3 Vertical displacement of the second-stage at the vertical position of a 4 K GM cryocooler, SHI model of RDK408. (Reprint from *Cryogenics*, 44, Tomaru, T. et al., Vibration analysis of cryocoolers, 309–317. Copyright 2004, with permission from Elsevier Publisher.)

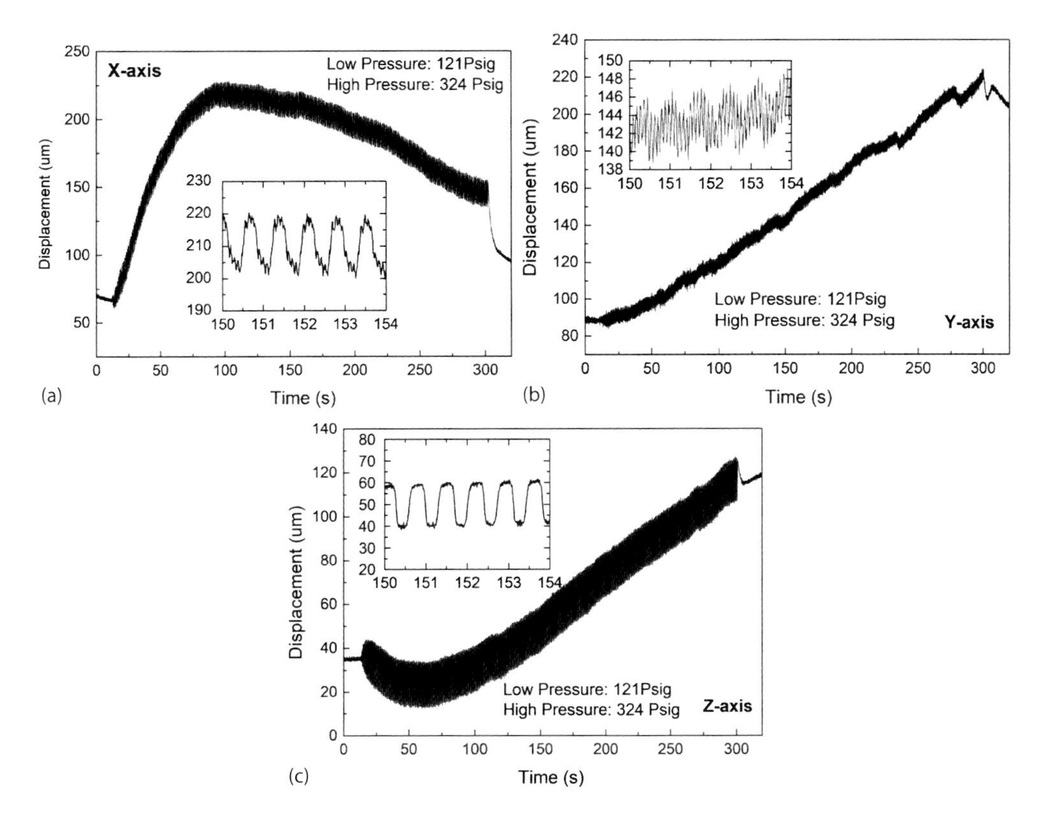

FIGURE 4.4 Displacements on three axes of the second stage of the 4 K PT cryocooler: (a) *X*-axis. (b) *Y*-axis. (c) *Z*-axis.

on the operating temperature of the length gauge. Figure 4.4 shows displacements of the second stage of the 4 K PT cryocooler in three axes, *X*, *Y*, and *Z*. The cooling station position drifts during the cold-head cool down because of thermal contraction of the cold head. High-resolution displacement curves are given in the figures. The displacement amplitudes at the horizontal direction, *X*, and the vertical direction, *Z*, are both around ±10 μm. The displacement amplitude at the horizontal position Y is ±4 μm. The second-stage displacement of the 4 K PT cryocooler is smaller than that of the 4 K GM cryocooler, but they are in the same order of magnitude.

The remarkable vibration differences for the two 4 K cryocoolers are the mounting forces and cryostat vibrations. The measured mounting forces from the GM and PT cryocoolers are given in Figure 4.5. The mounting force from the GM cryocooler is ~40 times higher than that from the PT cryocooler. The GM vibration curve has peak values due to the motion of the displacer. The PT vibration curve is similar to the sinusoidal wave of the oscillating pressure.

Figure 4.6 compares the accelerations on the top of the cryostat generated by the 4 K GM cryocooler (SHI model RDK408) and a 4 K PT cryocooler. The PT cryocooler in this comparison is a smaller model, RP-062B, made by SHI cryogenics with 0.5 W at 4.2 K, not a 1 W class model compared in Figure 4.5. The acceleration with the 4 K GM cryocooler is two orders of magnitude higher than that with the 4 K PT cryocooler.

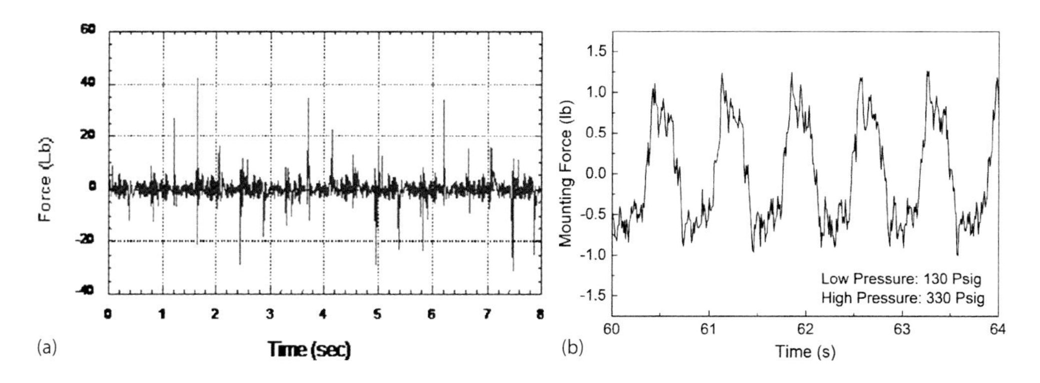

FIGURE 4.5 Mounting force from GM and PT cryocoolers: (a) GM cryocooler. (b) PT cryocooler.

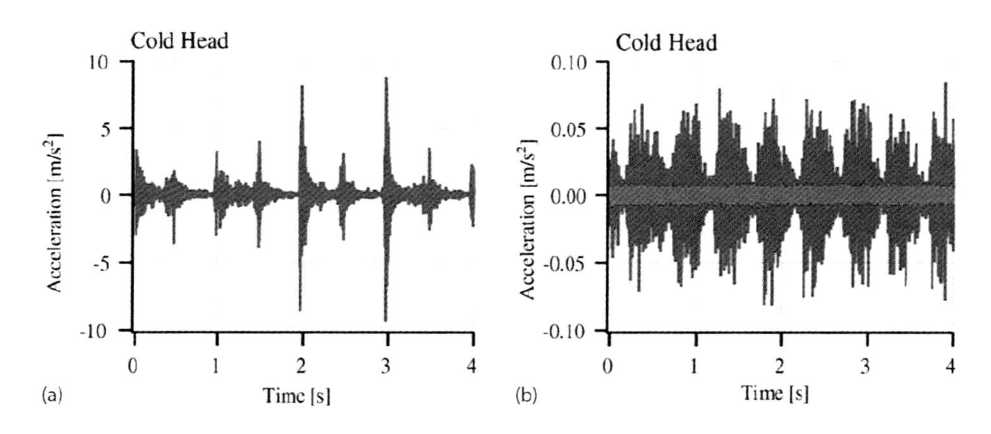

FIGURE 4.6 Acceleration on the top of the cryostat: (a) GM cryocooler. (b) 0.5 W class PT cryocooler (SHI model RP-062B, 0.5 W@4.2 K). (Reprint from *Cryogenics*, 44, Tomaru, T. et al., Vibration analysis of cryocoolers, 309–317. Copyright 2004, with permission from Elsevier Publisher.)

Vibrations of the 4 K cryocoolers are big concerns for many applications. More details of vibration reduction are introduced in Chapter 5 of this book.

4.3 Temperature Oscillations and Damping

Refrigeration in a regenerative cryocooler is achieved by periodic expansion of the working fluid (helium). For an adiabatic process, the temperature variation can be expressed as:

$$\Delta T = \frac{\alpha_v T}{\rho C_p} \Delta P \tag{4.5}$$

where α_v is the coefficient of volume expansion: $\alpha_v = 1/V(dV/dT)_P$, ρ is the density, C_p is the specific heat, and ΔP is the pressure variation. The temperature of the helium gas in the cold head fluctuates because of the oscillating of pressure and gas volume. A numerical simulation of 4 K GM and 4 K PT cryocoolers show that the gas temperature oscillations at the 4 K stage are about a few Kelvin in a refrigeration cycle [4]. The temperature oscillations on the outside of the 4 K cooling stages are damped by the stage heat capacity and are normally a few hundred miliKelvin.

Some applications require temperature stabilities from ~100 μK to a few mK. Several technologies for temperature damping on the 4 K cooling station have been developed. Three typical designs of temperature damping are shown in Figure 4.7a–c.

Figure 4.7a is a temperature damping plate made of a rare earth material, such as $HoCu_2$ or Er_3Ni. The rare earth plate provides a high heat capacity near 4 K. The rare earth damper is simple and easy for system integration. However, the cooling capacity loss caused by the rare earth damper could be as high as 20%–50% at 4.2 K because of poor thermal conductivity of the rare earth materials. The temperature damping mechanism with the rare earth plate is to combine its high specific heat near 4 K and low thermal penetration because of its low thermal conductivity. Equation 4.6 can be used to estimate the temperature stability with the damper of different thickness [5]. Where ΔT is the peak-to-peak temperature oscillation on the damper, $\Delta T\ (0)$ is the peak-to-peak temperature oscillation on the second stage, d the thickness of the rare earth damper, f cryocooler operating frequency, $\alpha = \lambda / (\rho c_p)$ is the thermal diffusivity, ρ density, λ thermal conductivity, and c_p specific heat.

$$\Delta T = \Delta T(0)\exp\left(-d\cdot\left(\frac{\pi f}{\alpha}\right)^{1/2}\right) \tag{4.6}$$

A copper helium pot damper in Figure 4.7b must be designed long enough to reduce the heat conduction effect along the copper wall of the pot [6]. Conduction fins are used in the copper pot to enhance heat transfer between the copper surface and liquid helium. The temperature damping replies on the high heat capacity of liquid helium. An SS helium gas supply tube is used to connect the damper pot to a room temperature gas reservoir. The cooling capacity difference is negligible for the 4 K cryocooler with and without the copper helium pot damper.

An SS helium pot damper in Figure 4.7c has a thin-wall SS tube connecting the copper heat exchangers on the top and bottom of the damper [7]. The SS helium pot damper is compact and relies on the helium phase change to transfer cooling from the second

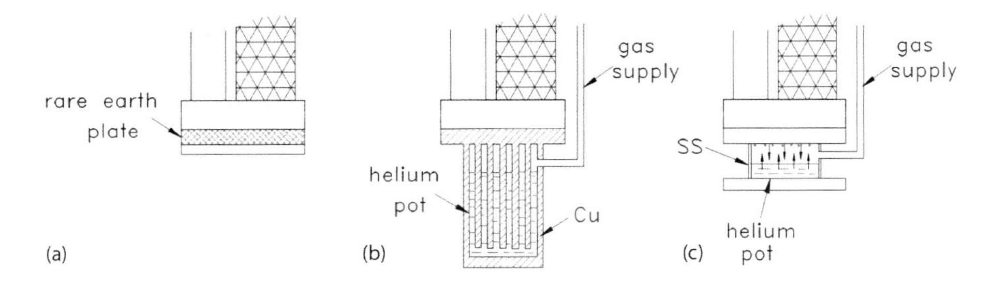

FIGURE 4.7 Three typical temperature dampers used in 4 K cryocoolers: (a) Rare earth damper. (b) Copper helium pot damper. (c) SS helium pot damper.

stage to the cooling plate. There is a cooling capacity loss of ~5% associated with the heat transfer loss in the helium pot based on the measured performance of these products.

Figure 4.8 is a photo of a rare earth damper plate produced at Cryomech, Inc., which has a center layer of rare earth 4.4-mm thick and copper layers 1-mm thick on both ends. The damper plate has the same diameter as the second-stage cooling station of the cold head. Figure 4.9 shows the temperature oscillations on the second-stage cooling station and the rare earth damper in a 4 K PT cryocooler, Cryomech model PT407. The 4 K PT cryocooler has a heat load of 0.5 W on the temperature damper during the temperature measurement. The temperature oscillations after the temperature damping

FIGURE 4.8 A rare earth damper made of $HoCu_2$ material. (Courtesy of website of Cryomech, Inc., at: http://www.cryomech.com/.)

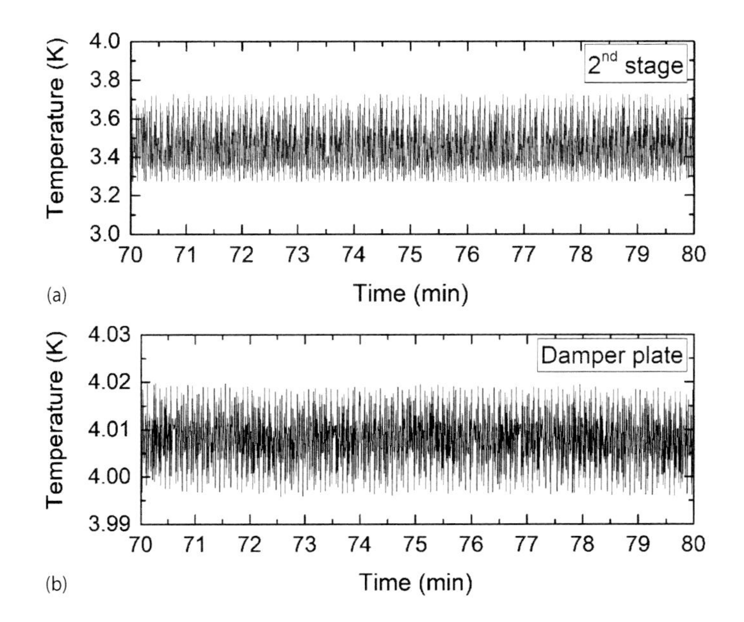

FIGURE 4.9 Temperature oscillations in a 4 K PT cryocooler (model PT407) with the rare earth material damper: (a) Temperature on the second-stage cooling station. (b) Temperature on the damper.

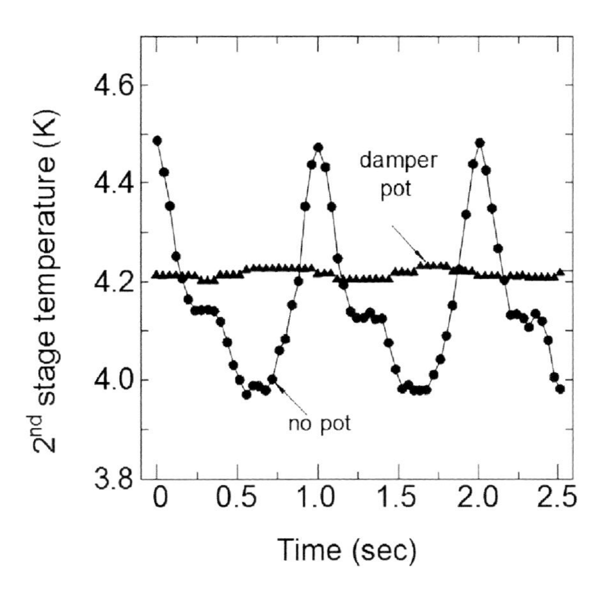

FIGURE 4.10 Temperature oscillations of a 4 K GM cryocooler with and without the copper pot damper. The second stage has a heat load of 0.6 W. (With kind permission from Springer Science+Business Media: *Cryocooler*, Temperature stabilization on cold stage of 4 K GM cryocooler, 9, 1997, Plenum Press, New York, 765–771, Li, R., et al.)

is reduced to ±12 mK while compared to the temperature oscillations of ± 220 mK on the second stage. However, the temperature on the damper is ~0.5 K higher than that on the second-stage cooling station, which indicates a high heat transfer loss through the damper.

Temperature stability with and without the copper helium pot damper in a 4 K GM cryocooler [6] are compared in Figure 4.10. The 4 K GM cryocooler has a peak-to-peak temperature oscillation of ± 270 mK without the temperature damping. The peak-to-peak temperature oscillation is reduced to ± 27 mK by using the damper. Adding the copper helium pot damper has almost no influence on the cooling capacity on the 4 K stage of the GM cryocooler.

Figure 4.11 shows a photo of a 4 K PT cryocooler, Cryomech model PT405 (0.5 W at 4.2 K), with the SS helium pot damper. An SS tube connects the helium pot to an external gas reservoir for supplying helium gas. Helium gas from an external gas reservoir is precooled by a heat exchanger on the first stage and the spiral precooling tube thermally attached on the second-stage regenerator before entering the helium pot to be liquefied. Precooling the helium gas accelerates the cool-down speed of the cold head with the helium pot damper. More details about the precooling tube will be introduced in Section 4.7.

Temperature stability of the 4 K cooling stations with the damper is given in Figure 4.12. A heat load of 0.45 W is applied on the damper station. The temperature differential between the second-stage cooling station and the damper station is 58 mK, which indicates a very small heat transfer loss in the SS helium pot damper. The temperature stability on the damper station is ±2.5 mK, and that on the second-stage cooling station is ±70 mK with the damper. The temperature oscillation on the second-stage cooling station is also reduced by the large heat capacity of the damper.

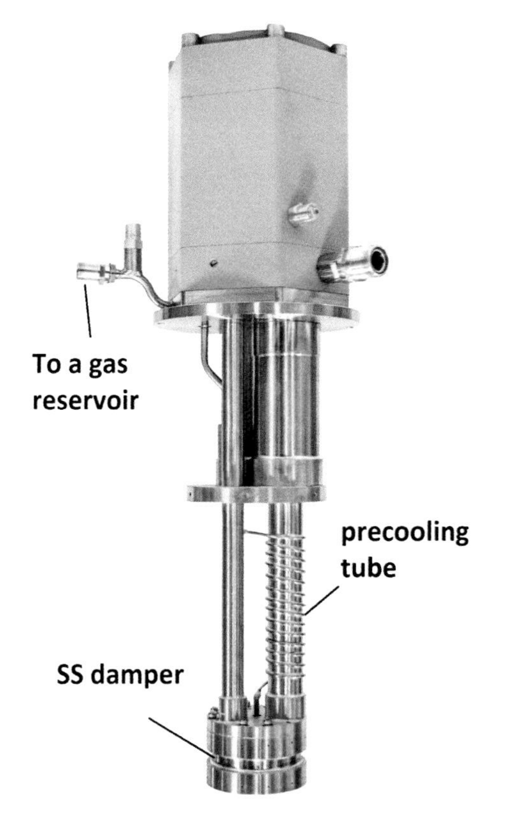

FIGURE 4.11 A 4 K PT cryocooler (model PT405) with a SS helium damper pot. (Courtesy of Website of Cryomech, Inc., at: http://www.cryomech.com/.)

4.4 Magnetic Distortion

Small magnetic field distortions from the 4 K cryocoolers were found in NMR, MRI, superconducting quantum interface devices (SQUIDs) systems, and other sensitive devices. The amplitude of the magnetic distortion near the second-stage regenerator of a 4 K PT cryocooler, Cryomech model PT410, was found to be ~20 nT compared to that of ~200 nT from a 4 K GM cryocooler, SHI model SRDK408. The measurement of the magnetic distortion is near the second-stage cooling station.

The rare earth regenerative materials used in the second-stage regenerator have high specific heat at low temperatures because of the magnetic phase transition. The rare earth regenerative materials are also called "magnetic regenerative materials." The magnetic field distortion is generated by the magnetization fluctuation of the rare earth materials due to the temperature variation during a cycle.

Figure 4.13 shows the variations of magnetization of the rare earth materials of $HoCu_2$ and Er_3Ni at different temperatures. The rare earth materials were put into a very sensitive solenoid to measure their magnetization at external magnetic fields of 0.1 Oe and 1 Oe [1]. The magnetization of the rare earth materials is related to the material temperature and external magnetic field outside of the cold head. The magnetic susceptibility (dM/dH) and temperature dependence of magnetization (dM/dT) of $HoCu_2$ are

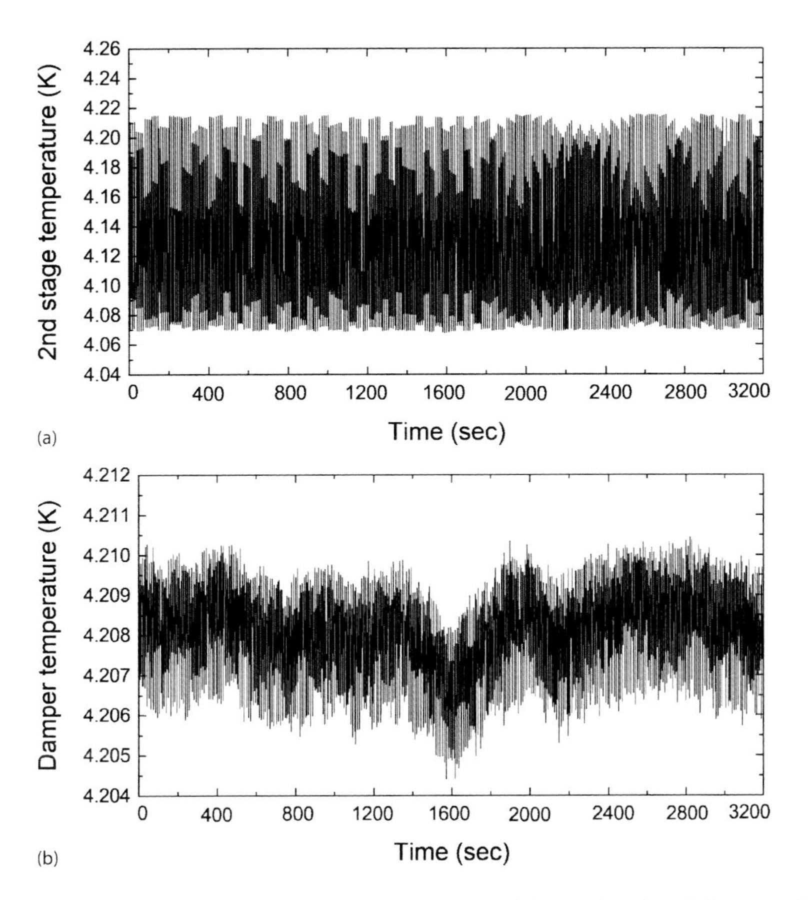

FIGURE 4.12 Temperature stability of a 4 K PT cryocooler (model PT405) with SS helium pot damper: (a) Temperature on the second-stage cooling station. (b) Temperature on the damper station.

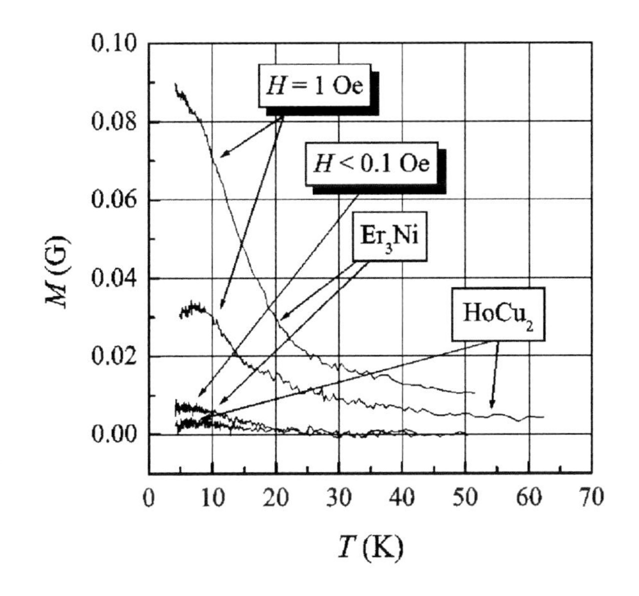

FIGURE 4.13 Magnetization of the rare earth materials of $HoCu_2$ and Er_3Ni in different magnetic fields. H, external magnetic field; M, magnetization of materials; T, temperature of materials.

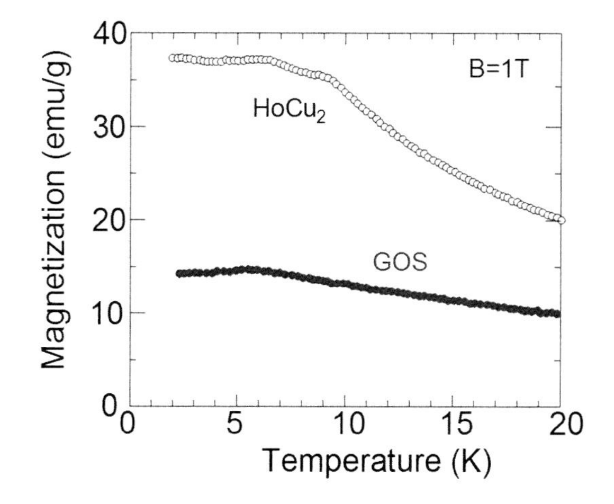

FIGURE 4.14 Magnetization of GOS and $HoCu_2$ as a function of temperature in an external magnetic field of 1 T. (With kind permission from Springer Science+Business Media: *Cryocooler*, A new ceramic magnetic regenerator material for 4K cryocoolers, Kluwer Academic/Plenum Publishers, New York, 12, 2003, 473–482, Numazawa, T. et al.)

smaller than those of Er_3Ni. $HoCu_2$ generates less magnetic distortion introduced by its magnetism fluctuation.

The newer material of Gd_2O_2S (GOS) has even better characteristics of magnetization as a regenerative material. Figure 4.14 compares the magnetization of GOS and $HoCu_2$ as a function of temperature in the 1 T magnetic field [8]. However, the temperature dependence of magnetization as well as the magnetic susceptibility of GOS is much less than those of $HoCu_2$. Therefore, the latest 4 K cryocoolers using GOS and $HoCu_2$ in the second-stage regenerator have less magnetic distortions than the older systems with the second-stage regenerator using $HoCu_2$ or $HoCu_2$ and Er_3Ni.

The rare earth regenerative materials inside of the second-stage regenerator have relatively high periodic temperature oscillations in an operation cycle because of their low specific heat compared to that of the helium. Calculated temperature variations of the second-stage regenerative materials in a cycle for a 4 K PT cryocooler [9] and a 4 K GM cryocooler [10] are given in Figure 4.15a and b. The periodic temperature oscillation of the regenerative material of Er_3Ni can be more than 10 K, which results in magnetic field fluctuation due to the temperature dependence of magnetization (dM/dT) of the rare earth material. The second-stage regenerator in the 4 K GM cryocooler normally has higher temperature oscillation than that in the 4 K PT cryocooler.

The magnetic distortions generated from a 4 K PT cryocooler and a 4 K GM cryocooler are schematically illustrated in Figure 4.16. For the 4 K GM cryocooler, the magnetic field distortions are generated not only by the regenerator temperature oscillation, but also by the motion of the displacer. The intensity of the magnetic field at the position of the magnetic regenerative material fluctuates with the reciprocation of the displacer. The magnetic field distortion generated by the motion of rare earth materials is approximately ten times higher than that by the periodic temperature oscillation.

External magnetic shielding covering the portion of the rare earth materials in the second-stage regenerator is required for some applications that are sensitive to the magnetic distortion. μ-metal is a favorite material used for the magnetic shielding.

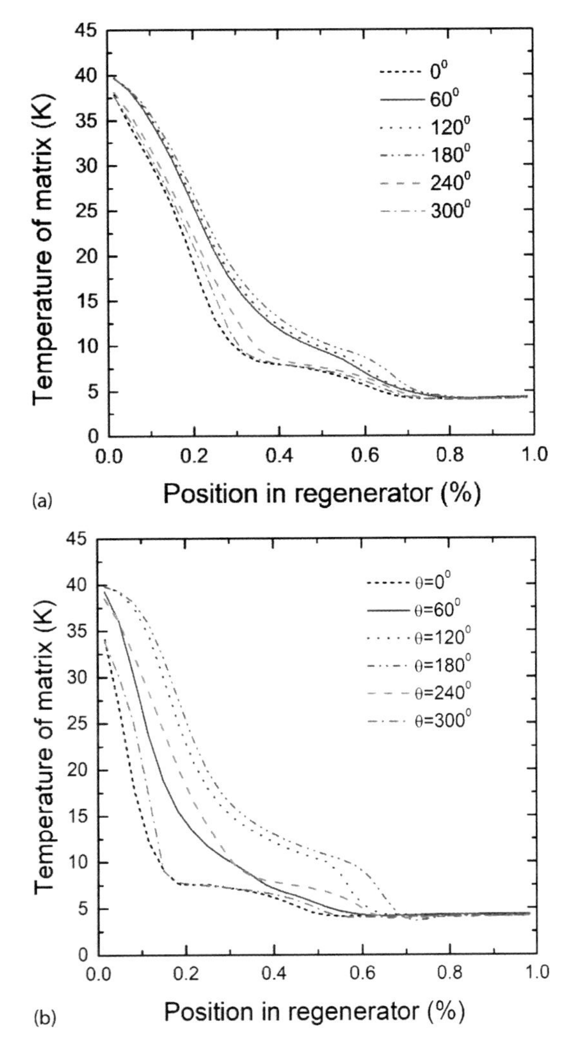

FIGURE 4.15 Calculated temperature distributions of the second-stage regenerative material in a cycle. The second-stage regenerator is filled with Er_3Ni. θ is the angle in a cycle, which is divided into 360°: (a) 4 K PT cryocooler (Data from Wang, C., *Cryogenics*, 37, 215–220, 1997.). (b) 4 K GM cryocooler (Data from Wang, C., Hafner, H. U. and Heiden, C., Performance and internal process of a 4K GM cooler, *Advances in Cryogenic Engineering*, 43B, Plenum Press, New York, 1775–1782, 1998.)

4.5 Cooling Performance Degradations in High Magnetic Fields

4 K cryocoolers have been widely used for cooling superconducting magnets, such as magnets in MRI systems. In these systems, 4 K cryocoolers are installed close to the superconducting magnets with high magnetic fields. Here the 4 K cryocoolers should be designed to be able to operate in a strong magnetic field. Two components in a 4 K cold head—cold-head motor and the second-stage regenerative materials—can be affected by an external magnetic field.

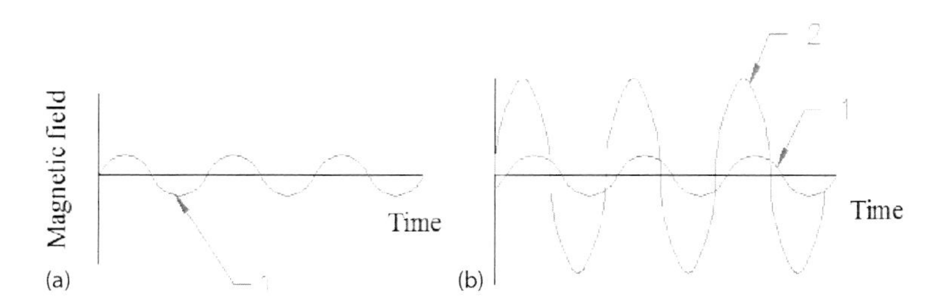

FIGURE 4.16 Schematic illustration of magnetic field fluctuation generated by the cryocoolers: (1) Generated by temperature oscillation. (2) Generated by motion of the GM displacer. (a) Pulse tube. (b) GM.

The cold-head motor for driving the rotary valve (pulse tube) or the rotary valve and Scotch–Yoke (GM) is either an AC synchronous motor or DC stepper motor, which has a permanent magnet on the rotor. The strong external magnetic field can reduce the magnetic strength of the permanent magnet and then cause the cold-head motor to stall. Most manufacturers recommend the external magnetic field around the cold-head motor should be ≤500 Gauss.

The cooling capacity of 4 K cryocoolers can be affected by reduced specific heat of magnetic regenerator materials in high magnetic fields. Early development of 4 K GM cryocoolers [11] and 4 K PT cryocoolers [12] employed $ErNi_{0.9}Co_{0.1}$ in the second-stage regenerator due to its large peak of specific heat between 4 K and 8 K. Figure 4.17a–c shows variations of the specific heat of $ErNi_{0.9}Co_{0.1}$, $HoCu_2$ and GOS in high magnetic fields [8,11]. $ErNi_{0.9}Co_{0.1}$ is a ferromagnetic material, and its peak value of specific heat degrades significantly in a high magnetic field. $HoCu_2$ and GOS are antiferromagnetic materials; magnetic fields of under 1 T have a small impact on the specific heat of $HoCu_2$ and almost no effect on that of GOS.

Figure 4.18 shows effects of the external magnetic field on the cooling capacity of a GM cryocooler, SHI model SRDK-408, with the rare earth material of $ErNi_{0.9}Co_{0.1}$, Er_3Ni and $HoCu_2$ [11]. The second-stage cooling performance of the cryocooler with $ErNi_{0.9}Co_{0.1}$ degraded dramatically with increasing external magnetic field. However, the cryocooler performance with Er_3Ni and $HoCu_2$ has much less degradation in the magnetic field. The cryocooler with $HoCu_2$ has the best performance in the comparison test under the magnetic field of <1 T. Therefore, in the early 4 K GM cryocoolers, manufactured by SHI, lead, and $HoCu_2$ were used in the second-stage regenerator. The early 4 K PT cryocooler manufactured by Cryomech, Inc. (before GOS was introduced) employed lead, Er_3Ni, and $HoCu_2$ in the second-stage regenerator. $ErNi_{0.9}Co_{0.1}$ has not been used in commercial 4 K cryocoolers.

Figure 4.19 compares cooling performance of two 4 K GM cryocoolers (SHI SRDK-408) packed with the $HoCu_2$ or $HoCu_2$ and GOS in the second-stage regenerator, respectively, under a magnetic field of <2 Tesla [13]. A constant heat load of 1 W is applied on the second stages of the cryocooler during the testing. The external magnetic field (axial magnetic field) is parallel to the cryocooler displacer. In magnetic fields of <0.6 T, no obvious reduction of cooling capacity is observed for both 4 K cryocoolers. But when the magnetic field increases above 0.6 T, the second-stage temperature of the cryocooler with $HoCu_2$ [type (a) in the figure] starts increasing. On the other hand, the cryocooler with $HoCu_2$ and GOS [type (b) in the figure] has a slower temperature rise with increasing the magnetic field of above 0.6 T, which only increases by ~0.4 K under a magnetic field

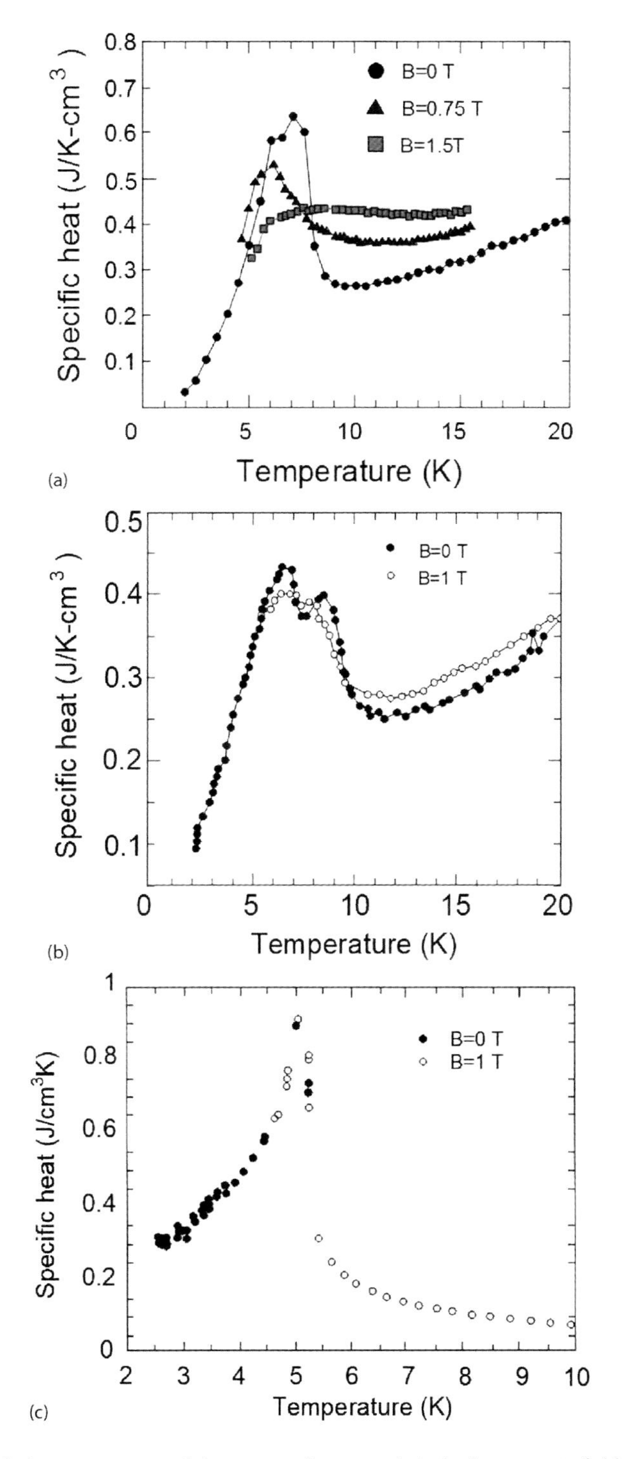

FIGURE 4.17 Specific heat variations of the rare earth materials in high magnetic fields. B is external magnetic field: (a) $ErNi_{0.9}Co_{0.1}$, (b) $HoCu_2$. ([a,b] Data from Onishi, A. et al., *J. Cryogenic Society of Japan*, 34, 196–199, 1999 with permission from Cryogenic Society of Japan.) (c) GOS. (With kind permission from Springer Science+Business Media: *Cryocooler*, A new ceramic magnetic regenerator material for 4K cryocoolers, 12 2003, Kluwer Academic/Plenum Publishers, New York, 473–482, Numazawa, T. et al.)

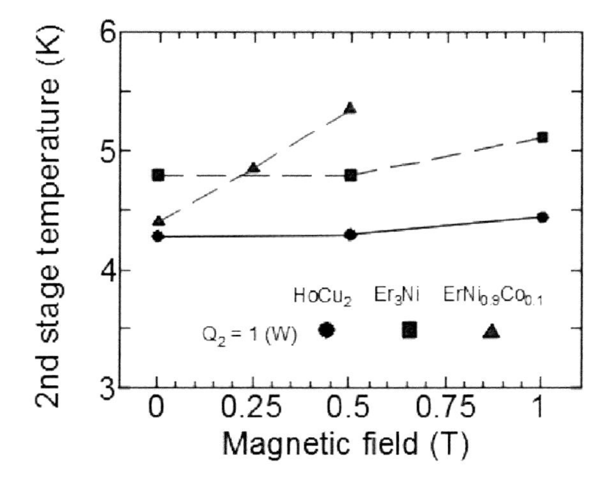

FIGURE 4.18 Effect of cooling performances of a 4 K GM cryocooler with different rare earth materials. (Data from Onishi, A. et al., *J. Cryogenic Society of Japan*, 34, 196–199, 1999. with permission from Cryogenic Society of Japan.)

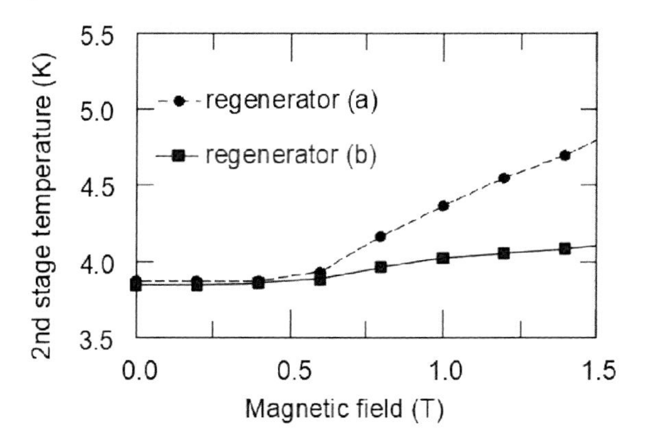

FIGURE 4.19 The second-stage cooling performance of 4 K GM cryocoolers with two types of the second-stage regenerators under various-strength axial magnetic fields: (a) $HoCu_2$ in the second-stage regenerator. (b) $HoCu_2$ and GOS in the second-stage regenerator. (Modified from Moriea, T. et al., *Phys. Procedia*, 67, 474–478, 2015.)

of 2 T. GOS regenerative material has very good characteristics and (see Figure 4.17c) performs better in high magnetic fields.

A radial external magnet field has two impacts on a 4 K GM cryocooler: (1) the second-stage displacer made of SS is forced in a radial direction, resulting in the increased wear on one side between the displacer and cold-head cylinder; and (2) the cooling capacity of the second stage is reduced because of specific heat reduction of regenerator materials. It is estimated that the side force applied on the radial direction of the second stage is ~9.8 N when the type (a) regenerator works in a radial magnetic field of 0.45 T [13]. Uneven wear on the second-stage displacer and cold-head cylinder had been found in some investigations, which shortened the life time of the cryocooler. A frictional heat loss could be generated due to the forced contact between the displacer and cylinder. The radial magnetic field should be carefully considered when integrating the 4 K GM cryocooler into a magnet system.

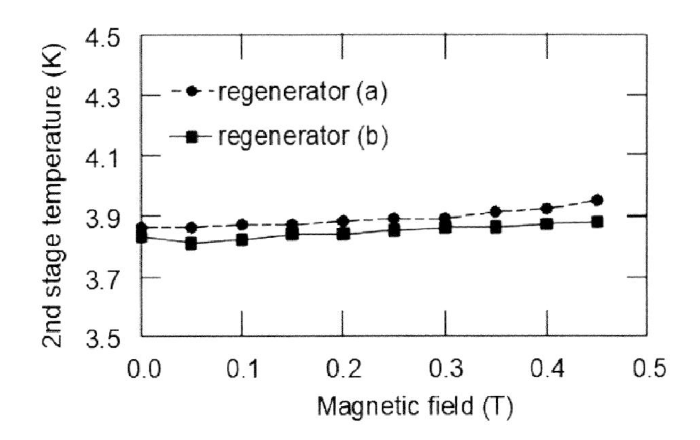

FIGURE 4.20 The second-stage cooling performance of 4 K GM cryocoolers with two types of the second-stage regenerators under various strengths of radial magnetic fields: (a) $HoCu_2$ in the second-stage regenerator. (b) $HoCu_2$ and GOS in the second-stage regenerator. (Modified Moriea, T. et al., *Phys. Procedia*, 67, 474–478, 2015.)

Figure 4.20 shows the second-stage temperatures with the two types of regenerators under different radial magnetic fields. The cryocooler can only be tested up to 0.45 T due to the restriction of the experimental equipment. Slight degradation of the cooling capacity is observed in the radial magnetic field of up to 0.45 T, which is close to the performance degradation in the axial magnetic field.

So far, performance variations of 4 K PT cryocoolers have not been quantified in high magnetic fields. However, 4 K PT cryocoolers are expected to have the same or better performance than the 4 K GM cryocooler in high magnetic fields because they have no moving parts at cryogenic temperatures. Many 4 K PT cryocoolers have been used in very high field superconducting magnets, such as high magnetic field MRIs (4–12 T) and Fourier Transform Mass Spectroscopy (FTMS, 7–15 T) [14]. They have demonstrated very good performances in these applications.

Magnetic shielding for reducing the magnetic field around the cold-head second stage of 4 K cryocoolers may be required in some high magnetic field applications.

4.6 Cold-Head Orientation

Cooling performance of the low-frequency PT cryocoolers is significantly affected by orientation because of the natural convection generated by gravity in the empty pulse tube. The large temperature differential of helium gas from the cold end to the warm end of the pulse tube results in large differences of helium gas density along the tube. To evaluate the natural convection effects on the performance of the pulse tube, the ratio of buoyancy force of natural convection over inertia force of forced convection can used [15]:

$$R = Gr/Re^2 = (g\beta\Delta td)/(u^2) \quad (4.7)$$

where Gr is the Grashof number, Re is the Reynolds number, g is the gravitational acceleration, β is the coefficient of thermal expansion, Δt is the temperature differential for the natural convection, d is the ID of the pulse tube, and u is the gas velocity of the forced flow.

A smaller ratio R of the buoyancy force over the inertia force results in lower orientation effects on the cryocooler performance. From Equation 4.7, apparently increasing the gas velocity u and selecting proper geometry of the pulse tube can reduce the natural convection effects on the cryocooler performance. Higher operating frequency will be very helpful in increasing the gas velocity u. High-frequency PT cryocoolers (Stirling type) with an operating frequency above 50 Hz can operate at any orientation because of high gas velocity in the pulse tube, in which the forced flow (Re) is dominant.

All low-frequency (GM type) PT cryocoolers with an operating frequency below 2.4 Hz have issues of orientation. The forced flow velocity is not big enough to overcome the impact of natural convection. The PT cold head must operate with the cold end down.

Figure 4.21 shows typical cooling performance of a 4 K PT cryocooler (Cryomech model PT410) at different orientations. Zero degree on the X-axis indicates the vertical position with the cold end down. The second-stage cooling station maintains the same temperature until the cold head is tilted to 50° off the vertical position. The temperature increases steeply as the angle increases to 60° and the second stage is not able to provide any cooling at 4.2 K. The first stage begins to lose the cooling capacity when starting to offset the cold head from the vertical position. The first-stage cooling capacity drops quickly with the tilting angle of >30°.

The second-stage performance is not as sensitive as the first stage to the orientations for two reasons. First, the second-stage pulse tube has a much larger ratio of length over diameter than that of the first stage. Second, there is a helium temperature plateau near the cold end of the second-stage pulse tube [4], which does not exist in the first-stage pulse tube. The natural convection loss in the portion of temperature plateau is small when tilting the pulse tube. Normally, the tilting angle of ≤30° for 4 K PT cryocooler is recommended by the manufacturers to prevent large performance degradation of 4 K PT cryocoolers.

In a 4 K GM cryocooler, the regenerators are the only components with a large temperature differential. The hydraulic diameters of the screen regenerator for the first-stage and the sphere regenerator for the second stage are too small to introduce natural convection. Therefore, the performance of the 4 K GM cryocooler is not as sensitive to the cold-head orientation as the 4 K PT cryocooler.

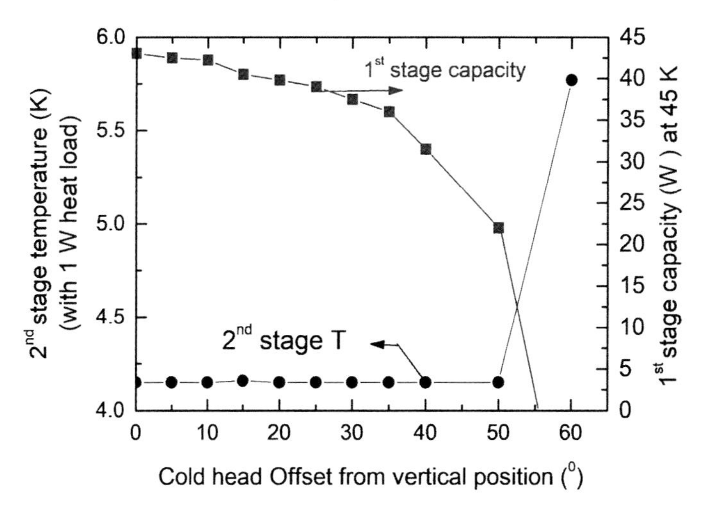

FIGURE 4.21 Performance of a 4 K PT cryocooler (model PT410) in different orientations.

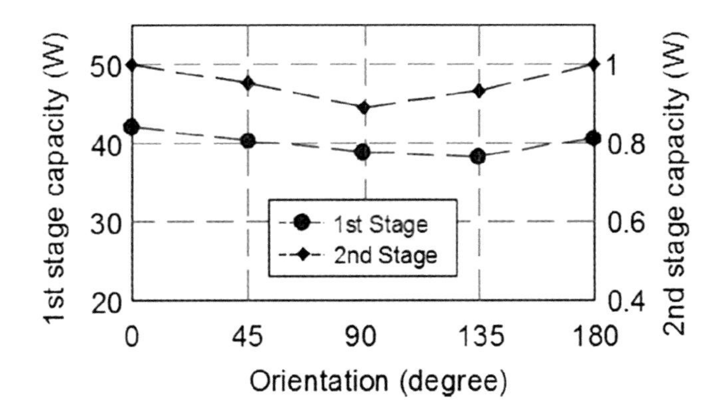

FIGURE 4.22 Performances of a 4 K GM cryocooler (SHI model RDK-408D) at different orientations. (Date from reference Ikeya, Y., Expanding market of 4 K GM cryocooler, *Presented at 24th IIR International Congress of Refrigeration*, 2015.)

In early development of 4 K GM cryocoolers with traditional displacer seals, some performance degradations have been observed when tilting the cold head. The performance degradation is mainly caused by the natural convection in the space between the second-stage displacer and cylinder. This convection loss is eliminated by using SHI spiral groove on the second-stage displacer. The forced gas flowing through the spiral groove suppresses the natural convection in that space.

Figure 4.22 shows orientation dependency of a SHI 4 K GM cryocooler (model RDK-408D). The largest performance degradation for both the first and second stages is ~10% at the cold-head tilting angle of 90°. The cold head recovers to the best performance when it is tilted to180°with the cold end up. The lower performance dependency of orientation enables the 4 K GM cryocooler to be installed at any orientation.

4.7 Intercept Cooling at Temperatures of 4–40 K from the Second Stage

This section introduces unique abilities of a 4 K PT cryocooler for providing intercept cooling at temperatures of 4–40 K from the second-stage pulse tube and regenerator. A 4 K GM cryocooler might be able to provide a small amount of cooling from the second-stage cylinder between the first stage and second stage because of the gas flowing through the spiral groove on the second-stage displacer. The intercept cooling from the 4 K GM cryocooler has not been studied so far and will not be discussed in this section.

The second law of thermodynamics shows that it is always desirable to remove heat in a cryocooler at the highest possible temperature since the increase in entropy carried by the refrigerant to the warm end is given by

$$\Delta\dot{S} = \dot{Q} / T \tag{4.8}$$

where $\Delta\dot{S}$ is the change in entropy flow, $\dot{Q}$ heat input, and T the temperature from which heat is being input. In the case of precooling gas or thermal sinking the instrumentation leads, it is desirable to have many cooling stages to remove heat down to a 4 K stage.

However, it is not practical since the additional stages will increase the system complexity and cost.

Intercept cooling between the first stage (~40 K) and second stage (~4 K), provided by a two-stage 4 K PT cryocooler, is very useful for some 4 K cooling applications. It was first used for precooling helium gas to be liquefied in a small helium liquefaction system with a 4 K PT cryocooler [17,18]. A small SS tube is wrapped around the second-stage regenerator to precool the helium gas to be liquefied. This precooling stage significantly increases the helium liquefaction rate.

The precooling tube installed on the second-stage regenerator has become a standard option (see a photo in Figure 4.29) and widely adapted in cryogen-free sub-4 K systems, in particular for all continuous flow-type systems such as cryogen-free 3He cryostats and dilution refrigerators where more efficient helium (including 3He) condensation rate is crucial for their performance [19,20]. More detailed discussion on the applications is in Chapters 8 and 11.

An intercept heat exchanger installed on the second-stage regenerator or the second-stage pulse tube has been used as a thermal sink for instrumentation leads at a temperature of ~10 K in some applications.

4.7.1 Analysis of the Energy Flow in the Second Stage

4.7.1.1 Intercept Cooling from the Regenerator

Intrinsic enthalpy loss of the second regenerator in a 4 K cryocooler is generated by the real gas property. This portion of enthalpy losses could provide the excess cooling from the second-stage regenerator. Figure 4.23 shows time-averaged energy flows in a two-stage PT cryocooler. In order to focus on the intrinsic enthalpy loss and exclude the rest of the enthalpy losses in the regenerator, an ideal second-stage regenerator is assumed in the discussion: (1) the specific heat of regenerator materials is infinite, (2) perfect heat transfer between gas and regenerator material, and (3) no regenerator dead volume. Assuming a perfect pressure wave with constant high and low pressures as well as a constant mass flow rate within each half cycle, thus:

$$P = \begin{cases} P_H,\ 0 \le t < 0.5\tau \\ P_L,\ 0.5\tau \le t \ge \tau \end{cases} \tag{4.9}$$

$$\dot{m} = \begin{cases} \dot{m}_a,\ 0 \le t < 0.5\tau \\ -\dot{m}_a,\ 0.5\tau \le t \ge \tau \end{cases} \tag{4.10}$$

Enthalpy flow in the regenerator is

$$\langle \dot{H} \rangle = \frac{1}{\tau} \oint \dot{m} h \, dt = 0.5 m_a (h_{pH} - h_{pL}) \tag{4.11}$$

where, P is gas pressure, P_H high pressure, P_L low pressure, $\dot{m}$ mass flow rate, $\dot{m}_a$ amplitude of mass flow rate, t time, τ time of a cycle, $\dot{H}$ enthalpy flow, h enthalpy of gas, h_{pH} enthalpy of high-pressure gas, and h_{pL} enthalpy of low-pressure gas.

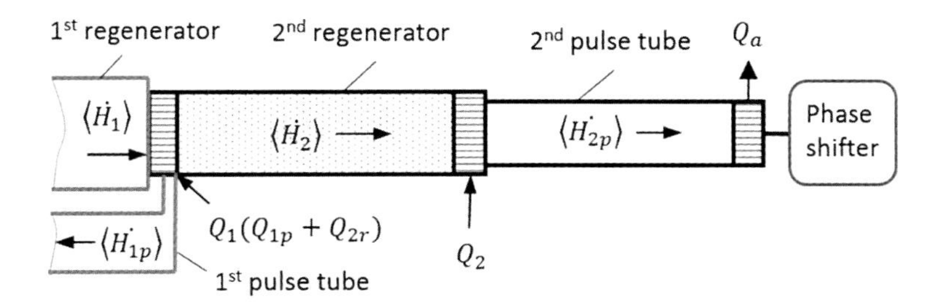

FIGURE 4.23 Time-averaged heat and enthalpy flows in a two-stage PT cryocooler.

For the first-stage regenerator from 300 K to ~45 K, helium gas inside the regenerator can be treated as an ideal gas, $\dot{H}_1 = 0$ because of $h_{pH} = h_{pL}$. For the second regenerator from ~45 K to 4 K, helium inside of the regenerator is considered as a real gas, $\dot{H}_2 > 0$ since $h_{pH} > h_{pL}$. The intrinsic enthalpy loss (to the second stage) normally contributes to the cooling capacity of the first stage. The first-stage cooling capacity can be expressed as

$$Q_1 = \left(\left\langle \dot{H}_{1p} \right\rangle - \left\langle \dot{H}_1 \right\rangle\right) + \left\langle \dot{H}_2 \right\rangle \tag{4.12}$$

where $\left\langle \dot{H}_1 \right\rangle = 0$ for idea gas.

$$Q_1 = \left\langle \dot{H}_{1p} \right\rangle + \left\langle \dot{H}_2 \right\rangle \tag{4.13}$$

The intrinsic enthalpy loss $\left\langle \dot{H}_2 \right\rangle$ can be used for the intercept cooling from the second-stage regenerator at a cost of reducing or eliminating the intrinsic enthalpy flow at the inlet of the second-stage regenerator. Figure 4.24 shows the energy flows in the second stage of the 4 K PT cryocooler with a heat load applied on the regenerator. The intercept cooling Q_c can be a distributed cooling along the regenerator or at an intermediate cooling on a certain location of the regenerator. The applied heat load causes differences of enthalpy flows, $\left\langle \dot{H}_{21} \right\rangle$ and $\left\langle \dot{H}_{22} \right\rangle$ at the two ends of the regenerator. At the cold end:

$$\left\langle \dot{H}_{22} \right\rangle = \left\langle \dot{H}_{21} \right\rangle + Q_c \tag{4.14}$$

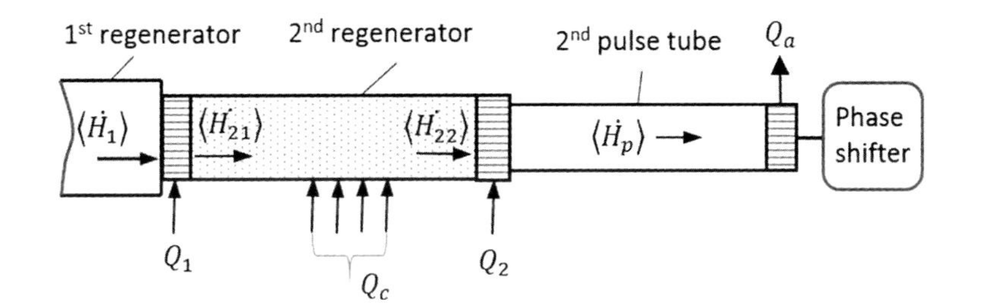

FIGURE 4.24 Energy flows in the second-stage of the PT cryocooler with a heat load on the regenerator.

where $\langle \dot{H}_{22} \rangle$ is equal to $\langle \dot{H}_2 \rangle$ in Equation 4.12. Theoretically, the maximum intercept cooling capacity, which can be provided without affecting the second stage capacity, is:

$$Q_c = \langle \dot{H}_{22} \rangle \tag{4.15}$$

where $\langle \dot{H}_2 \rangle = 0$. Zhu et al. [21] performed a simplified modeling for an ideal 4 K regenerator and showed the intrinsic enthalpy loss could be approximately four times the gross cooling capacity at the 4 K stage.

A numerical analysis, which considers the real gas property, heat transfer, and properties of regenerator materials, has been conducted to analyze the intercept cooling effect of energy flows in the second stage of a 4 K PT cryocooler (Cryomech model PT405) [18]. The first-stage and second-stage cooling stations are at 60 K and 4.2 K, respectively. A distributed heat load of 2.04 W total is applied along the entire second-stage regenerator. The results are given in Table 4.2.

The applied heat load results in slightly increasing the enthalpy losses $\langle \dot{H}_{22} \rangle$ at the cold end of the regenerator by 0.24 W, which is the loss of cooling capacity at 4.2 K. The increased losses might be caused by the real regenerator operation and the higher intercept cooling capacity than the intrinsic enthalpy loss. The enthalpy flow at the inlet of the regenerator $\langle \dot{H}_{21} \rangle$ is reduced by 1.8 W, which indicates the intrinsic enthalpy loss has been used for the intercept cooling. The intercept cooling reduces the 4 K cooling capacity by ~12% of the intercept cooling load applied on the regenerator. This percentage can vary with the different first-stage temperatures and other factors.

4.7.1.2 Intercept Cooling from the Pulse Tube

There is heat transfer between the gas and the wall of the second-stage pulse tube in a 4 K PT cryocooler. Due to the real gas property, it has a non-linear temperature distribution along the second-stage pulse tube with a temperature plateau near the cold end [4]. The function of the pulse tube is to pump heat from a cooling station to a room temperature heat exchanger where the heat dissipates.

The second-stage pulse tube can pump heat added at the intercept location on the pulse tube at the cost of slightly impacting the second-stage performance. Figure 4.25 shows the energy flow when an intercept heat load is applied to the second-stage pulse tube. The enthalpy flow in the second-stage pulse tube is:

Table 4.2 Calculated Energy Flow in the Second Stage of a 4 K Pulse Tube Cryocooler

	$\langle \dot{H}_{21} \rangle$ (W)	$\langle \dot{H}_{22} \rangle$ (W)	$\langle \dot{H}_p \rangle$ (W)	Q_2 at 4.2 K (W)
$Q_c = 0$	15.94	15.93	16.54	0.60 (0.57[a])
$Q_c = 2.04$ W	14.14	16.17	16.54	0.37

[a] Reprinted from *Cryogenics*, 41, Wang C., Helium liquefaction with a 4K pulse tube cryocooler, 491–496, Copyright 2001, with permission from Elsevier.

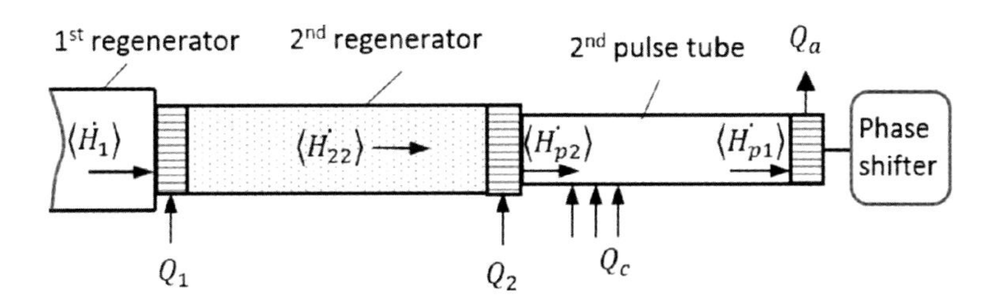

FIGURE 4.25 Energy flow in the second stage of the PT cryocooler with a heat load on the pulse tube.

$$\left\langle \dot{H}_{p1} \right\rangle = \left\langle \dot{H}_{p2} \right\rangle + Q_c \tag{4.16}$$

The impact of Q_c on $\left\langle \dot{H}_{p2} \right\rangle$ is relatively small based on the experimental results [22,23]. The heat ($Q_a = \left\langle \dot{H}_{p2} \right\rangle + Q_c$) is pumped to the warm-end heat exchanger located at room temperature. The heat pumping mechanism for the intercept cooling from the second-stage pulse tube has not been analyzed theoretically, but it has been confirmed in the following experiments.

4.7.2 Distributed Cooling from the Regenerator and Pulse Tube

Distributed cooling can be extracted along the entire length of the second-stage regenerator and pulse tube. Experiments were conducted on a 4 K PT cryocooler, Cryomech PT410 (1W at 4.2 K) to study the distributed cooling configuration [22]. Figure 4.26 is an experiment set up to measure the distributed cooling on the regenerator and the pulse tube, respectively. Manganin wires (a) and (b) are coiled evenly and thermally anchored along the second-stage regenerator (7) and the low-temperature portion (below the first stage) of the second-stage pulse tube (5). Cooling capacities of the distributed cooling are measured by using a DC power supply to apply a current through the manganin wires. During the testing, the first stage temperature is maintained at 45 K, and the second stage has a consistent heat load of 1 W.

Figure 4.27 shows the distributed cooling capacity from the second-stage regenerator as well as its effects on the first and second-stage performances. The distributed heat load increases the second-stage temperature slightly, but it has higher impact on the first stage cooling capacity. For example, applying the distributed heat load of 1 W along the second-stage regenerator increases the second-stage temperature from 3.9 to 4.0 K but reduces the first-stage cooling capacity at 45 K by ~3 W.

Figure 4.28 shows the effects of distributed cooling from the second-stage pulse tube. With the distributed heat load of ≤2 W on the pulse tube, the second-stage temperature has almost no change, and the first-stage cooling capacity has small degradation by ≤0.7 W at 45 K. Comparing to the results of the distributed cooling from the regenerator (Figure 4.27), the second-stage pulse tube shows higher capability to provide the distributed cooling. It is owning to the heat-pumping mechanism of the pulse tube, which pumps the input heat to the warm-end heat exchanger.

Figure 4.29 is a photo of a production model 4 K PT cryocooler with the precooling tube on the second-stage regenerator. Due to reliability concerns of installing a coiled tube on the second-stage pulse tube, a thin-walled tube, the precooling tube is thermally anchored on the regenerator only in the most applications. The commercial 4 K

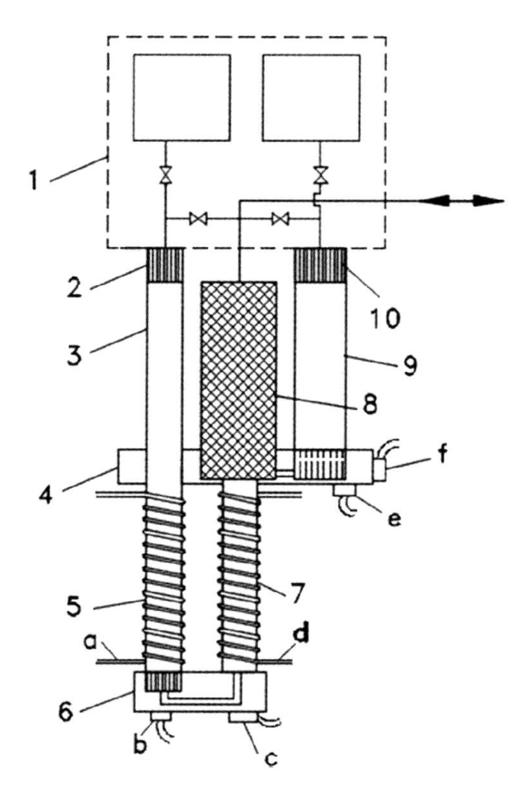

FIGURE 4.26 Schematic of extraction of the distributed cooling on the second-stage regenerator and the pulse tube in a 4 K PT cryocooler: (1) Phase shifters. (2) Hot heat exchanger of the second stage. (3) Warm temperature portion of the second-stage pulse tube. (4) First-stage cooling station. (5) Low-temperature portion of the second-stage pulse tube. (6) Second-stage cooling station. (7) Second-stage regenerator. (8) First-stage regenerator. (9) First-stage pulse tube. (10) Hot heat exchanger of the first stage. (a) Manganin wire heater. (b) Temperature sensor. (c) Heater. (d) A manganin wire heater. (e) Temperature sensor. (f) Heater.

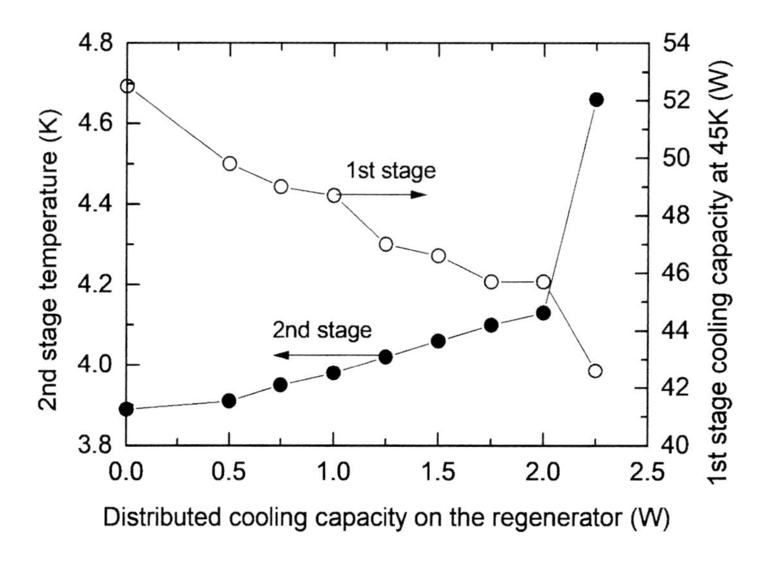

FIGURE 4.27 Effects of the distributed cooling from the second-stage regenerator.

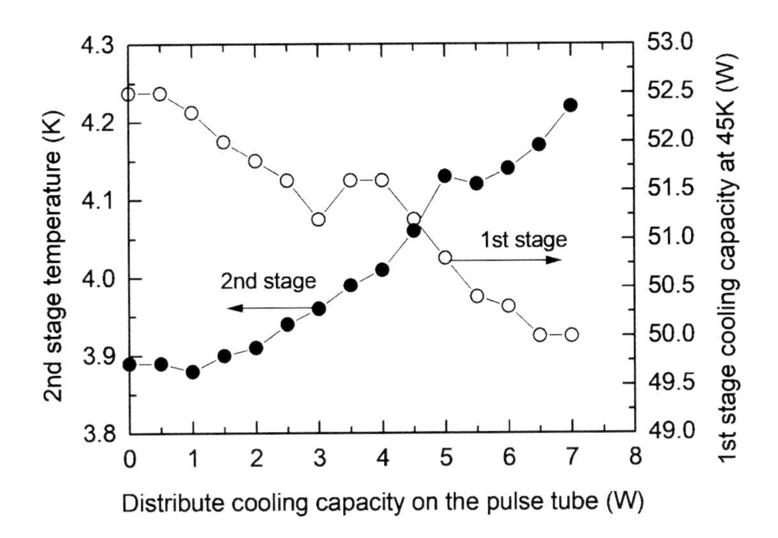

FIGURE 4.28 Effects of the distributed cooling from the second-stage pulse tube.

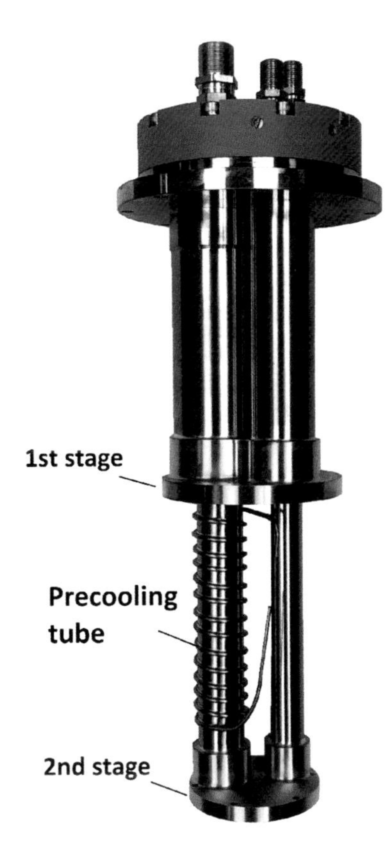

FIGURE 4.29 Standard option of the precooling tube on a commercial 4 K PT cryocooler. (Courtesy of Website of Cryomech, Inc., at: http://www.cryomech.com/.)

PT cryocooler with the precooling tube is a standard option used for precooling helium gas from ~45 K down to ~4 K in cryogen-free 1 K cryocoolers, cryogen-free dilution refrigerators, etc.

4.7.3 Intermediate Cooling

Intermediate cooling can be provided at a localized location on the second-stage regenerator and pulse tube. Location of the intermediate cooling is chosen based on the required cooling temperature and capacity. The experimental setup for studying the intermediate cooling on the second-stage regenerator and pulse tube is illustrated in Figure 4.30a. A 4 K PT cryocooler, Cryomech PT407 (0.7 W at 4.2 K) was used to conduct the measurement. The design and installation of the intermediate cooling heat exchanger are shown in Figure 4.30b.

The intermediate heat exchangers (7) and (15) are made of copper rings with a thickness of 12.7 mm and are thermally attached on the walls of the second-stage regenerator and the pulse tube [23]. The intermediate heat exchanger (7) on the second-stage pulse tube is called the "PT-intermediate heat exchanger" and the one on the second-stage regenerator (15) is called the "R-intermediate heat exchanger." The center of both intermediate heat exchangers is 51 mm away from the top of the second-stage cooling station (10). Two additional temperature sensors (5) and (9) are mounted on the wall of the second-stage pulse tube to measure the temperature distribution for analysis. They are located 20 mm away from the center of the PT-intermediate heat exchanger. During the measurement, the first stage is maintained at 55 K, and the second stage has a heat load of 0.7 W.

Figure 4.31 shows the cooling capacity of the R-intermediate heat exchanger and its effects on the first and second stages. The R-intermediate heat exchanger can provide a cooling capacity of 0.5 W at 12.3 K or 0.75 W at 13.3 K. Extracting the intermediate cooling capacity of 0.75 W increases the second-stage temperature by 0.08 K and lower the first-stage cooling capacity by 1.7 W at 55 K.

Figure 4.32 shows the cooling performance of the PT-intermediate heat exchanger and its effects on the first and second stages. With a heat load of 0.75 W on the PT-intermediate heat exchanger at 8.7 K, the second-stage temperature increases by 0.11 K, and the first-stage cooling capacity drops by 4.7 W. The first-stage capacity loss is higher than the loss (1.7 W) caused by the same amount of the intermediate cooling on the R-intermediate heat exchanger, and also higher than the loss (~0.2 W) introduced by the same amount of the distributed cooling on the pulse tube (see Figure 4.28).

To help understand the work mechanism of the PT-intermediate heat exchanger, the temperature distributions on the wall of the second-stage pulse tube with different intermediate heat loads are measured and given in Figure 4.33. The heat loads on the PT-intermediate heat exchanger changes the temperature distributions on the PT wall. With a heat load of ≥0.25 W, the intermediate heat exchanger temperature is higher than that at the upper spot, which is located 20 mm away toward the warm end.

The heat-transfer process with a heat load on the PT-intermediate heat exchanger is illustrated in Figure 4.34. The heat load Q_c applied on the intermediate heat exchanger

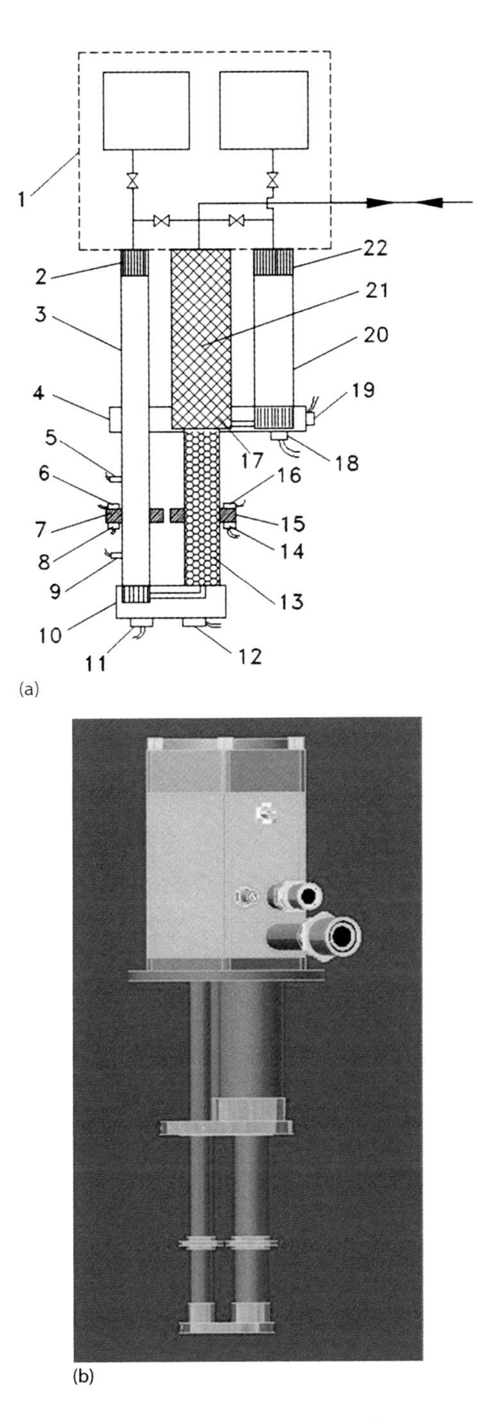

FIGURE 4.30 (a) Schematic of performance measurement setup of a 4 K PT cryocooler with the two intermediate heat exchangers. (b) Installation of intermediate heat exchangers on 4 K PT cryocooler, model PT407. (1) Phase shifters. (2) Second-stage hot heat exchanger. (3) Second-stage pulse tube. (4) First-stage cooling station. (5, 6) Temperature sensors. (7) PT-intermediate heat exchanger. (8) Heater. (9) Temperature sensor. (10) Second-stage cooling station. (11) Heater. (12) Temperature sensor. (13) Second-stage regenerator. (14) Heater. (15 R-intermediate heat exchanger. (16) Temperature sensor. (17) First-stage regenerator. (18) Heater. (19) Temperature sensor. (20) First-stage pulse tube. (21) First-stage regenerator. (22) First-stage hot heat exchanger.

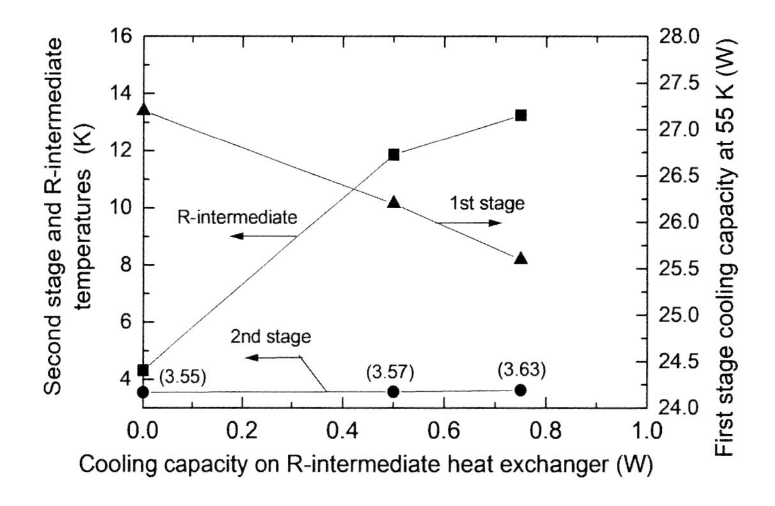

FIGURE 4.31 Cooling performances of the intermediate heat exchanger on the second-stage regenerator. (Reprint from *Cryocooler*, 15, Wang, C., Extracting cooling from the pulse tube and regenerator in a 4K pulse tube cryocooler, ICC Press, Boulder, CO, 177–184, Copyright 2009, with permission from Elsevier.)

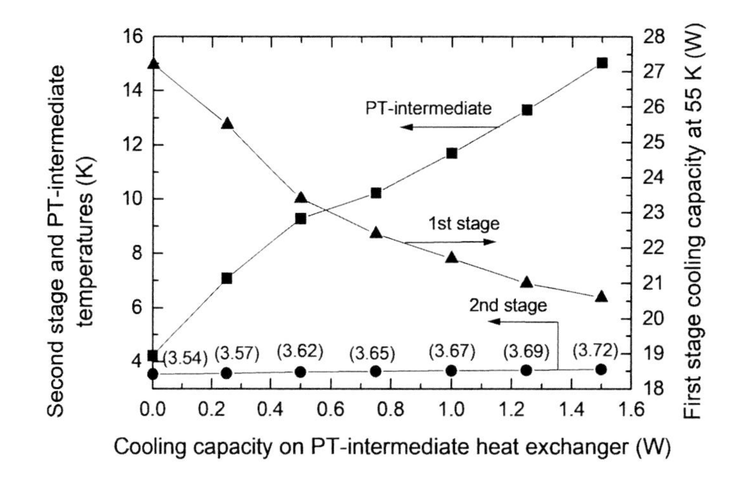

FIGURE 4.32 Cooling performance of the intermediate heat exchanger on the second-stage pulse tube. (Reprint from *Cryocooler*, 15, Wang, C., Extracting cooling from the pulse tube and regenerator in a 4K pulse tube cryocooler, ICC Press, Boulder, CO, 177–184, Copyright 2009, with permission from Elsevier.)

conducts radially through the wall of the pulse tube, Q_w, and also conducts axially to both the low temperature side, Q_l, and the high temperature side, Q_h. This results in a large area of the pulse tube with higher temperatures. Eventually, the helium gas inside of the pulse tube absorbs the heat through heat transfer between the wall and the helium gas. The cooling capacity on the intermediate heat exchanger is primarily extracted from the expansion gas inside of the pulse tube. A very small portion of the cooling capacity on the intermediate heat exchanger can be contributed by the thermal conduction between

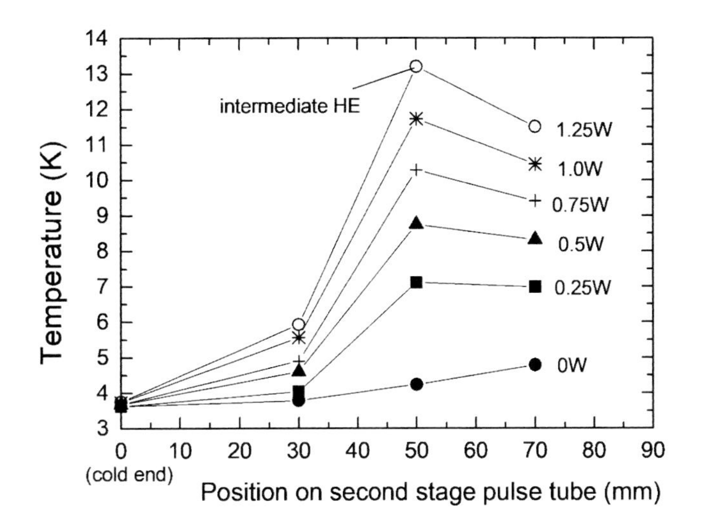

FIGURE 4.33 Temperature distribution on the pulse tube with different PT-intermediate heat loads. (Reprint from *Cryocooler*, 15, Wang, C., Extracting cooling from the pulse tube and regenerator in a 4K pulse tube cryocooler, ICC Press, Boulder, CO, 177–184, Copyright 2009, with permission from Elsevier.)

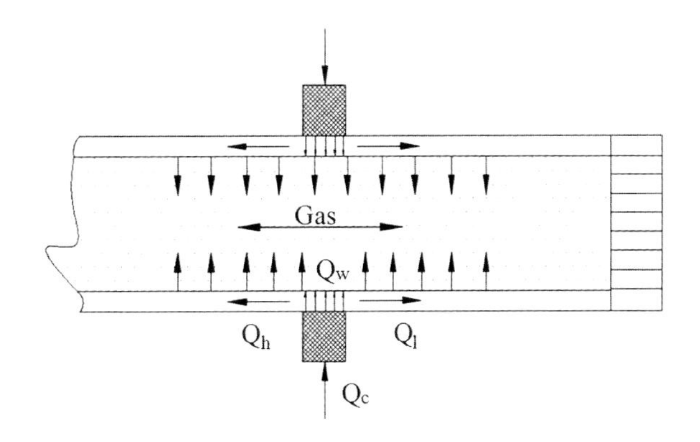

FIGURE 4.34 Heat transfer process of extracting intermediate cooling on the second-stage pulse tube. (Reprint from *Cryocooler*, 15, Wang, C., Extracting cooling from the pulse tube and regenerator in a 4K pulse tube cryocooler, ICC Press, Boulder, CO, 177–184, Copyright 2009, with permission from Elsevier.)

the second-stage cooling station and the PT-intermediate heat exchanger, which results in the reduction of the second-stage cooling capacity.

In summary, both the second-stage regenerator and pulse tube in 4 K PT cryocoolers can provide the intercept cooling. Comparing the intercept cooling on the regenerator and the pulse tube, the second-stage regenerator efficiently provides the intermediate cooling with less effects on the first and second cooling stages. Controversially, the second-stage pulse tube can efficiently provide the distributed cooling. The experimental results above can be used as references while choosing the intercept cooling. The performance will vary with different intercept locations and different 4 K PT cryocoolers.

Acknowledgments

Many thanks to Dr. Peter Kittel, retired from NASA Ames Research Center, and Mr. Jeff Raab, retired from Northrop Grumman, for reviewing and improving the manuscript. Thanks also goes to Andrey Olesh at Cryomech, Inc. for preparing some figures in this chapter.

References

1. Wang, C., "Characteristic of 4K pulse tube cryocooler in application", *Proceeding of Twentieth International Cryogenic Engineering Conference* (2005), Beijing, China, Elsevier, pp. 265–268.
2. Tomaru, T., Suzuki, T. and Haruyama, T. et al., "Vibration analysis of cryocoolers," *Cryogenics*, vol. 44 (2004), pp. 309–317.
3. Vibrations of PT410, (2010), Internal Report at Cryomech Inc.
4. Wang, C., "Numerical analysis of 4K pulse tube coolers: Part II. Performances and internal processes," *Cryogenics*, vol. 37 (1997), pp. 215–220.
5. Allweins, K., Qiu, L.M. and Thummes, G., "Damping of intrinsic temperature oscillation in a 4K pulse tube cryocooler by means of rare earth plates," *Advanced Cryogenic Engineering*, vol. 53A (2008), AIP, Melville, NY, pp. 109–116.
6. Li, R., Onishi, A., Satoh, T., and Kanazawa, Y., "Temperature stabilization on cold stage of 4K GM cryocooler," *Cryocooler*, vol. 9 (1997), Plenum Press, New York, pp. 765–771.
7. Website of Cryomech, Inc., at: http://www.cryomech.com/
8. Numazawa, T. et al., "A new ceramic magnetic regenerator material for 4K cryocoolers," *Cryocooler* 12 (2003), Kluwer Academic/Plenum Publishers, New York, pp. 473–482.
9. Wang, C., "Numerical analysis of 4K pulse tube coolers: Part II. Performances and Internal Processes," *Cryogenics*, vol. 37 (1997), pp. 215–220.
10. Wang, C., Hafner, H. U. and Heiden, C., "Performance and internal process of a 4K GM cooler," *Advances in Cryogenic Engineering*, vol. 43B (1998), Plenum Press, New York, pp. 1775–1782.
11. Onishi, A., Satoh, T., Hasebe, T., and et al., "Behavior of a 4K-GM cryocooler in a magnetic field," *Journal of the Cryogenic Society of Japan*, vol. 34 (1999), No. 5, pp. 196–199.
12. Wang, C., Thummes G. and Heiden C., "A two-stage pulse tube cooler operating below 4K," *Cryogenics*, vol. 37 (1997), p. 159.
13. Moriea, T., Shiraishib T. and Xu, M., "Experimental investigation of cooling capacity of 4K GM cryocoolers in magnetic fields," *Physics Procedia*, vol. 67 (2015), pp. 474–478.
14. Product catalog from Bruker BiospinSHI cryogenics group. Website: https://www.bruker.com/ Bruker BioSpin Co.
15. Wang, C. and Gifford, P.E., "A single-stage pulse tube cryocooler for horizontally cooling HTS MRI probe," *Advances in Cryogenic Engineering*, vol. 49B (2004), AIP, Melville, New York, pp. 1805–1811.
16. Ikeya, Y., "Expanding market of 4K GM cryocooler," *Presented at 24th IIR International Congress of Refrigeration* (2015).
17. Thummes, G., Wang, C. and Heiden, C., "Small scale 4He liquefaction using a two-stage 4K pulse tube cooler," *Cryogenics*, vol. 38 (1998), pp. 337–342.
18. Wang C., "Helium liquefaction with a 4K pulse tube cryocooler," *Cryogenics*, vol. 41 (2001), pp. 491–496.
19. Uhlig, K., "Condensation stage of a pulse tube pre-cooled dilution refrigerator," *Cryogenics* (2008), vol. 48, pp. 138–141.
20. Product catalogs of Janis Research Co., BlueFors Cryogenics Oy and Leiden Cryogenics BV.
21. Zhu, S., Ichikawa, M., Nogawa, M., and Inoue, T., "4K pulse tube refrigerator and excess cooling power," *Advances in Cryogenic Engineering* (2002), vol. 47A, AIP, Melville, New York, pp. 633–640.
22. Wang, C., "Extracting cooling from the pulse tube and regenerator in a 4K pulse tube cryocooler," *Cryocooler*, vol. 15 (2009), ICC Press, Boulder, CO, pp. 177–184.
23. Wang, C., "Intermediate cooling from regenerator and pulse tube in a 4K pulse tube cryocooler," *Cryogenics*, vol. 48 (2008), pp. 154–159.

5. Reduction of Vibration and Drift

Luke Mauritsen

5.1 Reduction of Vibration and Drift

Cryocoolers offer the promise of reducing our dependence on liquid cryogens and simplifying the experience at low temperatures. However, the cyclic mechanical oscillations introduced by the refrigeration cycle of a cryocooled system bring new challenges to many applications. Oscillation of pressure, rotation of motors, and cyclic motion of linkages create vibrational energy, which couples into the surroundings. Additionally, the cyclic nature of the refrigeration cycle creates temperature oscillations, which directly couple into the device or sample being cooled. It is important to have a common understanding of the terminology and a practical understanding the sources of cryocooler vibrations and drift in order to understand principles for the reduction of vibrations and drift for cryocooled applications.

5.1.1 Terminology

It will be helpful to establish a common terminology for the technical concepts presented in this chapter. There has been some confusion surrounding the use and meaning of terms below, which creates a barrier to a common practical understanding of the concepts and adds to the difficulty in proliferation of clear concepts to co-workers and future generations of students.

Vibration: A mechanical vibration is any motion that repeats itself after an interval of time [1]. Vibrations in a cryocooled system can be generated by a number of sources, which will be discussed later in this chapter. Examples of such sources of vibrations in a cryocooled system can be a rotary valve, head motor, helium compressor capsule, cooling fan, Scotch-Yoke and displacer mechanism, and gas pressure oscillations. These sources create the repeating motion, or energy, which causes vibrations in other parts of the entire system. Vibrations can be further described in terms of vibration displacement and accelerations, and it is important to know the difference between these terms.

Natural frequency: After an initial disturbance or input of energy, an object will vibrate at its own unique frequency. The frequency with which it oscillates without external forces is known as its natural frequency or resonant frequency [1]. Every individual component in a cryogenic system and an experimental setup has a natural frequency and can be "excited" by vibration sources, such as a cryocooler and compressor. A system of components may have several natural frequencies. It is important to understand the dominant natural frequencies of a system before attempting to control the displacement and acceleration of specific components.

Resonance: If the frequency of the external source of vibration coincides with any of the natural frequencies of the system, resonance occurs [1]. This is generally an undesirable condition since vibration displacement of a component or system will increase dramatically in a resonant condition. System design will usually require natural frequencies that are either far above or below the source excitation frequencies depending on the goal of the designed system. For instance, minimization of displacement of a component may require that it has a resonant frequency much higher than the source of the vibration, while minimization of acceleration may require an isolated component, which has a lower natural frequency than the source.

Displacement: The displacement of an object is the amplitude of oscillation from its equilibrium position. It is important to understand the total displacement of one object relative to another. For instance, a cooled sample attached directly to cryocooler may have a large displacement relative to the components on an optical table because of how the cryocooler is supported. Often, a large optical table is the best vibrational reference plane because of its large mass and large table space to build various elaborate experiments. In this case, a relevant way to describe vibrations is in terms of total displacements relative to the optical table. Components on an optical table typically have a small displacement relative to one another and to the optical table, while the table has a high displacement relative to the surrounding room.

Acceleration: Acceleration is the rate of change of velocity per unit time. An object that is oscillating at a high frequency and large displacement will have high accelerations. Often, vibrations are described in terms of acceleration and displacement and, depending on the application, one may be desirable over the other. An object may have a low displacement and a high acceleration, or alternatively, a high displacement and a low acceleration. For instance, effective isolation of vibrations from a source may require a design that increases displacement in order to reduce accelerations.

Vibration isolation: Vibration isolation is a broad term, which means to bring about a reduction in a vibratory effect [2]. A vibration isolator is used to reduce vibrations between a source and an object such that a desired response of the system is achieved. Cryogenic systems may incorporate both passive and active vibration isolation techniques. Passive vibration isolation incorporates both passive isolation (spring stiffness) and energy dissipation (damping). There is no such thing as a pure isolator or a pure damper. An appropriate spring between a vibration source and an object exhibits minimal energy dissipation but effectively transfers energy to a lower frequency range. Alternatively, an elastomer damper will dissipate energy while increasing transmissibility at high frequencies. In passive systems, the proportion of isolation and energy dissipation is carefully chosen for the frequencies necessary to isolate from. Active vibration isolation uses feedback and an actuator to reduce the vibratory effect.

Vibration damping: Vibration damping is the mechanism by which vibrational energy is converted into heat. In cryogenic systems, vibration isolation and vibration damping are often mistakenly used interchangeably. Instead, it is appropriate to think of vibration damping as one element of vibration isolation. An increase in vibration damping will increase the amount of energy dissipated but only to the degree that there exists relative velocity between one end of the damper to the other. This relative velocity also produces a force across the damper, which increases the high-frequency transmissibility of the isolator. For these reasons, it is important that damping elements are used to carefully produce the vibration isolation at the frequency range, which is desired.

Positional drift: Positional drift can be thought of as a motion on a thermal timescale, and it is necessary to define separately from the traditional understanding of vibration because the source of it is different. Positional drift is motion that results from expansion or contraction of a material due to a change in temperature of the material. Positional drift is often used interchangeably with "thermal drift" since it is understood that the origin of the drift is thermal. The more general term "thermal drift" is used to describe any change (electronic or mechanical) that is due to changes in temperature. In cryogenic systems, positional drift of components can be a result of ambient room temperature changes, deliberate temperature changes in the cryocooled sample, or undesired temperature oscillations in the cryogenic system.

5.1.2 Sources of Vibrations in Cryocoolers

Cryocoolers are mechanical devices that recirculate a refrigerant through a refrigeration cycle, producing continuous low temperatures. Two widely used cryocooler systems are based on the Gifford–McMahon (GM) and pulse tube (PT) refrigerator cycles.

A drawback of each of these cryocoolers is mechanical vibrations that have a number of sources. An understanding of the sources of these vibrations and practical considerations for addressing them is important.

A cryocooler system consists of a compressor, a cold head (expander), and helium hoses to transfer pressurized helium gas to the cold head from the compressor (Figure 5.1). Vibrations in the compressor are generated primarily by the compressor capsule. The helium hoses extend to the cold head and are a path for vibrations to be transferred from the compressor to the cold head and to anything they physically contact. The hoses vibrate at 50 to 80 Hz due to the capsule operating frequency, and they also exhibit a pressure oscillation at a frequency of approximately 1–3 Hz. This change in pressure creates an oscillating length change in the hose and can be significant enough to rock an optical table back and forth if the hose is not able to freely expand. Audible noise from the compressor varies by manufacturer and by size of cryocooler and may be distracting in a laboratory environment; however, some new cryocooler systems have incorporated quieter compressor technology, which creates significantly less audible noise to the level where operation even inside a laboratory is common. The compressor can be located in a separate room with long helium hoses extending to the cold head. Since the compressor is only supplying room temperature helium to the cold head, there are no significant efficiency losses due to longer helium hoses. Due to these compressor vibrations, as well as cold-head vibrations explained below, the cold head usually requires some type of isolation from the optical table or

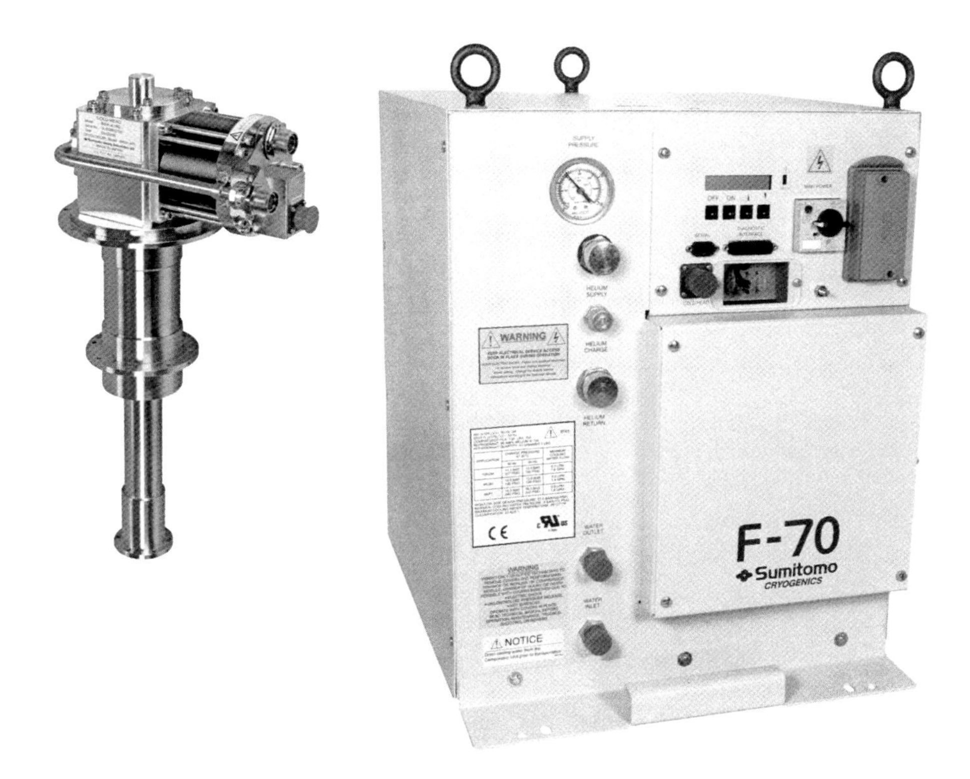

FIGURE 5.1 Image of a cryocooler and compressor. (Courtesy of SHI Cryogenics Group.)

experimental platform. This chapter provides an in-depth look at the vibrations of a GM cryocooler. Vibrations are also present with a PT cryocooler but are not discussed in this chapter.

5.1.2.1 Vibrations Sources from a Gifford–McMahon Cryocooler

Chapter 3 discusses the refrigeration cycle and mechanical operation of both the GM and PT cryocoolers. This section discusses vibration sources of the GM cryocooler. The primary vibration sources in a GM cold head are the motor, the mechanical components including the Scotch–Yoke, the first- and second-stage displacer linkages, and the pressure fluctuations created with each rotation of the rotary valve. Figure 5.2 shows the cross section of a GM cold head. As the motor rotates, a Scotch–Yoke connection drives the first- and second-stage displacers up and down. The motor also rotates a rotary valve, which allows a sudden exchange of high-pressure helium gas into and out of the cold-head displacers area. One source of vibration is the motor itself. Motor vibrations will be transmitted to the first and second stages of the cryocooler as well as the vacuum flange. Many GM cold heads use a *single-phase motor*, which is the cost-effective choice for a motor, but also inefficient and noisy. A *multi-phase motor*, which requires more sophisticated drive electronics, will significantly reduce motor noise. Some manufacturers are beginning to integrate these multiphase motors into cryocoolers for laboratory applications. Other sources of noise in a GM cold head are due to the sudden cyclical pressurization and depressurization in the first- and second-stage displacer area with each rotation of the rotary valve. There are mechanical linkages between the motor, the Scotch–Yoke, and the displacer(s), and each have some mechanical tolerance. Each time the displacer area is pressurized and a new volume of helium gas is forced through the displacers, this series of mechanical linkages are forced to one extreme in tension and then the other extreme in compression. This generates a series of mechanical impulses. Another source of vibration is directly from the stainless-steel tubes, which house the displacers. Each time these tubes experience a change in pressure, they undergo a mechanical strain and lengthen at the same frequency as the pressurization and

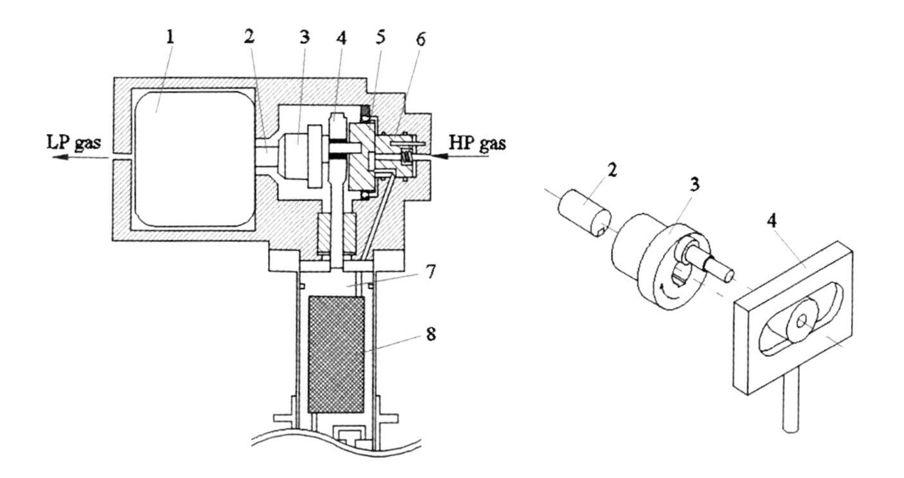

FIGURE 5.2 Image from Chapter 3 (courtesy Figure 3.13 in Chapter 3) showing a cross section of a Scotch–Yoke-driven GM cryocooler: (1) Cold-head motor. (2) Motor shaft. (3) Crank. (4) Scotch–Yoke. (5) Rotary valve. (6) Valve plate. (7) Displacer. (8) Regenerator.

depressurization of the cycle, which is usually 1–3 Hz. Each time these tubes experience a sudden pressure change, they resonate at their natural frequency, which is typically a hundred to a few hundred Hz.

5.1.3 Sources of Positional Drift

Positional drift, also sometimes called "thermal drift" in a cryogenic system is motion that results from expansion or contraction of a material due to a change in temperature of the material. This motion exists on a thermal timescale of seconds and minutes and should be understood and addressed separately from mechanical vibrations because the sources of and the solutions to positional drift are different than the sources of and the solutions to vibrations. The equation for linear thermal contraction (or expansion) is as shown below, where α is the coefficient of linear expansion, L_i is the initial length of material, and T_i and T_f are the initial and final temperatures of the material. dL represents the change in material length as a result of the change in temperature. Keep in mind that α is a temperature dependent coefficient, so if the temperature change is large, we must integrate over the range of α. A useful reference for thermal expansion of commonly used materials in cryogenics is found in Jack Ekin's book, *Experimental Techniques for Low-Temperature Measurements* [3].

$$dL = \alpha L_i \left(T_i - T_f\right) \tag{5.1}$$

Minimizing positional drift can be particularly important when your sample is to be aligned with measurement apparatus, such as a high numerical aperture optical lens assembly. The higher the spatial resolution and the lower depth of focus, for instance, the greater the need for sample stability relative to the optics over long timescales. Some microscopy applications require this relative stability to be less than a few hundred nanometers drift per hour. Section 5.7 highlights an application where high-performance optics are integrated into the cryogenic system in order to minimize drift. In cryogenic systems, there are a few common sources of positional drift.

1. Changing the temperature of the sample stage during temperature-dependent measurements can be one of the largest sources of unwanted positional drift. Each time the temperature is increased for a new set point, more heat is added to the sample stage, which adds heat to each component thermally connected to the sample stage. As the materials along the conductance path from the cryocooler stage to the sample change temperature, they also expand proportional to their thermal expansion coefficient, length, and temperature change. If the conductance path is long, the positional drift can be significant. If the temperature of any of the components supporting the sample rises above approximately 50 K, the coefficient of linear expansion increases dramatically. This is especially impactful in wide temperature measurements and for cooldown and warm up of the cryocooled system. During cooldown, since some materials supporting the sample will undergo a temperature change of nearly 300 K, the positional drift can typically cause hundreds of microns and sometimes even millimeters of change in the sample position, depending on the cryostat design.

2. An advantage of cryocoolers is that once a low temperature is reached, a relatively stable temperature can be maintained for long periods of time, even months or years. However, even when a cryogenic system is stabilized to fractions of a degree so that positional drift becomes negligible in the cryostat, fluctuations of laboratory temperature can be a significant contributor to positional drift of a cooled sample stage relative to the surrounding measurement apparatus. Since temperature variations in a typical laboratory can vary by more than 1°C–2°C over a day, this relative drift can be hundreds of nanometers per hour. In a stabilized cryogenic system, this becomes the limiting factor for high numerical aperture measurements.
3. If cryogens are used to provide cooling between the cryocooler and the sample stage, special consideration must be given to minimizing drift because the cryogen gas flow creates a highly dynamic environment, which creates thermal fluctuations. For example, in cryogenic systems where a cryogen is used as a gas vibration isolation stage, the dynamics of the gas flow can create a variation in sample position as the gas interacts with surrounding surfaces of cryostat construction materials at various temperatures. This can be especially significant if the temperatures of portions of the materials and gas are above about 50 K.
4. Although not a source of significant positional drift, for completeness it is worth mentioning the thermal oscillations at each of the cooling stages during steady-state operation, which result from the mechanical oscillations of the cryocooler. This is because for each instance of the refrigeration cycle, the gas expansion portion, which cools the cryocooler stage(s), is only a fraction of a second, after which, the cooled gas must exit the expansion chamber and be replaced with a new volume of gas. During this time, between cyclical expansion, the cryocooler stage temperature increases in proportion to the heat load that is acting on it. Typical thermal fluctuation amplitudes of a 4 K cryocooler are 150–200 mK on a timescale of 1–2 Hz determined by the frequency of the cryocooler operation. If the cryocooler has a higher heat load on it, fluctuations in temperature can be over 500 mK. For applications where a higher heat load is present, some cryocooled systems have variable speed technology that allows the cryocooler to run a faster cycle, reducing the time in between expansion cycles, which can reduce the amplitude of thermal fluctuations. For applications that require stable temperature, this is useful; however, these fluctuations do not significantly affect positional drift at low temperatures because most metals nearly stop contracting below about 50 K.

5.1.4 Measuring Vibrations and Positional Drift

The first step to understanding how to stabilize a cryogenic environment is to understand how to measure vibrations and positional drift. There are at least a few good techniques for this, and three of these techniques are discussed in this section. In choosing a measurement technique, the most important thing is to first *know what exactly you want to measure*. If what matters is relative displacement between components on an optical table, and you are measuring from subnanometer to micron range displacements with high accuracy and resolution, then a capacitive displacement sensor or an interferometer may be a good choice. If what matters is accelerations on a cryogenic platform or breadboard, then an accelerometer is a good choice. Whichever method you choose, it is very helpful to use a measurement technique that gives instantaneous results. This is because displacements that are small, such as nanometer scale displacements, are often not intuitively predictable. The ability to make many adjustments with instant feedback helps to build a more accurate understanding of the causes of the displacements.

5.1.4.1 Capacitive Linear Displacement Sensor

Capacitive linear displacement sensing is a non-contact method of using electrical capacitance to detect distance from a sensor to a conductive target. As the distance between the target and sensor change, the capacitance between the surfaces changes, and this is calculated electronically to output a linear voltage range corresponding to distance. Linear capacitive displacement sensors can resolve subnanometer displacements, work with any conductive surface, such as copper, aluminum, and steel, and they are relatively inexpensive compared to other subnanometer measurement techniques. The sensors are small, versatile, and robust, which makes them easy to use with confidence in the measurement. The image below shows the simple configuration of a linear capacitive measurement for a cryogenic system on a table. This is a relative measurement, which means the table is the vibrational reference frame, and any measurement is showing displacement of the cryogenic sample stage relative to the table. The sensor probe should be rigidly mounted to the table (Figure 5.3).

Capacitive sensing is not suitable where a large gap between the senor and target is required (an interferometer is better here). When measuring cryocooler vibrations and positional drift, the frequencies of the dominating displacements are usually less than a few hundred hertz. Cryocoolers can also be a source of high-frequency electromagnetic interference (EMI) noise, which can affect sensitive electronics. For these reasons, it is useful to use a low-pass filter in line with the measurement. This will filter out unwanted high frequencies, including any electrical noise, and will lower the noise floor. Some capacitive measurement system manufacturers will include low-pass filter options in their electronics. If it is necessary to build one, a single pole low-pass filter like the simple resistor-capacitor (RC) circuit shown in Figure 5.4 works well. Selecting the appropriate resistor and capacitor for a cutoff frequency of 1 kHz will significantly reduce higher frequency noise while allowing signals in the hundreds of hertz.

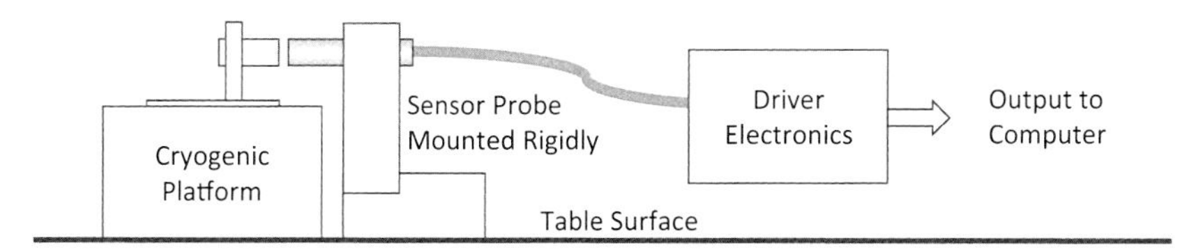

FIGURE 5.3 Capacitive linear displacement sensor arrangement for a cryogenic system measurement.

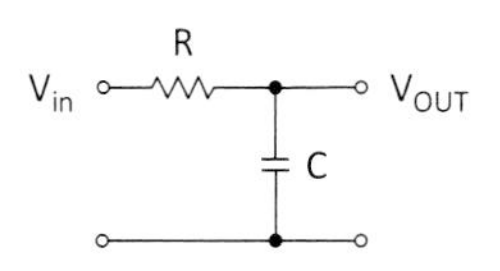

FIGURE 5.4 Single-pole low-pass filter RC circuit diagram.

Most measurements of vibrations on cryogenic systems can be made at room temperature. At room temperature, vacuum housings and radiation shields are not necessary, so the setup is relatively simple, and many adjustments can be made quickly while measuring. In many cases, this may be sufficient to understand vibrations of the system. 4 K GM systems typically exhibit lower vibrations at low temperature than at room temperature. This is because the viscosity of helium gas flowing through the displacers is lower, and the pressure differential across the cryocooler head is smaller at low temperatures. If vibrations are acceptable before cooldown, they are likely to be better at base temperature. Vibration displacement data are shown in Figure 5.5 at room temperature and at low temperature.

One reason for testing at low temperatures is to ensure that a bolted joint hasn't loosened or an epoxied joint hasn't fractured due to differential thermal contraction during cooldown. If displacement measurements at low temperatures are necessary, a hardware configuration that locates the sensor probe inside vacuum and mounted rigidly to the table can be made. It is important that the sensor probe is connected rigidly to the table while the vacuum flange is connected flexibly via bellows to ensure that displacement measurements represent the sample mount relative to the table and any vacuum flange movement does not affect the measurement. An example is shown in Figure 5.6.

Two commercial suppliers of capacitive linear-displacement sensors are shown below:

Lion Precision
7166 4th Street North
Oakdale, MN 55128
651-484-6544
toll free 800-250-9297
www.lionprecision.com

Micro-Epsilon
8120 Brownleigh Dr.
Raleigh, NC 27617
919 787 9707
919 787 9706
www.micro-epsilon.com

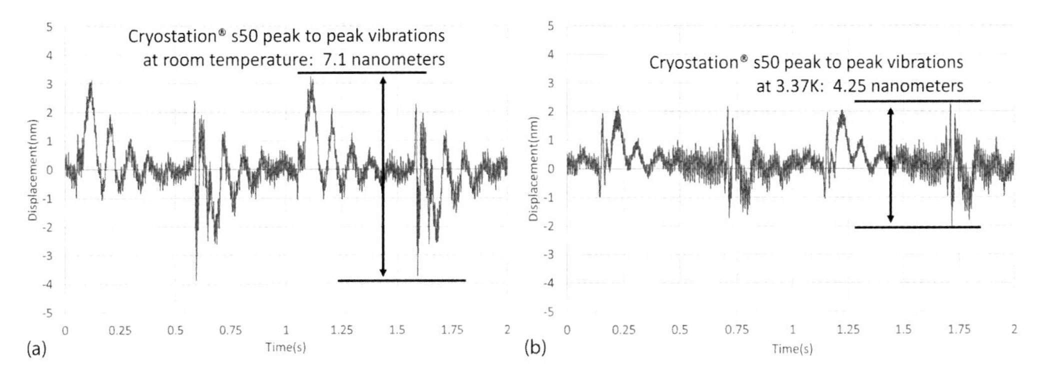

FIGURE 5.5 Vibration displacement measured at the sample mount of a Cryostation® s50, which uses a custom version of a Sumitomo Heavy Industries RDK-101D 4K cryocooler. Vibrations at room temperature are about 7.1 nm (a), and at low temperature are about 4.25 nm (b).

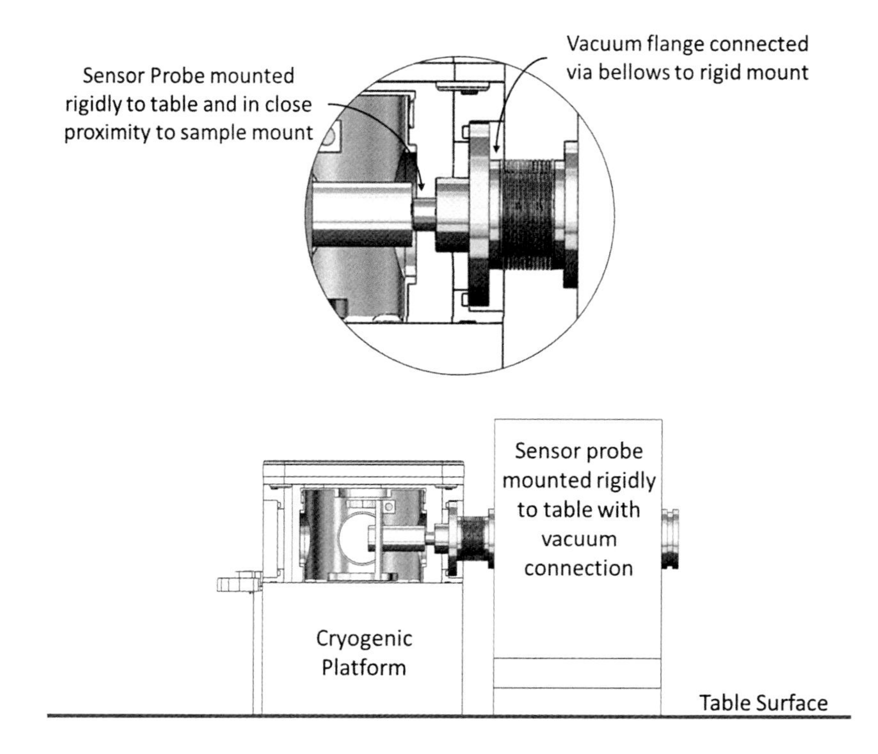

FIGURE 5.6 Capacitive linear displacement measurement arrangement for low-temperature measurement.

5.1.4.2 Accelerometer

An accelerometer is a sensor that outputs an electrical signal proportional to the applied acceleration. The transduction of the signal from acceleration to electrical output may be through piezoelectric, piezoresistive, or capacitive techniques. Whatever the technique, this section will give some basic guidelines on when an accelerometer is the appropriate measurement device to use and some considerations for how to choose the right product for testing a cryogenic system.

An accelerometer directly measures the rate of change in velocity of the platform it is attached to. Often represented in g's, this measurement provides insight into how much vibrational energy is present on a cryogenic platform. Most importantly, the power spectral density (PSD) of an accelerometer measurement provides information about which frequencies the vibrational energy is present at. See the Section 5.1.5 "Understanding Cryocooler Vibrations" for more about PSD. The accelerometer is the ideal tool for measurement of a cryogenic platform when there are devices, nanopositioners, optics, or probes that must be stable relative to each other on the platform. Vibrational energy present on a cryogenic platform will couple in to these elements and excite each of them at their unique resonant frequency. Since many of these elements, such as combinations of nanopositioners stacked on top of each other in series with a device, optic, or probe to position, may have resonant frequencies in the low hundreds of hertz, vibrational energy coupling into them may create large relative displacements from one element to another on a cryogenic platform.

An accelerometer is not the best method to measure vibrations of a cryogenic platform relative to other objects, such as other optics on an optical table. There are a few reasons for this. One is that the accelerometer will only measure the vibrations present on the object it is attached to and cannot measure relative to another object, which also has its own vibrational signature. Another reason is that accelerometers with low-frequency sensitivity are typically large in size, and therefore it is difficult or impossible to attach them to small objects. It is better to measure relative vibrations directly with a capacitive displacement sensor or an interferometer because these direct measurements will include resonant frequencies of both relative objects being measured.

There are many specifications to consider when choosing an accelerometer. A few important considerations for measuring cryogenic systems are mentioned here. Since cryogenic systems often have a dominant working frequency near 1–2 Hz, choosing the appropriate sensitivity is important in order to measure low frequencies. If isolation from 1 to 2 Hz working frequencies of the cryocooler is desired, a sensitivity that allows the measurement of subhertz frequencies will be important. One accelerometer may not work for all the frequencies that need to be measured since high-sensitivity accelerometers have a low internal resonant frequency and may not be able to measure above hundreds of hertz. Some companies offer cryogenic-rated accelerometers that can operate down to 4 K. While these may be necessary to use in some applications, the bandwidth of measurement is limited in these models, and the time necessary to make each measurement can be substantial due to cooldown, warmup, and the added necessity of appropriate electrical feedthroughs for signals. Most or all of the characterization of vibrational energy on a platform can be done at room temperature because a typical cryogenic system will decrease in vibration as it cools (see Figure 5.5 for comparison vibration data from room temperature to low temperature).

Accelerations on a cryogenic platform should be characterized for all three axes of measurement *X*, *Y*, and *Z* because each axis will have some unique content. Although three-axis accelerometers are available, the bandwidth of measurement is limited and most likely will not be adequate to measure subhertz frequencies. Whichever model is chosen, it is important to ensure that the accelerometer is fastened securely to the platform it is measuring. If it is not, the high-frequency end of the bandwidth it is able to measure will be limited. Ideally, the resonant frequency determined by the mass and mounting of the accelerometer is higher than the internal resonant frequency of the accelerometer specified by the manufacturer.

A few manufacturers and particular models of accelerometers are shown below:

Seismic accelerometer—Model 731A
Wilcoxon Sensing Technologies®
20511 Seneca Meadows Parkway
Germantown, MD 20876
Phone: 301-330-8811
Website: www.wilcoxon.com
Various cryogenic and room temperature accelerometers
PCB Piezotronics, Inc.
3425 Walden Avenue
Depew, NY 14043-2495

Toll-free (in the US): 800-828-8840
24-hour SensorLineSM: 716-684-0001
Website: www.pcb.com
Great resource website: http://www.pcb.com/Resources/Technical-Information/tech_signal.php

Cryogenic accelerometer—Model 876
Columbia Research Laboratories
1925 Mac Dade Blvd.
Woodlyn, PA 19094
Phone: 1-800-813-8471
Website: www.crlsensors.com

5.1.4.3 Laser Interferometer

Interferometry can offer some advantages over other displacement measurement techniques for measuring cryocooler vibrations but also has some limitations. A laser interferometer is a highly versatile measurement tool because it can operate at virtually any distance from the target, unlike the capacitive sensor, which must be in close proximity. An interferometer can measure objects through optical windows, which allows low-temperature measurements to become relatively easy. There are a variety of interferometric techniques for measuring displacement. A basic laser-based Michaelson interferometer used with a fringe counting method is a useful technique for measuring displacements, provided the displacement velocity is slow and amplitude is greater than half the wavelength of the laser [4]. These limitations restrict practical use for many applications. If the measurement of displacements less than a few hundred nanometers are desired, then more sophisticated interferometers exist but are substantially higher in cost. Other factors that may contribute to limitations for use are the need for a reflective surface on the target and consistent optical alignment with the target throughout the measurement [5].

5.1.5 Understanding Cryocooler Vibrations

5.1.5.1 Vibration Spectrum Analysis

To understand the vibrations of a cryogenic system, it is helpful to know both frequency content as well as amplitude of displacement. Understanding frequency content is important for avoiding resonances within the system, especially on the sample platform, and displacement is important because often what ultimately matters is how much the sample is moving relative to measurement apparatus, such as optics next to the cryogenic system.

Power Spectral Density (*PSD*) is a good way to look at vibration frequency content. Similar to a fast Fourier transform (FFT), in that it *shows the distribution of vibrations across a frequency spectrum*, the PSD goes further to normalize the frequency bin width so that bandwidth measurement choices, as well as any random vibrations, do not affect overall accuracy of the data when comparing datasets. Units of a PSD for acceleration may be shown as g^2/Hz or $g/(Hz)^{1/2}$. The measurement that is being quantified (in this case accelerations g^2) is represented as power, or the mean-square value so that it is always represented as a positive value even if voltage output shows negative values because the measurement is taken near zero. This mean-square value is then divided by the frequency bandwidth *so that it becomes independent of chosen measurement bandwidth.* Either way of displaying units may be chosen, but it is important when comparing datasets that the same units are used in each. While the PSD is a useful way to analyze frequency content across a spectrum and between datasets, it is important to understand that the PSD is not sufficient in showing peak-to-peak amplitude of displacement.

When a cryogenic system is part of a measurement, such as an optical experiment with high-numerical aperture optics mounted next to the cryogenic system, what may become the most important metric for predicting successful performance of spatial resolution is the amplitude of displacement of the sample relative to the optics. While a PSD will show amplitude distributed across frequency, a plot of displacement versus time will show peak-to-peak displacement of all frequencies combined. Measuring peak-to-peak vibrations in this way provides a simple and straightforward way to compare performance of cryogenic systems and configurations. It is important to measure at least out to 1 kHz in bandwidth and at least down to 1 Hz to have an accurate representation of the combined frequency content, which will contribute to overall peak-to-peak displacement amplitude. The next section will further describe why most mechanical resonances that dominate displacement amplitude on the sample platform are in the hundreds of hertz. It is also important that the cryogenic system being tested exhibits a repetitive cycle and therefore has no significant contribution to the amplitude that is random.

5.1.5.2 Frequency Content in a Pulse

For any particular cryogenic system, its vibrations are a function of frequency. Understanding the frequency content of vibrations measured in a pulse of the cryocooler is important because this will help in understanding how to mitigate vibrations at the sample mount and surroundings. We know from previous sections that the typical mechanical operating frequency of a 4 K cryocooler is about 1–2 Hz. This obvious and dominant operating frequency would lead many to believe that the operating frequency is the primary frequency to be concerned about when isolating a sample mount from cryocooler vibrations. A closer look at the accelerations created by each pulse of the cryocooler shows a broad spectrum of vibration generated. This is a key point to understand because this broad spectrum of vibrational energy is able to excite the resonances of support structures, cryogenic sample mounts, and any device or actuator on the sample platform. To show the vibrational energy generated by the cryocooler and how this energy transmits to a sample environment, an accelerometer was placed in each of the locations shown in Figure 5.7. One placement was on top of the cryocooler cold head (custom SHI RDK-101D), and the other placement was on the cryogenic sample platform. The cryocooler cold head is supported by a support structure and isolated using vacuum bellows and thermal links from the sample platform. The sample platform is supported by a cryogenic sample support.

The graphs in Figure 5.8 show the vibrational energy generated by the cryocooler measured by an accelerometer placed on top of the cold head as shown in Figure 5.7. As we learned from Section 5.2, the vibration generated by a cryocooler is dominated by the pulse of gas as a rotary valve allows a sudden exchange of high-pressure helium into and out of the cold-head displacers area. On the left, accelerations resulting from pulses of high-pressure helium gas into and out of the cold head are shown. The operating

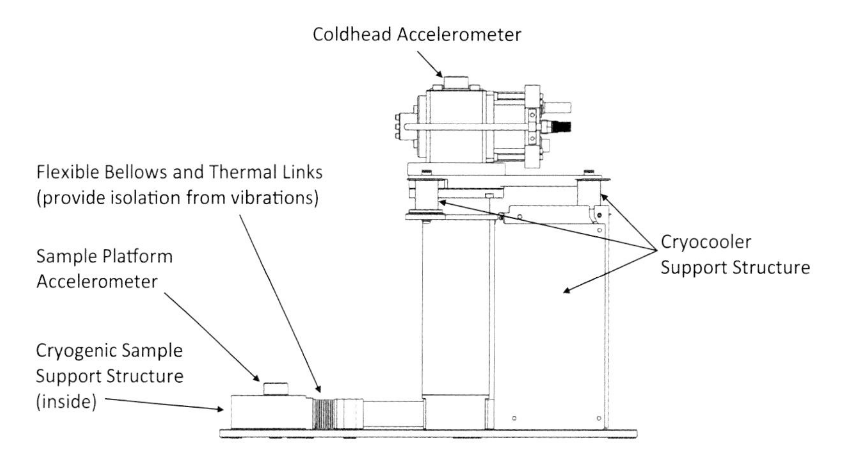

FIGURE 5.7 Accelerometers were placed on a GM cryocooler cold head (custom SHI RDK-101D) and the isolated sample platform locations. This helps to understand the spectrum of vibrational energy generated by the cold head and transmitted to the sample space.

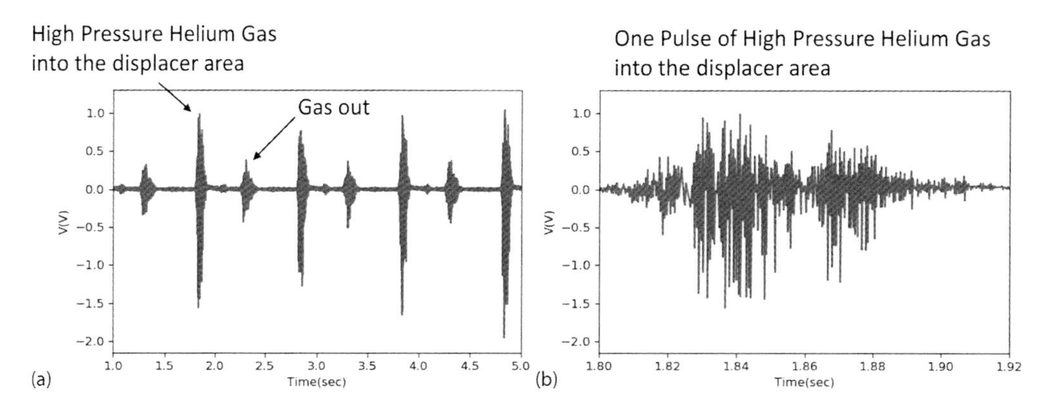

FIGURE 5.8 (a) Cryocooler accelerations measured on the cold head showing each pulse of high-pressure helium gas into and out of the displacer area at a 1 Hz operating frequency. (b) One pulse of high-pressure helium gas into the displacer area.

frequency of this cryocooler is 1 Hz, so every second contains one larger pulse resulting from gas expanding into the displacer area and followed by one smaller pulse resulting from gas out. A common misunderstanding around cryocoolers is that the dominant ~1 Hz operating frequency is the primary frequency to isolate from; however, a closer look will reveal there is more to understand. On the right, the accelerations generated from one pulse of gas into the displacer area shows that there is a large amount of higher frequency content contained within a pulse. This is an important clue that there exists much more vibrational energy content than just at 1 Hz. Looking at the PSD will help us understand just how broad the vibrational energy is that is generated with each pulse at the cryocooler cold head.

The PSD of accelerations generated by the cold head are shown in Figure 5.9 from subhertz to 50 Hz on the left, and from 50 to 500 Hz on the right. In both plots, the blue represents the accelerations at the cold head and the red shows a baseline noise level

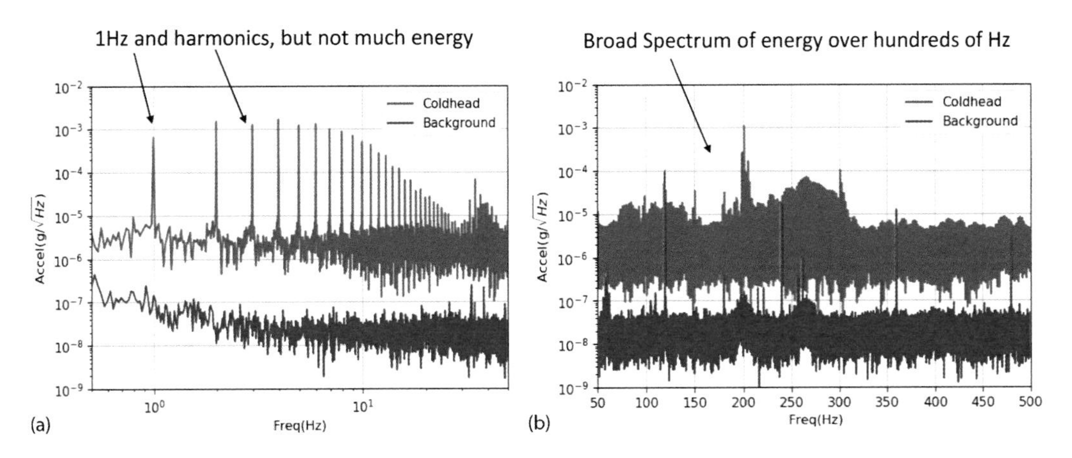

FIGURE 5.9 The power spectral density plot on the left (a) shows accelerations from subhertz to 50 Hz for the cryocooler cold head (blue) and for the baseline noise level with the cryocooler off (red). We see the operating frequency of 1 Hz; however, we see in the plot on the right (b) there is broad vibrational energy out into the hundreds of hertz significantly above the noise floor.

with the cryocooler off. The operating frequency of the cryocooler shows the peak at 1 Hz and the multiple harmonics. However, the spectral content from 10 Hz out into the hundreds of hertz remains significantly above the noise floor. This is perhaps the most important point of this chapter in that this spectrum shows us there is significant energy generated from the cryocooler cold head, which is in the hundreds of hertz frequencies. *Whatever method is used to isolate from the cryocooler vibrations, it must isolate not only the 1 Hz but out past hundreds of hertz as well.* The range from 50 to 500 Hz as shown in the plot on the left is significant because this is the frequency range where many devices, positioners, sample mounts, and cryogenic sample assemblies have their resonances. Whatever object is placed on a sample mount, if it has a resonance in the hundreds of hertz, and the isolation scheme does not adequately isolate hundreds of hertz, then the object will be excited at resonance. Once an object is excited at resonance, the displacement of vibrations increases dramatically, and this displacement of the sample relative to the surrounding optics or probing device is what limits the stability of sensitive cryogenic measurements. This is one primary reason why isolating from cryocooler vibrations is so difficult.

Referring back to Figure 5.7 and the accelerometer placed on the sample platform, the accelerations measured here represent vibrational energy that has transmitted through the bellows, thermal links, and through both the cryocooler support structure as well as the cryogenic sample support structure to the sample platform. Figure 5.10 shows the PSD of the accelerations on the sample platform. This graph shows two dominant resonances and almost no energy at 1 Hz. This plot helps make the point clear that there is very little energy at 1 Hz and substantial energy from tens to hundreds of hertz. Whatever structures exist in this frequency range of tens to hundreds of hertz will be excited. There are many ways to isolate from cryocooler vibrations and many ways to construct a sample support, and whichever structures are built, they will probably have resonances in this range. For this particular system design, there exists a resonance at 35 Hz, which is the cryocooler support structure that supports the cold head. The resonance is relatively low because it is designed to provide isolation of the cold-head vibrations from both

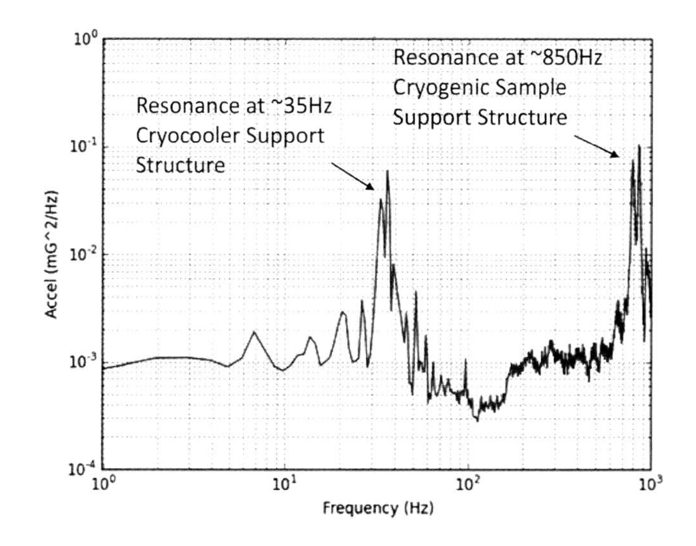

FIGURE 5.10 Power spectral density of acceleration on the sample platform shows two dominant resonances of the cryocooler support structure and the cryogenic sample support structure, but there is almost no energy at 1 Hz.

the sample as well as the table it is mounted on. The second resonance is the cryogenic sample support structure, which supports the sample platform. This resonance is about 850 Hz. The frequency space between 35 and 850 Hz is significantly lower; however, there is some energy present. Any device, positioner, probe, or other cryogenic assembly placed on this sample platform will have a resonance that will be excited, and many of these assemblies have resonances in the hundreds of hertz. Ideally, anything placed on the sample platform will not have a resonance close to 35 or 850 Hz but somewhere in between or above 850 Hz.

5.1.6 Isolating from Cryocooler Vibrations

5.1.6.1 Thermal Links

Vibrations are present in every type of cryocooler system, and therefore many applications require isolation from the cryocooler. There are many techniques for isolating cryocooler vibrations, and each technique has practical considerations that are helpful to understand because each technique has trade-offs. One of the primary components of any vibration-isolated cryocooled system is the mechanism by which heat is transferred away by the cryocooler without inducing vibrations and disturbing the application. This is commonly accomplished with flexible thermal links. Pure metals, such as copper and aluminum, can be good material choices for low-temperature thermal links. At room temperature, heat transfer is determined by both phonon activity of lattice vibrations as well as electron transport, but at low temperatures thermal conductivity of metals is dominated by electron transport. Because electron transport is dependent upon the amount of impurities in a metal, which scatter electrons, a pure metal will have high thermal conductivity at low temperatures. In fact, the thermal conductivity of a pure metal can be accurately estimated from 4 K to room temperature by measuring the electrical resistivity at room temperature and again at 4 K to calculate a "residual resistivity ratio" (RRR) and then correlating to thermal conductivity tables available at NIST.

Oxygen-free high-purity copper is a common material for thermal links because it has relatively low impurity content. Bulk copper can be purchased with very low impurity content and very high RRR values above 300; however, wire and foil are limited to an RRR value of about 100, which corresponds to a thermal conductivity of about 400–600 W/mK at 4 K. This is because the forming processes necessary to draw or roll copper wire or foil into its final form introduces impurities and increases grain boundary dislocations, even if it began as a higher purity, higher thermal conductivity bulk material.

Once the thermal conductivity of the material is known, it is just geometry and temperature that will determine heat flow. It is helpful to think of the thermal performance of a link in terms of conductance. This is derived from Fourier's heat conduction equation where Q is the amount of heat flow in Watts, k is thermal conductivity in W/mK, A is the cross-sectional area of the material, and dT/dx is the temperature gradient along the length of the material.

$$Q = \frac{kAdT}{dx} \tag{5.2}$$

Writing in terms of conductance where the total length of the link is L, we write the heat equation as heat flow Q per degree Kelvin:

$$\text{Conductance} = Q \text{ per degree Kelvin} = \frac{kA}{L} \tag{5.3}$$

Writing the heat equation in terms of conductance is a useful way to show the thermal effectiveness of a thermal link because it considers only geometry and material property. Conductance has units of Watts per Kelvin, which means that if a heat load of 1 Watt is cooled by a thermal link with a temperature rise of 2 K from one end of the links to the other, the link has a conductance of 0.5 W/K. It is important to remember that thermal conductivity k is a temperature-dependent value and it must be appropriately chosen for a particular temperature range. Once a thermal link has been constructed, its conductance can be empirically tested with a relatively simple setup like the thermal link conductance test station shown in Figure 5.11. In this test station, a material can be clamped between a high-temperature and low-temperature clamping point. The high-temperature clamping point has a heater embedded in the mount, and the mount is simply supported with a loop of Nylon string (unwaxed Nylon dental floss) to ensure that heat generated from the heater has virtually one path to flow and that is through the material under test (thermal link). The-low temperature clamping point is connected directly to the second-stage of the cold head. The clamps should be high-force clamping joints with at least an M3 size bolt to ensure good thermal contact at each end of the thermal link.

High-accuracy thermometers should be attached either to the thermal link near each clamping joint as shown in Figure 5.12 or to the clamping joint itself. It is useful to know the distance between thermometer attachment points, or at least to ensure that when

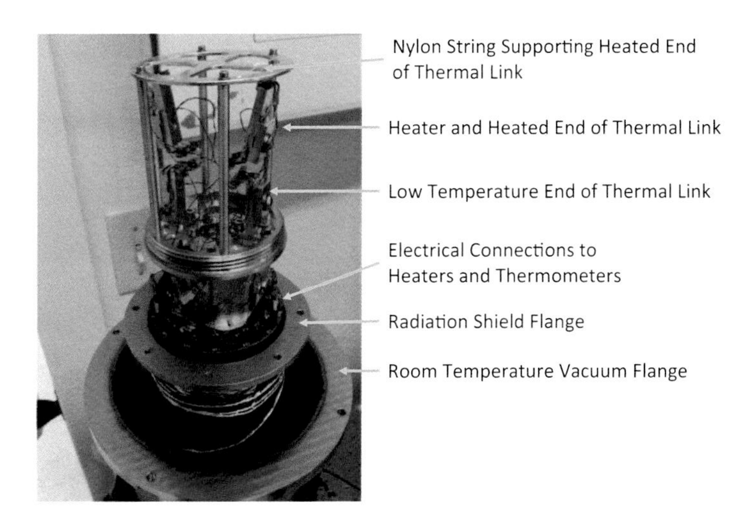

FIGURE 5.11 Thermal link conductance test station courtesy of Montana Instruments. Thermal link conductance can be measured by clamping a thermal link between a low-temperature end and a heated end and measuring the delta T across the link for each level of heater power applied.

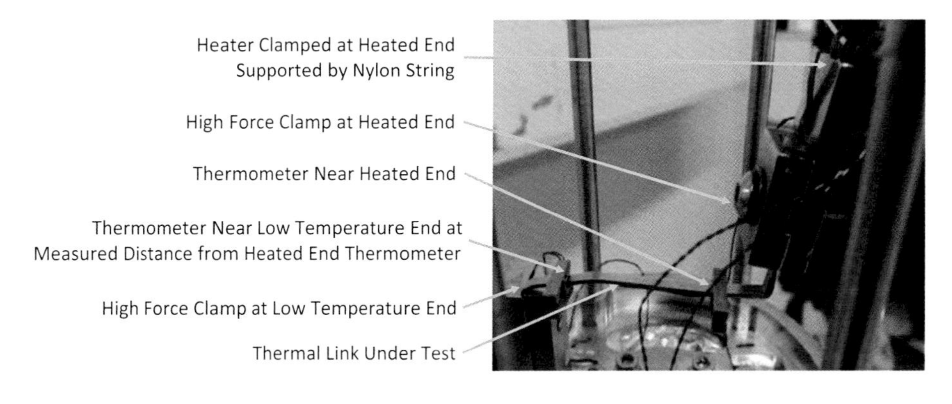

FIGURE 5.12 Thermal link to be tested in the conductance test station courtesy of Montana Instruments. The thermal link is clamped with high force between a low-temperature end and a heated end supported by nylon string. Thermometers are fastened to the thermal link with high-force clips at a measured distance from one another.

testing and comparing various materials that the distance between thermometer attachment points is the same for each test.

Care should be taken so that any additional heat load on the thermal link under test other than heat applied from the high-temperature clamp heater is negligible. For this purpose, a 4 K radiation shield is used to keep the entire test area isothermal and shielded from any radiative loads. Once the station is cooled down to base temperature, heat can be applied to the heater in the high-temperature clamp. The amount of heat applied should be carefully measured because this is one source for error in the measurement. This can be done with a four-point measurement and a high-quality power source, such as a Keithley 2400 Source Meter. Thermometers should also be read with a high-quality four-point measurement device, such as a Lakeshore Temperature Controller. It is advisable to measure conductance of a thermal link in the temperature range where you are interested in using it. Remember that thermal conductivity k is a temperature-dependent material property, and especially at low temperatures, the slope of k as temperature changes can be quite steep. A useful resource for thermal link testing apparatus and procedure is a research paper done by Fermi National Accelerator Laboratory and Technology Applications, Inc. titled "Thermal conductance characterization of a pressed copper rope strap between 0.13 and 10 K" [6].

When considering a thermal link for isolating vibrations in a cryogenic system, the two primary things that matter are thermal conductance and force transmitted across the link. We have already described thermal conductance and will now discuss force transfer. A thermal link is only effective if it adequately isolates vibrational energy from the source to the components, which it is removing heat from. Isolating a rigid cryogenic sample mount from a vibrating component is effectively done if the force transfer across the thermal link is reduced to an acceptable level. The force transfer determines how much force is exerted from the vibrating source through the thermal link onto the cryogenic structure supporting the sample. This force, (along with the stiffness of the cryogenic structure) will determine how much displacement the sample mount experiences. Although the equations are oversimplified and don't account for friction between wires or straps, some intuition for various designs can be gained by using a simple beam deflection equation. Force can be estimated by multiplying the thermal link stiffness k by the amount it is deflected as in the simple spring deflection equation below. If we idealize the thermal link as a fix-free cantilever beam, stiffness k can be calculated and used as a comparison metric for various materials, lengths, and number and diameter of wires. E is the elastic modulus of the thermal link material, L is the effective length of the link chosen for comparison, and I is the area moment of inertia of the thermal link. In the two-area moment of inertia equations below, I_{round} represents a bundle of small wires where N is the number of wires and D is the diameter of the wire used, and I_{rect} represents one rectangular strap (thin foil) where b represents the width and h represents the thickness of the strap.

$$F = kx \tag{5.4}$$

$$k = \frac{3EI}{L^3} \tag{5.5}$$

$$I_{\text{round}} = \frac{N\pi D^4}{64} \tag{5.6}$$

$$I_{\text{rect}} = \frac{bh^3}{12} \tag{5.7}$$

Comparing stiffness or force transfer of various designs is useful to build intuition, but since other factors, such as friction between wires, have a significant effect on link stiffness, it is useful to test stiffness empirically. Copper is continually building up oxide layers on its surface while it is in the presence of oxygen, which not only increases friction between moving wires or straps but also creates oxide bonds between closely spaced or touching wires or straps. These subtleties of thermal links show the importance of empirical testing of thermal link designs.

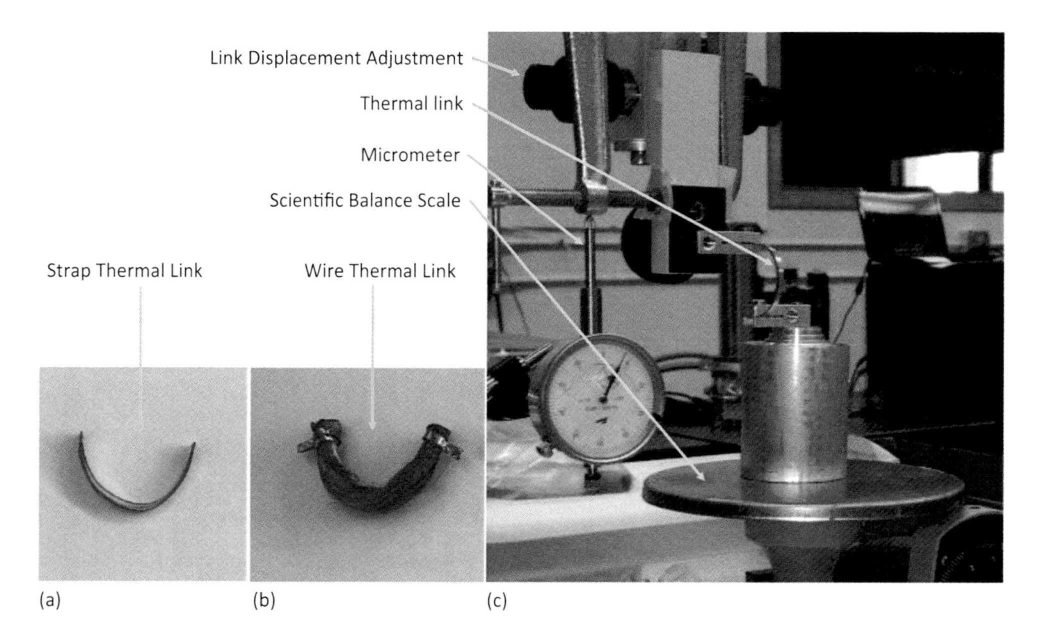

FIGURE 5.13 Ready for stiffness testing: (a) Copper strap. (b) Example wire bundle thermal link. (c) Thermal link stiffness test fixture. (Courtesy of Montana Instruments.)

Testing thermal link stiffness can be done with a relatively simple setup, such as shown in Figure 5.13. Each axis of stiffness is tested independently by measuring the force applied on one end of the thermal link using a weight scale, as the other end is precisely displaced and measured with a dial micrometer. The middle image of Figure 5.13 is an example thermal link consisting of 5,250 wires. It is useful to wrap each end of the link tightly with wire or strap to bind the wires together and provide something for the stiffness test fixture to clamp on to. Both ends of the link should be clamped or bolted rigidly similar to how they would be in use. It is important to test the link in the orientation or shape that it is intended to be used in, and also test within the relative displacement range it is expected to work in. For example, if the vibrations to be isolated from are on the order of millimeters, test the links with millimeter displacements, and if the vibrations are on the order of microns, test the links with similar micron displacements. The reason this is important is because there are a number of factors contributing to the stiffness of a thermal link, including bending of the wires or strips, friction between wires or strips, and bonding between wires or strips from oxidation. Each of these factors contribute differently, depending upon the displacement of the thermal link. Many people have been fooled into thinking that a thermal link, which feels "soft" as it is flexed manually by hand is, in actuality, relatively stiff on the micron or submicron scale, and this can be due to friction and oxidation bonding between wires. The only way to know this is to test within the displacement scale it is intended to be used.

5.1.6.2 Commercially Available Thermal Links

Thermal links for use at cryogenic temperatures have been made available by some commercial companies. Two options are shown for commercial cryogenic thermal links (Figures 5.14 and 5.15).

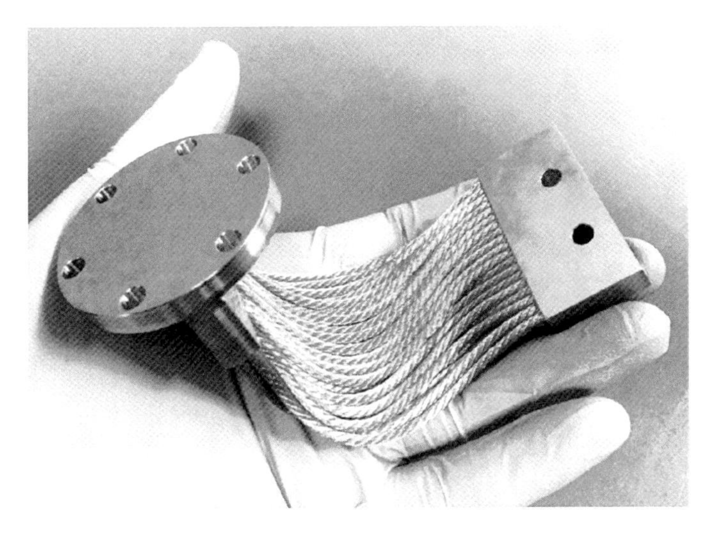

FIGURE 5.14 Image shows one of a variety of copper thermal straps for vibration isolation and conductance at low temperatures. This strap shown can be directly attached to the second stage of a cryocooler. (Courtesy of Technology Applications, Inc.)

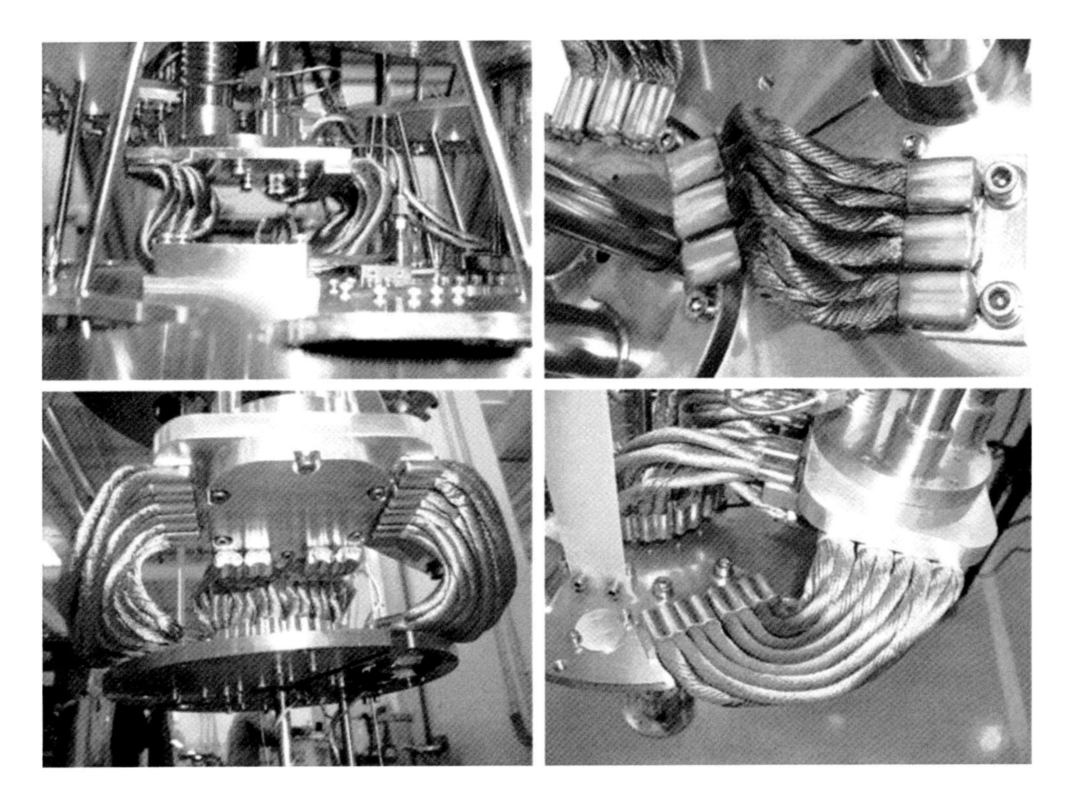

FIGURE 5.15 Images show examples of commercially available flexible links using wire braids from New England Wire Technologies connected to portions of the cryocooled system. These copper braids made by New England Wire Technologies contain of approximately 10,374 #46 AWG (0.00157″ diameter) OFHC copper wires. Both ends of these copper braids can be pressed or fused onto copper mounting plates to form a flexible thermal link that can be bolted into place. (Courtesy of Janis Research Company, LLC.)

Option #1: Commercial flexible thermal link
Technology Applications, Inc.
5303 Spine Rd., Suite 101
Boulder, CO 80301
Phone: 303-443-2262 | FAX: 303-443-1821
www.techapps.com

Option #2: Commercial OFHC wire braid
New England Wire Technologies
130 N. Main Street
Lisbon, NH 03585
Phone: 603-838-6624
Email: sales@newenglandwire.com

5.1.7 Importance of Vibrational and Thermal Stability in Selected Applications

5.1.7.1 Optical Application: Characterization of Low-Dimensional Materials over a Wide Temperature Range [7]

Optical characterization of materials from cryogenic temperatures to room temperatures and above requires attention to each of the vibration and drift considerations covered in this chapter. From 2D materials, carbon nanotubes, quantum dots, and single molecules to quantum sensors and devices, high spatial resolution as well as high spectral resolution requirements are necessary for observing fundamental phenomena, such as phase transitions, molecular thermal activities, and crystal structure changes. Precise control of temperature and mechanical stability over a wide temperature range makes these and related applications possible because the region of interest does not move for each temperature set point, remains in focus during each measurement, and quick stabilization at each new temperature point allows many data points to be collected within a reasonable time frame.

At the time of this writing, one interesting example of these applications is the Raman and Photoluminescence characterization of monolayer tungsten diselenide, WSe_2. Figure 5.16 shows an image of WSe_2 and a region of interest where temperature-dependent Raman spectra were measured from 5 to 350 K. This particular region indicated by the circle is an area in the material sample with good homogeneity and a single layer. Any significant deviation from this precise area during the entire measurement set over the wide temperature range would adversely affect the characterization of the material. WSe_2 is a representative material of atomically thin transition metal dichalcogenides, and because of its chemical and mechanical stability it represents a good platform for studying quantum effects and could one day be used in a variety of applications, such as a nanoscale field effect transistor and in solar cell photovoltaics.

Two-dimensional materials have characteristically low absorption and a low conversion efficiency for spectroscopy, so low laser power and long integration times become necessary to increase signal to noise. This increases the need for stability over long time periods for each measurement. The rich photonic and electronic temperature-dependent properties of materials like WSe_2 can be explored with scientific instrumentation,

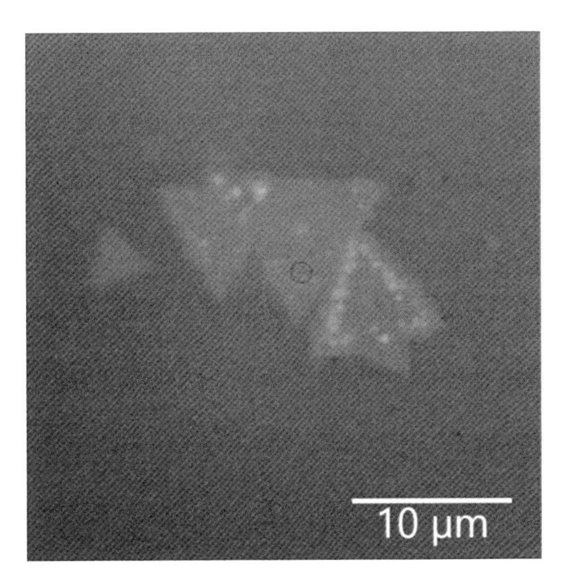

FIGURE 5.16 Image of WSe_2, an atomically thin transition metal dichalcogenide. Temperature-dependent Raman spectra from 5 to 350 K were measured at the region of interest identified by the circle. (Courtesy of Montana Instruments.)

which allows access to wide temperatures with nanometer-level vibration and drift stability. Figure 5.17a shows Raman intensity, and Figure 5.17b shows peak position and full width at half maximum (FWHM) peak narrowing as sample temperature changes from 350 to 5 K. Each measurement was taken at 20 K increments, and since stabilization time for each measurement was about 1 minute or less, the entire dataset was captured in less than 1 hour.

The scientific instrumentation that enabled this measurement of WSe_2 and many other optical applications over a wide temperature range, shown in Figure 5.18, is the Montana Instruments CRYOSTATION® s100 with integrated CRYO-OPTIC® and Agile Temperature Sample Mount ATSM®. This instrument combines three key technology developments in the field of low temperature instrumentation which remove significant barriers to stability when studying materials and devices over a wide temperature range.

A few key advances that enable optical applications are described for each of the three major elements of this system. The Cryostation serves as a general platform for a variety of applications and options. The cryocooler cold head is moved away from the sample space and isolated vibrationally from both the sample space as well as the table it rests on. This allows vibration isolation from the sample space and the optics on the table. Thermal links and thermal damping are integrated into the system below the sample space so that the sample space can easily be configured with hardware and devices for excitation, manipulation, and detection without affecting the performance of the system. The sample housing bolts directly to the table on which it is placed, and the cryogenic sample mount is also rigidly coupled to the table via the sample housing so that the sample remains rigidly coupled to the table just like every other optic on the table. This reduces the vibration of the sample relative to the optics on the table.

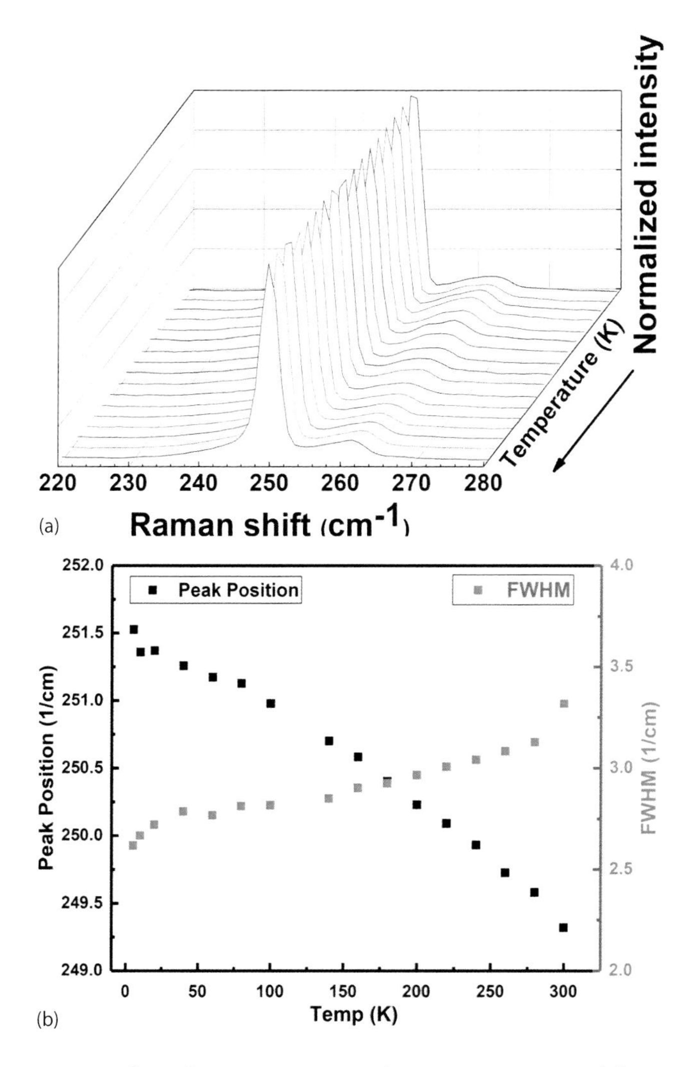

FIGURE 5.17 Temperature-dependent Raman spectra for WSe_2 was measured from 5 to 350 K. Raman signal intensity (a) and peak position and full-width half-max FWHM narrowing (b) are shown as temperature decreases. (Courtesy of Montana Instruments.) *Data was taken with a Montana Instruments Cryostation s100 - Cryo-Optic with Agile Temperature Sample Mount (ATSM) coupled to a Princeton Instruments IsoPlane SCT 320 spectrograph with a 532 nm cube set and PIXIS CCD detector.*

Electronics allow control of the cryocooler and compressor-operating frequencies for faster cooldowns and minimum vibrations at temperature (Figure 5.19).

The Cryo-Optic assembly is an integrated sample platform that combines vacuum-compatible high-numerical aperture (NA) optics, with cryogenic sample positioning, vibration, and drift stability. High NA optics with high-performance color correction are held at room temperature on a precise temperature-stabilized platform adjacent to the cryogenic sample space. This separation allows wide temperature operation of the sample without any adverse effect to the optics and imaging capability. A radiation shield separates the cryogenic sample and the optics with the option of a small aperture or thin window along the optical axis. The sample mount is fixed to a positioning stage, which

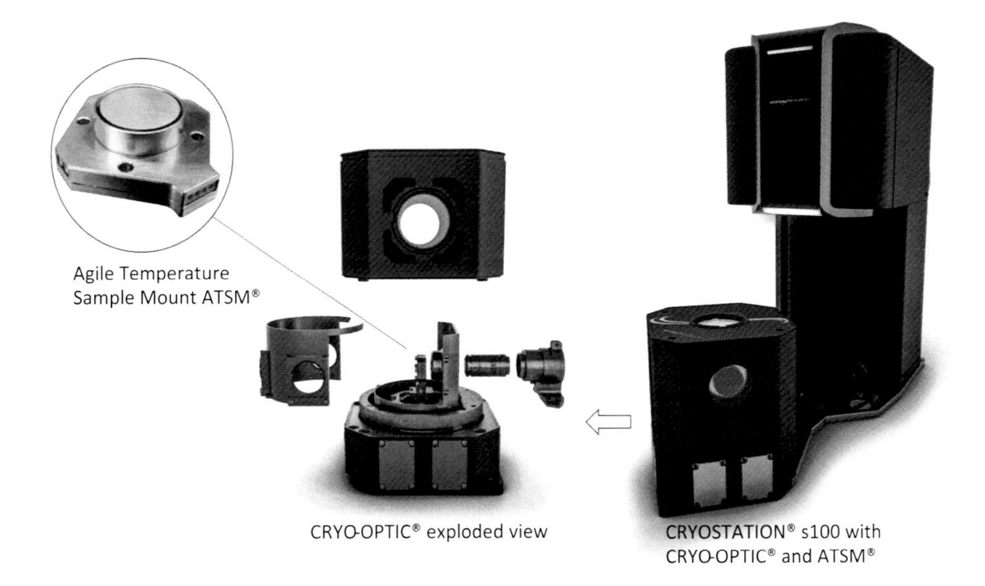

FIGURE 5.18 Key technologies have been developed for optical applications, such as Raman and Photoluminescence spectroscopy. Three of these technologies are the Cryostation s100, Cryo-Optic, and the ATSM. (Courtesy of Montana Instruments.)

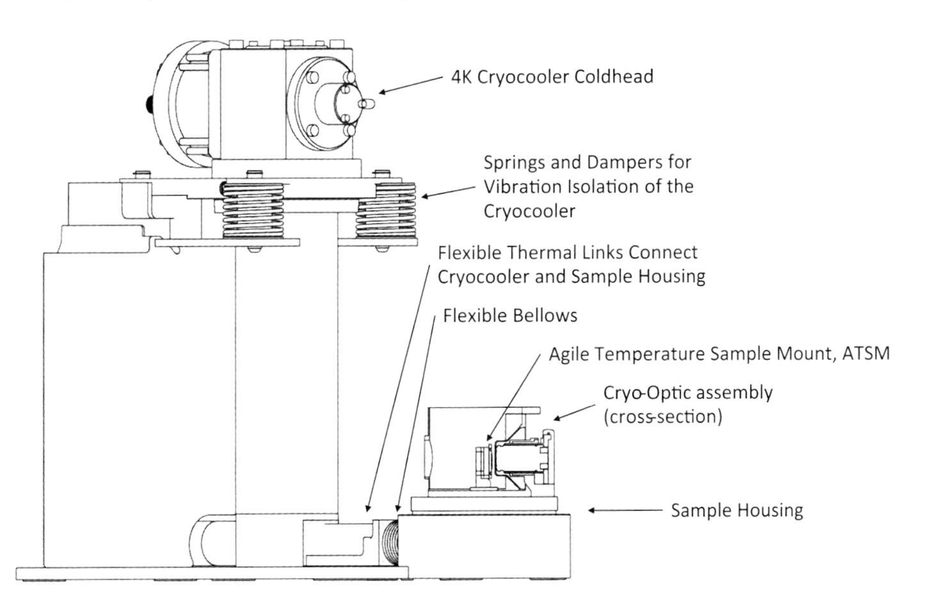

FIGURE 5.19 Line drawing of the Cryostation s100 with Cryo-Optic and ATSM installed in the sample space. The cryocooler cold head is vibrationally isolated from the sample space the table is placed on. The sample space is rigidly coupled to the table. (Courtesy of Montana Instruments.)

allows movement of the sample along all three principle axes. The Cryostation design allows the sample space to be isolated from vibrations, and the system has no detectable drift at each temperature set point. Some adjustment in focus is necessary for temperature set point changes above 50 K in sample temperature, but no adjustment perpendicular to the optical axis is necessary from 4 K to room temperature (Figure 5.20).

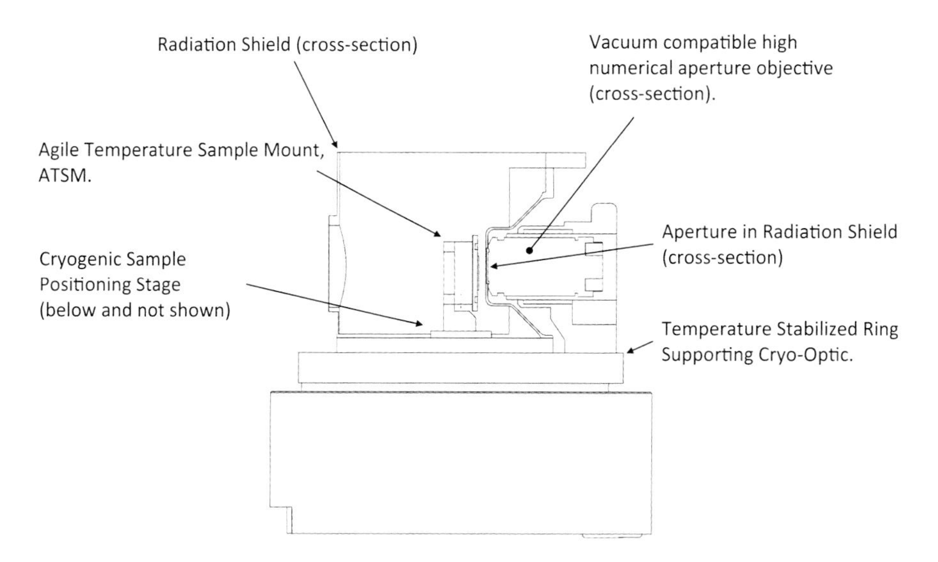

FIGURE 5.20 A cross section of the Cryo-Optic assembly supported by a temperature stabilized ring and the ATSM positioned close to the vacuum compatible high numerical aperture objective. The radiation shield surrounds the ATSM and allows light to pass through the aperture. (Courtesy of Montana Instruments.)

The Agile Temperature Sample Mount (ATSM) is a small device for sample mounting, which allows wide temperature changes within minutes or seconds. It is integrated with an active heater and thermometry, which can bring the sample from a base temperature below 4–350 K within less than 10 minutes, and again down to base temperature in less than 10 minutes as Figure 5.21 shows. Set-point change and stabilization of smaller steps can happen within seconds, which is very useful for wide temperature measurements for each feature of interest, which could otherwise become prohibitively time consuming. This agile temperature operation is possible because the ATSM creates

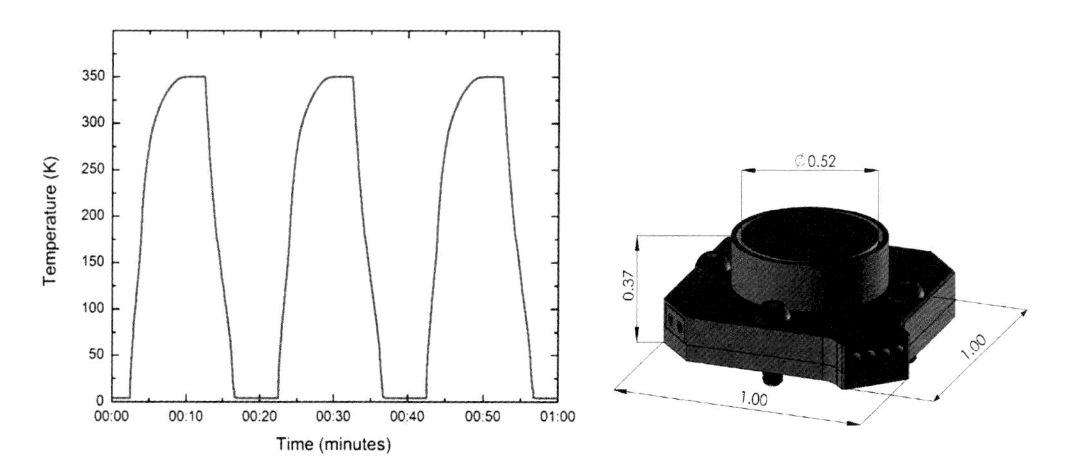

FIGURE 5.21 The Agile Temperature Sample Mount ATSM is a small device (right), which can be mounted on positioning stages at cryogenic temperatures and allows fast sample temperature changes from 350 to 4 K in less than 10 minutes (left). (Courtesy of Montana Instruments.)

a thermally isolated sample mount with small mass, which can quickly change temperature and stabilize while the rest of the cryogenic system temperature remains virtually unchanged. The ATSM can be mounted on positioning stages for alignment with optics and positioning or scanning of the sample spatially.

Special thanks to Tim Johnson, David Schiff, Will Baker, Craig Wall, and the team at Montana Instruments Corporation for help with technical figures, data, and supporting information to make this chapter possible.

References

1. Rao, S.S. 2004. *Mechanical Vibrations*, 4th ed. Pearson Education, Inc.
2. Crede, C.E. 1957. *Vibration and Shock Isolation*. John Wiley & Sons.
3. Ekins, J. 2007. *Experimental Techniques for Low-Temperature Measurements*. Oxford, UK: Oxford University Press.
4. Pineda, G.S., and Argote, L.F. 2000. *Vibration Measurement Using Laser Interferometry*. SPIE, Vol. 3831.
5. Berkovic, G., and Shafir, E. 2012. Optical Sensing Group, Applied Physics Division, Soreq NRC, Yavne 81800, Israel. *Optical Methods for Distance and Displacement Measurements*. Optical Society of America.
6. Dhuley, R.C., Ruschman, M., Link, J.T., and Eyre, J. 2017. Thermal conductance characterization of a pressed copper rope strap between 0.13 K and 10 K. *Cryogenics*, 86: 17–21.
7. DeGrave, J. 2018. *Optical Characterization of Low-Dimensional Materials*. Montana Instruments Applications Lab. https://www.montanainstruments.com/help/TG101.html (accessed December 2018).

6. Enabling Technologies

Zuyu Zhao

6.1 Introduction

The LIQUID HELIUM-FREE systems discussed in this book are often referred to as "dry" systems or "cryogen-free" systems, while the NON-LIQUID HELIUM-FREE are often called "wet" systems. This chapter will discuss some technologies often used in (but not limited) to dry systems, such as cryogen liquefaction from the Joule–Thomson effect and heat transfer through heat switch.

"Dry" systems do not use liquid helium as an "open-cycle coolant" where helium will be vented to the atmosphere, or sent to the recovery system, from the cryostat. However, "dry" systems do use helium as refrigerant in a "closed" loop. Examples discussed in this book include commercial cryocoolers using 4He, 3He cryostats using 3He, and dilution refrigerator system using 3He/4He mixture. All these coolants are in the gaseous phase at room temperature, and they are liquefied during operation.

It is a common practice for "wet" systems to employ a so-called 1 K pot to condense the coolant. The 1 K pot is filled with liquid helium (4He) and then cooled down by reducing the saturated vapor pressure inside the 1 K pot. Alternative approaches are needed for the "dry" systems because they *usually* do not have a 1 K pot. A device called the Joule–Thomson condenser based on the Joule–Thomson effect has become the indispensable part in the recirculating flow-type dry systems for cryogen liquefaction.

6.2 Joule–Thomson Experiment

6.2.1 Joule–Thomson Process and Joule–Thomson Effect

The Joule–Thomson process refers to a fluid (gas or liquid) that is forced through a throttling process adiabatically, and then expands from higher pressure to lower pressure while no heat is exchanged between the fluid and its environment. The Joule–Thomson effect, discovered by Joule and Thomson in 1852, describes the temperature change caused by this thermodynamic process.

Notes:

1. Fluid for our discussion in this book is often in the gaseous phase, and we will call it "gas" in most of our future discussions.
2. Temperature change can be that the temperature increases, decreases, or remains unchanged, depending of the specific gas and its thermodynamic status, such as temperatures and pressures.

3. Joule–Thomson expansion is an adiabatic process, that is, the fluid is kept thermally insulated from its environment so that it does not exchange heat with the environment.
4. This is a throttling process, that is, the fluid is forced to pass from a region at higher pressure to a region at lower pressure through a "throttling device," which can be a "needle valve," a "porous plug," an "impedance," etc.
5. The throttling process is inherently an irreversible process.

6.2.2 Experimental Setup

An early Joule–Thomson experiment setup is as shown in Figure 6.1. It included a high-pressure gas bottle (air was used at the beginning) with a pressure regulator; a "bath" filled with cold water to precool the inlet gas; a Joule–Thomson chamber divided by a porous plug, including two (2) thermocouple thermometers located on either side of the porous plug, and they were used to measure the temperature of the gas before and after expansion; and a manometer to measure the gas pressure before expansion.

An idealized physical illustration of the Joule–Thomson experiment is illustrated in Figure 6.2 for our discussion [2–4].

The gas with volume V_1 is first located at the left chamber under pressure P_1 applied by the piston #1. There is no gas in the right chamber at the beginning. A porous plug is located in between these two chambers.

A piston #2 inside the right chamber applies a lower pressure P_2 against the porous plug from the right-hand side.

Since P_1 is greater than P_2, the gas is forced to move toward the right by piston #1. After all the gas passes the porous plug, it occupies volume V_2 in the right chamber under pressure P_2.

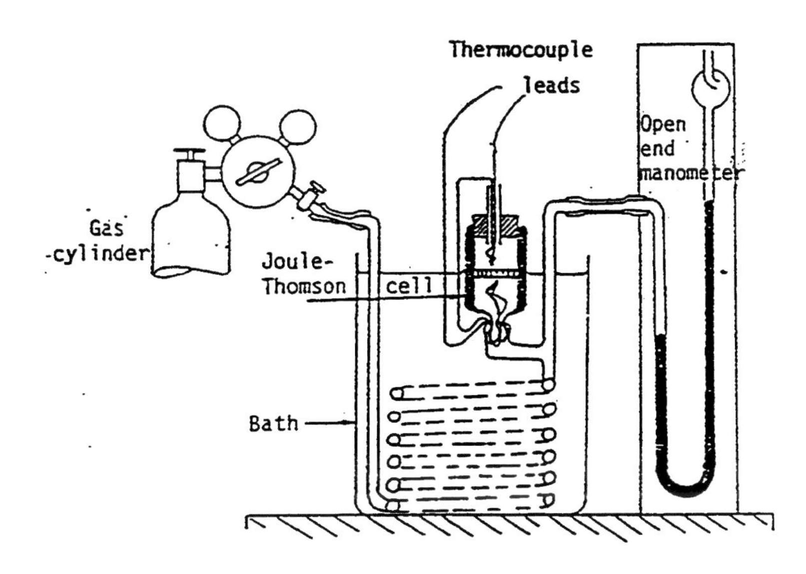

FIGURE 6.1 The first Joule–Thomson experimental setup. (From Perry, R.H. and Green, D.W., *Perry's Chemical Engineers' Handbook*, McGraw-Hill, New York, 1984 [1].)

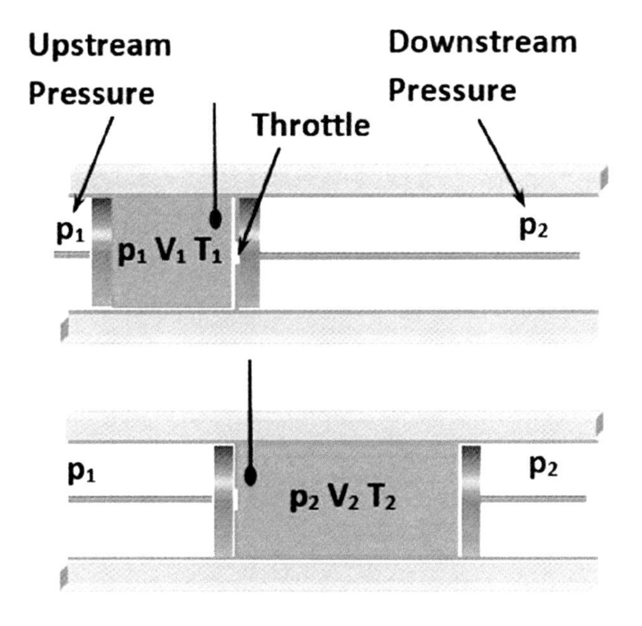

FIGURE 6.2 Ideal Joule–Thomson process.

This is an adiabatic process, and no heat exchanges between the gas and its environment, that is, $dQ = 0$ per the first law of thermodynamic (Eq. 1.6).

Assume:

U_1: the internal energy of the gas when it is located inside the left chamber.
$W_1 = P_1V_1$: the work done toward gas by the environment inside the left chamber.
U_2: the internal energy of the gas inside the right chamber;
$W_2 = -P_2V_2$: the work done toward the gas by the environment inside the right chamber.

(Note: since the gas is pushing the piston in the right chamber, the work done toward the gas by the environment inside the right chamber is negative)

Since this is an adiabatic process, the change of the internal energy of the substance, $(U_1 - U_2)$, should be equal to the ***total*** work done to the gas by the environment before and after the expansion, $\left[P_1V_1 + (-P_2V_2)\right] = (P_1V_1 - P_2V_2)$ (i.e. Q = 0 in equation 1.7).

We have $U_2 - U_1 = P_1V_1 - P_2V_2$ (6.1a)

or

$$H_1 = U_1 + P_1V_1 = U_2 + P_2V_2 = H_2 \quad (6.1b)$$

We conclude that the enthalpy of the gas remains unchanged during the Joule–Thomson process, that is, this is an **isenthalpic process**.

Discussions:

1. The throttling process happened inside the porous plug is an irreversible process.
2. The gas inside the porous plug is in a thermodynamically non-equilibrium state.

However, enthalpy is a state function of the equilibrium states per our discussion in Chapter #1; we have implied that the gas inside both chambers are in the thermodynamically equilibrium states when we call the Joule–Thomson process an isenthalpic process.

In this case, we should ignore the details happening inside the porous plug but focus on the "equilibrium" states "far" away from the porous pug when we discuss the Joule–Thomson effect.

3. When we are setting up the real system, care should be taken to minimize (if not completely isolate) the thermal link between the real throttling device and its surroundings, that is, to keep the throttling process as "**isenthalpic**" as possible.

6.2.3 Does the Joule–Thomson Process Cause Warming or Cooling?

The potential energy in the **real** gases originates from the interactions between the gas molecules, either attractive or repulsive forces. The gas molecules also move with kinetic energy, which is measured by "temperature." The gas temperature drops (rises) when its kinetic energy is reduced (increased).

The Joule–Thomson effect in real gas originates from a "fight" between potential energy and kinetic energy inside the gas. Whether the gas cools or warms during the Joule–Thomson process depends on how the energy is transferred between the potential energy and kinetic energy of the gas molecules.

In general, at low temperatures and/or low density (i.e., low pressure) the **attractive force** predominates, that is, the intermolecular attraction is the most important interaction. When the cold gas **expands adiabatically**, it takes energy to pull the molecules farther apart and increase the average distance between molecules. The only source of energy is the kinetic energy of the gas itself since the process is adiabatic. When the kinetic energy is reduced, the gas *cools*.

On the other hand, the predominant interaction between the molecular gas is the **repulsive force** at high enough temperatures and/or high density (i.e., high pressure). Energy is released from the potential energy between gas molecules, and it will only turn into, and increase, the kinetic energy of the gas itself during an **adiabatic expansion**. The gas *warms* because this is an adiabatic process.

The overall situation is complicated. The sign of the temperature change (ΔT) during the Joule–Thomson expansion depends on initial T and P of each different gas. Details will be discussed in the following sections.

6.2.3.1 Joule–Thomson Coefficient and Joule–Thomson Inversion Curve

Figure 6.3 illustrates a group of Joule–Thomson process data in a P–T diagram where each curve represents a separate isenthalpic process with constant enthalpy.

Recall that enthalpy is a function of T and P. Point D is located on the T-axis where $P = 0$.

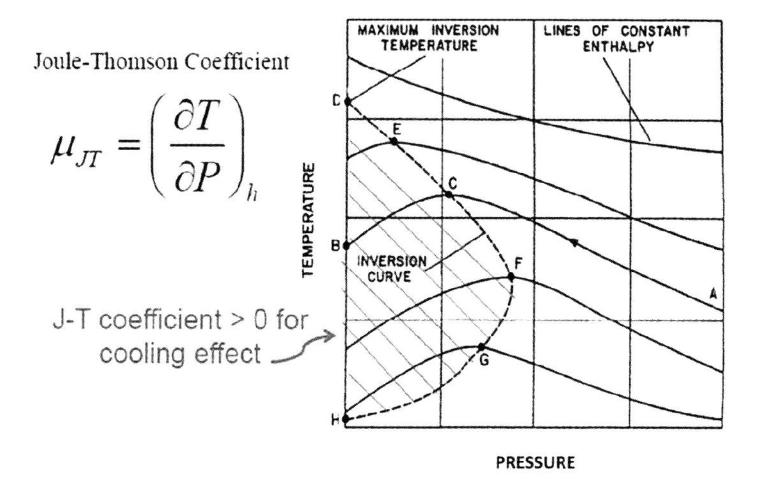

FIGURE 6.3 Joule–Thomson inversion curve.

Let us call the temperature at point D as T_D. Then all isenthalpic curves below T_D starts from a positive slope, and then transitions to a negative slope, with a turning point where the slope is equal to zero.

The slope in the curve of the isenthalpic process equals to

$$\mu_{JT} = (\partial \boldsymbol{T} / \partial \boldsymbol{P})_H \tag{6.2}$$

and it is called the Joule–Thomson coefficient. Obviously, when the Joule–Thomson coefficient is positive ($\mu_{JT} > 0$), the temperature of the gas should increase (decreases) when its pressure increases (decreases), as long as the states of the gas stay on the same isenthalpic curve.

Take curve B-C-A as an example. A–C has a negative slope while C–B has a positive slope. The slope is zero at C. The gas cools when it experiences a Joule–Thomson expansion from state C to state B, while it warms if experiences a Joule–Thomson expansion from state A to state C.

C is the slope turning point. A new curve is plotted in a dashed line from the locus of this point where the Joule–Thomson coefficient is zero. This new curve is called the *Joule–Thomson inversion curve* (or inversion curve).

All isenthalpic curves within the inversion curve have positive slope ($\mu_{JT} > 0$), that is, the gases cool when they experience a Joule–Thomson expansion.

All isenthalpic curves outside the inversion curve have negative slopes ($\mu_{JT} < 0$), that is, the gases warm when they experience a Joule–Thomson expansion.

Obviously, no gas can experience a Joule–Thomson expansion from point D because the pressure at point D is zero. T_D indicates a temperature limit called Maximum Inversion temperature, above which no cooling would take place during Joule–Thomson expansion.

6.2.3.2 Measurement of Joule–Thomson Coefficient

In order to make the Joule–Thomson effect practically useful, we need to relate μ_{JT} to some physical quantities that are experimentally measurable.

Recall that enthalpy is a function of T and P.

We have:

$$dH = (\partial H / \partial P)_T dP + (\partial H / \partial T)_p dT = (\partial H / \partial P)_T dP + C_p dT \quad (6.3a)$$

$$dH = 0 \text{ in the Joule–Thomson process} \quad (6.3b)$$

$$(\partial H / \partial P)_T dP = -C_p dT \quad (6.3c)$$

$$\text{Combine } dH = dU + PdV + VdP = TdS + VdP = 0 \quad (6.4a)$$

$$\text{with the Maxwell equation } (\partial S / \partial P)_T = -(\partial V / \partial T)_P = -\alpha \quad (6.4b)$$

$$\text{we get } (\partial H / \partial P)_T = -\alpha T + V \quad (6.5)$$

$$\mu_{JT} = [\alpha T - V] / C_P \quad (6.6)$$

where

V is the volume of the gas,

C_P is the heat capacity at constant pressure, and

α is the coefficient of thermal expansion.

The value of μ_{JT} is typically expressed in °C/bar or K/Pa.

The **Joule–Thomson coefficient** is now expressed with all measurable quantities in Eq. 6.6, and it can be estimated from the measured values of V, C_P, and α.

Recall the equation of state of one mole of classical ideal gas, PV = RT,

$$\alpha = (\partial V / \partial T)_P = R / P \quad (6.7)$$

Equation 6.6 becomes

$$\mu_{JT} = [\alpha T - V] / C_P = [RT / P - V] = 0 \quad (6.8)$$

The Joule–Thomson coefficient of the classical ideal gas is always zero. It implies that the temperature of the classical ideal gas will remain unchanged during the Joule–Thomson coefficient expansion [5, 6].

Hasan and Şişman [7] studied the behavior of the purely quantum nature of gas particles at low temperatures and derived the Joule–Thomson coefficients of monatomic Bose and Fermi-type quantum ideal gases. They concluded that Joule–Thomson coefficient of a Bose gas is always positive, while that of a Fermi gas is always negative.

When the quantum effects become negligible at high temperature, the Joule–Thomson coefficient of both diluted Bose and Fermi gases approach zero similar to that of the classical ideal gas.

6.2.3.3 Joule–Thomson Inversion Temperature

The temperatures on the **inversion curve** are called **Joule–Thomson inversion temperatures** (or **inversion temperatures**).

Generally speaking, gases always have two inversion temperatures for a specific pressure, and they are called the *Upper Inversion Temperature* and the *Lower Inversion Temperature*, respectively.

Point D in Figure 6.3 is located at the *Maximum Inversion Temperature*, $T_{D.}$ No cooling will take place above T_D during the Joule–Thomson expansion.

The maximum inversion temperatures of various types of gases were vigorously studied in the sixteenth to seventeenth centuries when people were focused on gas liquefactions under high pressures, including Onnes when he was planning to liquefy helium.

Table 6.1 has listed the maximum inversion temperatures of some commonly used cryogenic gases.

Many gases, such as N_2 and O_2, will cool upon expansion at room temperature, but He, H_2, and Ne will warm upon expansion at room temperature [8].

http://www.phys.ufl.edu/courses/phy4550-6555c/spring11/expansion%20cooling.pdf [9]

https://www.coursehero.com/sitemap/schools/552-University-of-Florida/courses/718116-PHY4550 [10]

Attention should be paid to the *Lower Inversion Temperature* of gases near the *T*-axis.

The lower end of the inversion curves of **gases** generally does not intercept the *T*-axis because most gases would turn into liquid or solid before the lower end of the inversion curve intercepts the *T*-axis. An inversion curve for **liquids** may start from there.

6.2.3.4 Inversion Curves and Inversion Temperatures of Helium

Since all systems discussed in this book use 4He as the coolant, a summary of the inversion curves and inversion temperatures of helium would be helpful to the readers when they design their own cryogen-free cryogenic systems [8] (Figure 6.4 and Table 6.2).

Table 6.1 Maximum Inversion Temperature

Elements	Maximum Inversion Temperature (K)
Helium	45
Hydrogen	205
Neon	250
	Room Temperature
Air	603
Nitrogen	621
Carbon monoxide	652
Oxygen	761
Argon	794
Methane	939
Carbon dioxide	1,500
Ammonia	1,994

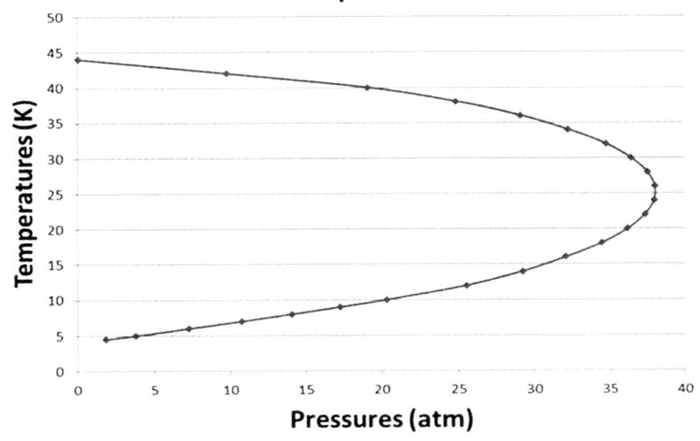

FIGURE 6.4 Inversion curve of 4He.

Table 6.2 Inversion Temperature of 4He

Helium-4 Inversion Temperature (K)	Pressure (atm)	Helium-4 Inversion Temperature (K)	Pressure (atm)
4.5	1.821	22	37.33
5	3.768	24	37.93
6	7.266	26	37.98
7	10.74	28	37.48
8	14.10	30	36.40
9	17.31	32	34.71
10	20.36	34	32.22
12	25.57	36	29.13
14	29.29	38	24.89
16	32.07	40	19.11
18	34.44	42	9.8
20	36.18	44	0.03

6.2.3.5 Reduced Inversion Curves

Hendricks et al. [11] at NASA have studied the equations of state (PVT relations) of methane, oxygen, argon, carbon dioxide, carbon monoxide, and helium.

An approximation for the critical isenthalpic Joule–Thomson coefficient $\mu_c = T_c / 6P_c$ for the critical μ_c was developed, where T_c and P_c are the critical temperatures and pressures of the substance.

The fluids helium, hydrogen, and neon were not included in the generalization because of their quantum effects. Carbon dioxide was also excluded from the generalization because of its high acentric factor per the authors.

Readers can refer to the following reference for more details: [11, 12].

6.2.4 Concept of Viscosity and Impedance

6.2.4.1 Viscosity

Fixed impedance is one of the commonly used devices in the throttling process.

This section is to introduce its working principle, fabrication, and measurement.

6.2.4.1.1 Definition and Newton's Law of Viscosity

Dynamic viscosity (or **viscosity**) is a measure of the resistance of a fluid flow from the shearing force. It describes the internal friction of a moving fluid and plays an important role for the correct impedance design.

Consider two parallel surfaces with overlapped area, A, as shown in Figure 6.5. The lower surface is fixed while the upper surface moves under a shear force, $\boldsymbol{F}$, along a shear direction with velocity, $\boldsymbol{u}$.

The space between these two surfaces is filled with fluid.

The top layer of the fluid, that is, the layer next to the upper plate, starts moving with velocity, $\boldsymbol{u}$, while the velocity of the lowest layer of the fluid, that is, the layer seated on the lower plate, remains zero.

Therefore, a velocity gradient, $\boldsymbol{du / dy}$, is developed within the fluid between these two plates due to the viscous feature of the fluid.

Note that $\boldsymbol{u}$ is not necessarily a linear function of $\boldsymbol{y}$, that is, $\boldsymbol{du / dy}$ may not be a constant.

The shear stress, $\boldsymbol{F / A}$, is found proportional to the velocity gradient,

$$F / A = \eta(\Delta u / \Delta y) \tag{6.9}$$

where η is the proportionality factor, and it is called the coefficient of **dynamic viscosity**. It is a function of material, temperature, pressure, the position, $\boldsymbol{y}$, etc.

Equation (6.9) is called Newton's law of viscosity: *The shear stress between adjacent fluid layers is proportional to the velocity gradient between the two layers.* (Note the directions of the shear stress applied on each layer of fluid.)

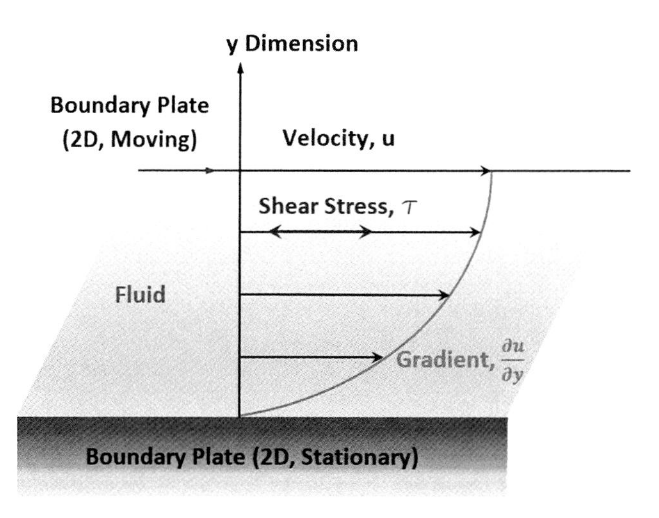

FIGURE 6.5 Dynamic viscosity.

Fluid that obeys Newton's law of viscosity is called Newtonian fluid.

Viscosity is often discussed in the cgs system with the unit of Poise (symbol: P), named after the French physician Jean Louis Marie Poiseuille (1799–1869).

One Poise is equivalent to *dyne-second per square centimeter.* It is the viscosity of a fluid in which a tangential force of 1 dyne per square centimeter maintains a difference in velocity of 1 cm per second between two parallel planes 1 cm apart.

Unit of Pa·s. is often used where 1 P = 0.1 Pa·s.

The unit of viscosity in the MKS system is Newton-second per square meter.

$$1.0\text{ poise} = 1.0\,\text{dyne s/cm}^2 = 1.0\,\text{g}/(\text{cm s}) = 0.1\,\text{Pa}\cdot\text{s} = 0.1\,\text{N s/m}^2$$

The preceding discussion has implied that the fluid has to develop a ***laminar flow***, instead of a ***turbulent flow***, during the viscosity measurements. The substance moves in imaginary thin layers in the laminar flow, and each molecule stays in the same layer instead of moving from one layer to another. The fluid flows with an orderly structure, and a definitive velocity gradient exists inside the fluid. Turbulent flow may be developed when the applied shear stress is too high where there is no recognizable structure and no layers can be observed.

An alternative quantity, $\boldsymbol{k} = \eta / \rho$, is defined as "kinematic viscosity," where ρ is the density of the fluid. Readers can refer to [13] for details. Our discussion will be limited to the **dynamic viscosity** [14].

6.2.4.1.2 Estimation of Gas Viscosities

1. *Pressure dependence of gas viscosity*

   Viscosity in gases is mainly originated from the molecular diffusion that transports momentum between layers of flow. Maxwell [15] published a paper in 1866 to discuss viscosities of air and other gases with the kinetic theory of gases. His calculations showed that the gas viscosity was proportional to the density, the mean-free path, and the mean velocity of the atoms. Since the mean-free path is inversely proportional to the density, that is, the density will increase but the mean-free path will decrease when the pressure increases, the gaseous viscosity is insensitive to pressure, and it almost remains constant under a fixed temperature within a wide range of pressures.

   NASA has used the torsional crystal parameters to measure the viscosity of helium gas at room temperature up to 131 × 10^6 Newton per square meter (19,000 psi). Changes of only a few percent in viscosity was observed [16].

   Interested readers can refer to [17, 18] for more extensive discussions.

2. *Temperature dependence of gas viscosity*

   Qualitatively speaking, molecules in gas are very far apart from each other. When the temperature increases, the molecules move faster and collide more often, resulting in greater resistance, or friction. The larger friction slows down motion of the fluid layers in Figure 6.5 and results in higher viscosity between the molecules per equation Eq. 6.9 The gas viscosity is a function of temperature.

   The viscosity of the *ideal* gas can be expressed by Eq. 6.10 [19]:

$$\eta = \frac{\sqrt{3\boldsymbol{mkT}}}{4\pi \boldsymbol{r}^2} \tag{6.10}$$

where

- m is the mass of the gas molecule,
- k is the Boltzmann constant,
- T is the temperature, and
- r is the radius of the gas molecule.

The viscosity for an ideal gas is independent of pressure but depends on the temperature only.

It increases (decreases) when the temperature increases (decreases).

Equation 6.10 applies for real gases for most circumstances near the conditions we live in, including that in the systems to be discussed in Chapters 8 through 11.

A question: Can we estimate the viscosity of real gas at different temperatures?

The answer is YES!

Sutherland [20] employed the kinetics theory of the ideal gas and studied the gas viscosity by applying an idealized intermolecular-force potential in between molecules. He has generated a relationship between the dynamic viscosity, η, and the absolute temperature, $\boldsymbol{T}$, of an ideal gas, called Sutherland's law:

$$\eta = \eta_{ref}(T / T_{ref})^{3/2}\left[\frac{(T_{ref} + S)}{(T + S)}\right] \tag{6.11}$$

where

- η is the gas dynamic viscosity (Pa s or kg m^{-1} s^{-1}) to be estimated,
- $\boldsymbol{T}$ is the gas temperature (K),
- η_{ref} is the gas viscosity at the reference temperature (Pa.s or kg m^{-1} s^{-1}),
- T_{ref} is the reference temperature (K), and
- $\boldsymbol{S}$ is the Sutherland temperature of the gas (K) at the reference temperature.

Although Sutherland's law provides a semi-empirical relationship, it has provided a good fit to the experimental data of real gases within a reasonable temperature range.

Sutherland's law has implied that the gas viscosity is not a function of pressure. Or the influence of pressure is negligible compared to that of temperature.

Listed in Table 6.3 are the reference temperatures, viscosities at the reference temperature, and the Sutherland temperatures for some commonly used gases.

Table 6.3 Reference Temperatures, Viscosities at the Reference Temperature, and the Sutherland Temperatures for Some Commonly Used Gases

Gas	S(K)	T_{ref} (K)	η_{ref} (microPa.s)
He	79.4	273	18.7
N_2	111	300.55	17.81
H_2	72	293.85	8.76
O_2	127	292.25	20.18
Air	120	291.15	18.27

3. *Temperature dependence of helium gas and helium gas mixture*

Extensive efforts including both experimental measurements and first-principle analysis have been carried out to study the viscosity of helium. For an example, Bich et al. [21] presented the zero density (dilute gas) values of the viscosity of Helium-4 as a function of temperature from 5.0 K up to 5,000 K. Viscosity values of Helium-4 above 20 K under 0.101325 MPa (one atmospheric pressure) were also presented. Interested readers should refer to Table 6.4 and reference [21] for data in the full temperature range.

Hurly and Moldover estimated the helium–helium interatomic potential based on the *ab initio* results from the quantum mechanical calculations of the interaction energy of pairs of helium atoms and the experimentally measured results. From this potential, they calculated the viscosities for diluted (*zero density*) 4He, 3He gas, and 3He/4/He mixture in a wide temperature range [22].

Table 6.4 4He Gas Viscosity at Low Temperatures

Temperature (K)	Viscosity Under Zero Density (microPa-s)	Viscosity at 0.101325 MPa (microPa-s)
5	1.224	
6	1.433	
7	1.619	
8	1.788	
9	1.948	
10	2.097	
12	2.378	
14	2.640	
16	2.886	
18	3.120	
20	3.344	3.357
25	3.872	3.885
30	4.360	4.371
35	4.818	4.827
40	5.252	5.260
45	5.666	5.673
50	6.066	6.071
60	6.832	6.836
70	7.556	7.556
80	8.240	8.240
90	8.894	8.894
100	9.529	9.529

Source: Bich, E. et al., *J. Phys. Chem. Ref. Data*, 19, 1289, 1990.

Table 6.5 Zero-density Ab Initio Temperature Dependence of 4He Gas Viscosity

Temperature (K)	Viscosity (microPa-s)	Temperature (K)	Viscosity (microPa-s)	Temperature (K)	Viscosity (microPa-s)
1.0	0.3279	16	2.895	90	8.920
2.0	0.4573	18	3.131	100	9.558
3.0	0.7123	20	3.356	120	10.775
4.0	0.9815	25	3.884	140	11.932
5.0	1.222	30	4.373	160	13.039
6.0	1.433	35	4.833	180	14.105
7.0	1.620	40	5.269	200	15.137
8.0	1.791	45	5.686	225	16.386
9.0	1.951	50	6.087	250	17.596
10	2.102	60	6.851	275	18.772
12	2.385	70	7.572	300	19.919
14	2.648	80	8.260	325	21.040

Table 6.6 Zero-Density Ab Initio Temperature Dependence of 3He Gas Viscosity

Temperature (K)	Viscosity (microPa-s)	Temperature (K)	Viscosity (microPa-s)	Temperature (K)	Viscosity (microPa-s)
1.0	0.5561	16	2.584	90	7.762
2.0	0.9850	18	2.783	100	8.313
3.0	1.165	20	2.974	120	9.367
4.0	1.261	25	3.423	140	10.369
5.0	1.355	30	3.841	160	11.328
6.0	1.461	35	4.235	180	12.253
7.0	1.576	40	4.610	200	13.148
8.0	1.695	45	4.696	225	14.231
9.0	1.814	50	5.314	250	15.280
10	1.931	60	5.973	275	16.301
12	2.159	70	6.596	300	17.296
14	2.377	80	7.910	325	18.268

Listed in Tables 6.5 and 6.6 are selected data of the zero-density viscosity values of both 4He and 3He gas respectively from [22].

Interested readers should refer to [22] for data in the full temperature range. The viscosity values of 3He/4He mixture gas/vapor not only depend on the temperature but also depend on the ratio of 3He and 4He.

Table 6.7 Zero-Density Ab Initio Temperature Dependence of Equimolar Mixture of 3He/4He Gas/Vapor Viscosity

Temperature (K)	Viscosity (microPa-s)	Temperature (K)	Viscosity (microPa-s)	Temperature (K)	Viscosity (microPa-s)
1.0	0.4147	16	2.751	90	8.366
2.0	0.6555	18	2.968	100	8.963
3.0	0.8884	20	3.177	120	10.102
4.0	1.093	25	3.666	140	11.185
5.0	1.276	30	4.121	160	12.221
6.0	1.443	35	4.548	180	13.220
7.0	1.599	40	4.955	200	14.187
8.0	1.747	45	5.344	225	15.357
9.0	1.889	50	5.718	250	16.490
10	2.024	60	6.431	275	17.592
12	2.281	70	7.105	300	18.667
14	2.522	80	7.748	325	19.717

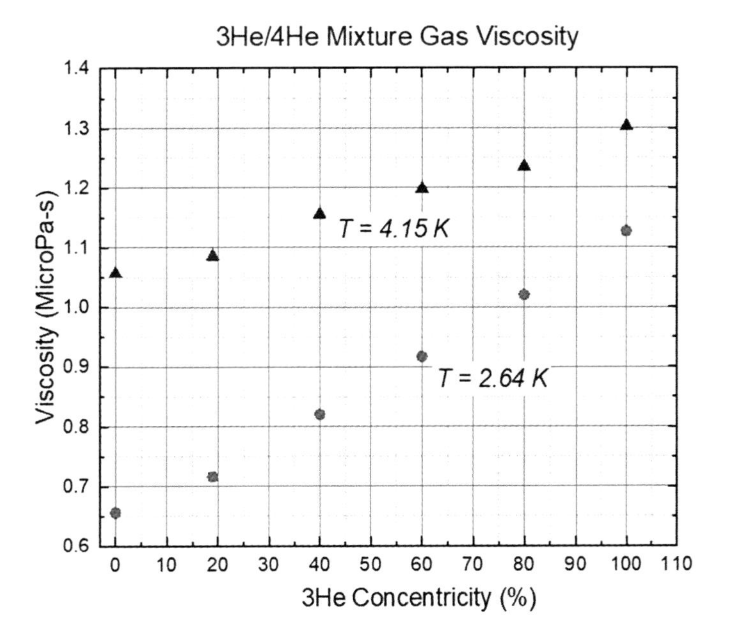

FIGURE 6.6 3He/4He mixture gas/vapor viscosity with different mole ratio.

Listed in Table 6.7 are selected data of the zero-density viscosity values of equimolar 3He/4He mixture [22].

Zhang et al. [23] presented the dilute mixture of 3He/4Helium-3 gas with different ratios at 2.64 and 4.15 K, respectively. Obviously, Helium-3 plays an important role in the viscosity of dilute mixture gas (Figure 6.6).

6.2.4.1.3 Brief Discussion of Viscosities of Liquid

The temperature dependence of liquid viscosity is different from that of gas. The liquid viscosity **increases** with decreasing temperature, except when the superfluid transition takes place in liquid helium.

Several models for predicting liquid viscosities are developed for a qualitative description of the observed behavior [24]. Since the molecules inside the liquid are very close to each other, the viscosity inside the liquid mainly originates from the resistance when the liquid molecules are pushing each other and make room when they move. Molecules at higher temperature move faster due to higher kinetic energy. At the same time, the mean-free path between liquid molecules decreases at higher temperatures. The preceding changes make it easier for a molecule to push itself through other molecules when it moves and results in a decrease of liquid viscosity at higher temperatures. The gas viscosity will increase with temperature. According to the kinetic theory of gases, viscosity should be proportional to the square root of the absolute temperature. In practice, it increases more rapidly. In a liquid, there will be a molecular interchange similar to those developed in a gas, but there are additional substantial attractive, cohesive forces between the molecules of a liquid (which are much closer together than those of a gas). Both cohesion and molecular interchange contribute to liquid viscosity.

Liquid helium is a "quantum liquid" at low temperatures. Their viscosities are illustrated in Figures 6.7 and 6.8, respectively.

a. *Liquid 4He*

   The viscosity of liquid Helium-4 follows the similar temperature dependence as the preceding discussion at higher temperatures, but 4He liquid becomes a "superfluid" at 2.17 K and below. Its viscosity drops dramatically below 2.17 K as shown in Figure 6.7.

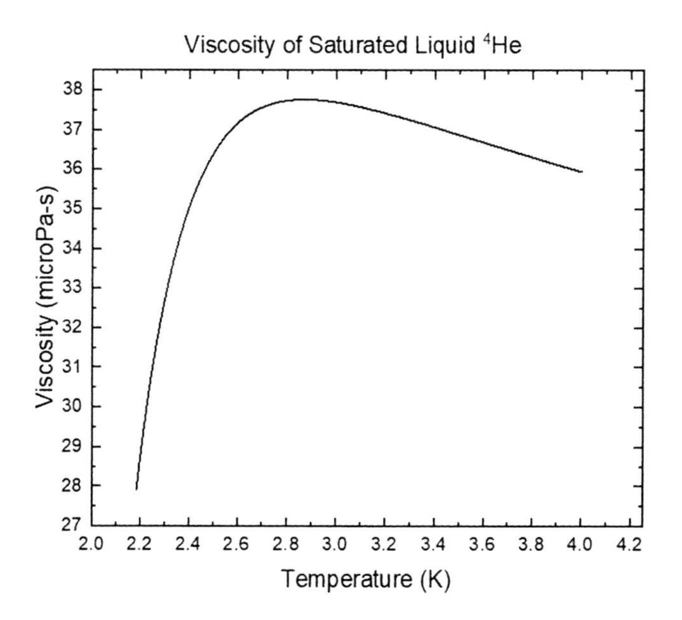

FIGURE 6.7 Viscosity of saturated liquid 4He. (From Angus, S. et al., *International Thermodynamic Tables of the Fluid State Helium-4*, Elsevier, Burlington, MA, 2016.)

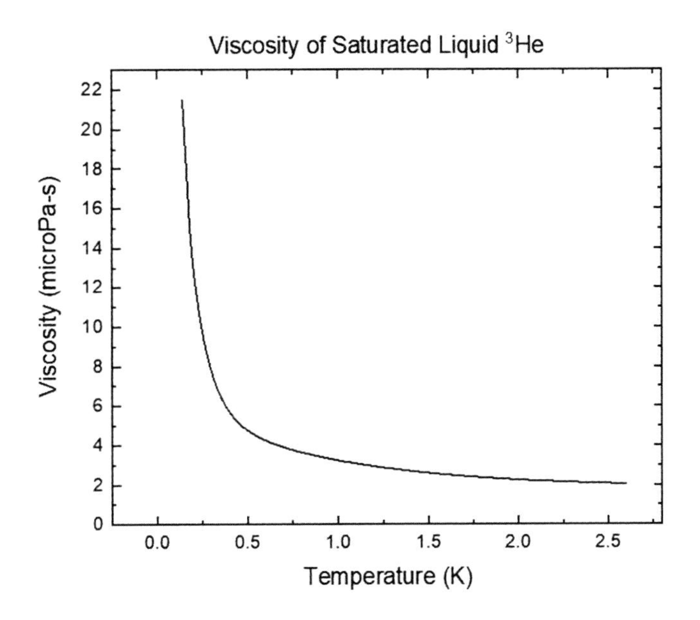

FIGURE 6.8 Viscosity of saturated liquid 3He. (From Vollhardt, D. and Wolfle, P., *The Superfluid Phases of Helium* 3, Courier Corporation, 2003.)

b. *Liquid 3He*

The viscosity of liquid Helium-3 increases when its temperature decreases. In particular, it increases dramatically below 1.0 K and causes obvious viscous heating when liquid Helium-3 flows fast through small channels. Proper heat exchangers should be designed to minimize the viscous heating for sub-K operation when 3He serves as the coolant.

Liquid Helium-3 becomes a "superfluid" at 2.46 mK around 34 atm, where its viscosity drops dramatically [26] (Figure 9.1).

6.2.4.2 Impedance

6.2.4.2.1 Definition When fluid (liquid or gas) moves through a small path (e.g., a thin capillary) by certain volume per unit time (**V / t**), the impedance, *Z*, of this small path is defined by Eq. 6.12.

$$Z = \frac{(P_{high} - P_{low})}{[\eta \times (V/t)]} \tag{6.12}$$

where

Z is the Impedance in 1/cm^3,

P_{high} is the capillary inlet pressure in Pascal,

P_{low} is the capillary exhaust pressure in Pascal,

$(P_{high} - P_{low})$ is the pressure drop along the capillary in Pascal,

η is the viscosity in Poise, and

V/t is the volume of the fluid that comes out of the capillary per unit time in cm^3 /second (CGS system is often used in impedance calculation).

Impedance depends on the geometrical dimensions of the "path." Since the most commonly used "path" is a capillary with a circular cross sections in our discussion, the impedance depends on the inner diameters and lengths of the capillaries, where the inner diameter is usually taken to be uniform.

For the practical system design and fabrications, the impedance of each capillary should be measured with a specific fluid. Different fluids behave differently when they pass through the same "path" because the viscosities for different fluids are different. At the same time, the same fluid behaves differently at different temperatures because the viscosity of the fluid is a function of temperature [27].

6.2.4.2.2 Fabrication and Measurement

Impedance is an "inevitable" device in all continuous flow cryogenic systems below 2 K. Readers can refer to Chapters 8 through 11 for discussions of its applications. This section will take the fixed impedance used in cryogen-free dilution refrigerators as the example and focus on fabrication and parameters.

The typical values of the fixed impedance employed in cryogen-free dilution refrigerator systems are in the range of 10^{10} (1/cm^3) ~ 10^{11} (1/cm^3) depending on the mixture flow rates needed for different systems.

The fixed impedance can be made of CuNi capillaries with approximately 0.5 mm outer diameter and 0.2 mm inner diameter. Since the inner diameter is so small, it is not unusual that the inner diameter varies for different batches of commercial products.

In order to make the device shorter and therefore easier to handle, a non-annealed manganin wire with approximately 0.15 mm diameter can be installed inside the capillary. Non-annealed manganin wire is stiffer and will be less likely to kink when it is inserted into the CuNi capillary.

The impedance can be measured with a setup as illustrated in Figure 6.9 with the following procedures.

- Fill a syringe with water.
- The inlet end of the impedance is connected to a helium gas bottle. The inlet pressure is regulated at a value higher than 1.0 atm, for example, 900 torr for a real example of measurement; the pressure difference across the capillary is (900–760 torr) = 18,662 Pa.

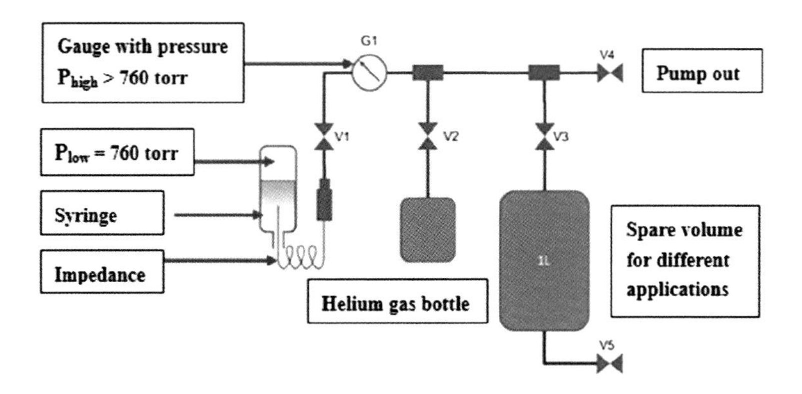

FIGURE 6.9 Impedance measurements setup illustration.

- The exhaust end of the impedance is connected to the syringe.
- Helium gas from the high-pressure bottle will pass through the capillary and bubble into the syringe. Count the total volume of the gas collected inside the syringe within a certain period of time, for example, 0.095 cm^3 per second.

The helium gas viscosity is around 20 microPa-s at room temperature per Table 6.5

We get the impedance of the example device as $\mathbf{1.5 \times 10^{10}}$ **(1/cm³)**.

6.2.4.2.3 Impedance Blockage and Unblocking Tricks

Residual air, nitrogen, or hydrogen gas will solidify at low temperatures and block the impedance when the system cools down. In this case, the user normally needs to warm up (at least part of) the system to unblock the impedance, which takes a lot of time, effort, and the cost of cryogens.

One of the approaches to unblock the impedance quickly is to solder a small copper tube (say, 0.062″ diameter) onto the inlet end of the impedance. Coil the copper capillary on a copper cylinder and then attach it with solder as shown in Figure 6.10. Install a heater and thermometer on the copper cylinder so that the user can heat the copper cylinder at a preset temperature to unblock the impedance rather than heating up the whole system.

This approach can give the user some idea regarding the source of the blockage. For example, if the impedance is unblocked at 20 K, the blockage is most likely caused by hydrogen. If it is unblocked at 80 K, it is likely a nitrogen block. The impedance needs to be warmed up to room temperature in order to get it unblocked if the block is caused by water.

This copper cylinder should be thermally isolated from the rest of the cryostat so that it can be heated up quickly.

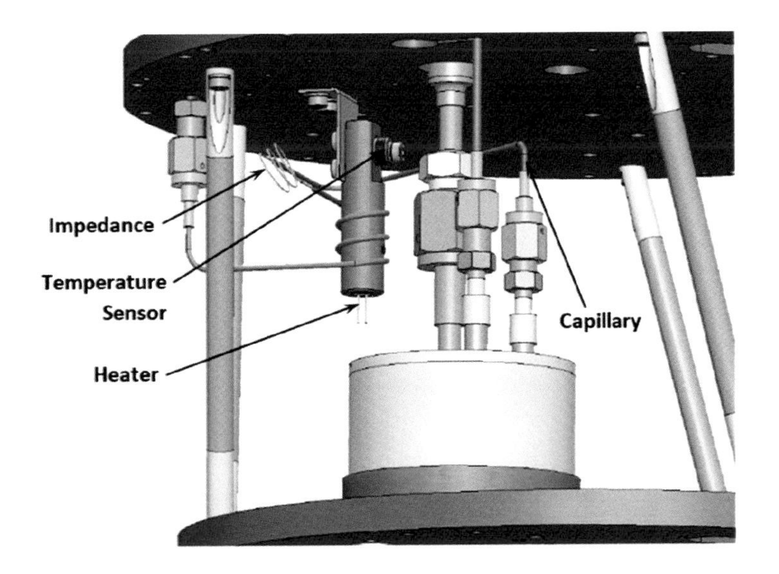

FIGURE 6.10 Unblock the impedance.

6.2.4.2.4 Turn the Fixed Impedance into a "Controllable" One

Due to the thin and long path of the impedance capillary, the temperature of the fluid becomes very close (if not the same) to that of the impedance. Since the viscosity of the fluid is a function of temperature, the flow rate of the fluid, either gas or liquid, can be controlled by varying its temperature, or practically by the temperature of the impedance.

This process can be implemented with the same setup as shown in Figure 6.10.

6.2.5 Adjustable JT Valve

Commonly used throttling device include the fixed impedance and the adjustable JT valve.

The fixed impedance has been described in Section 6.2.4.2. This section will introduce the adjustable JT valve.

The Joule-Thomson process can be implemented with an adjustable JT valve, often called a "*needle valve.*" A typical example of a needle valve is as shown in Figure 6.11.

The needle valve assembly is composed of the following major parts:

Needle assembly: It includes a stainless-steel needle-shaped plunger typically with 1.7° taper. Male threads are made above the tapered plunger. The needle assembly is attached to the bottom of an operating shaft for operation, and it can be moved up and down with the operation shaft.

Seat: This is a receptacle device with female threads that match the male threads on the needle assembly. Identical taper is fabricated inside the seat that matches the taper on the plunger.

An inlet orifice for the fluid with typical diameter of 0.062″ is located approximately in the middle of the seat.

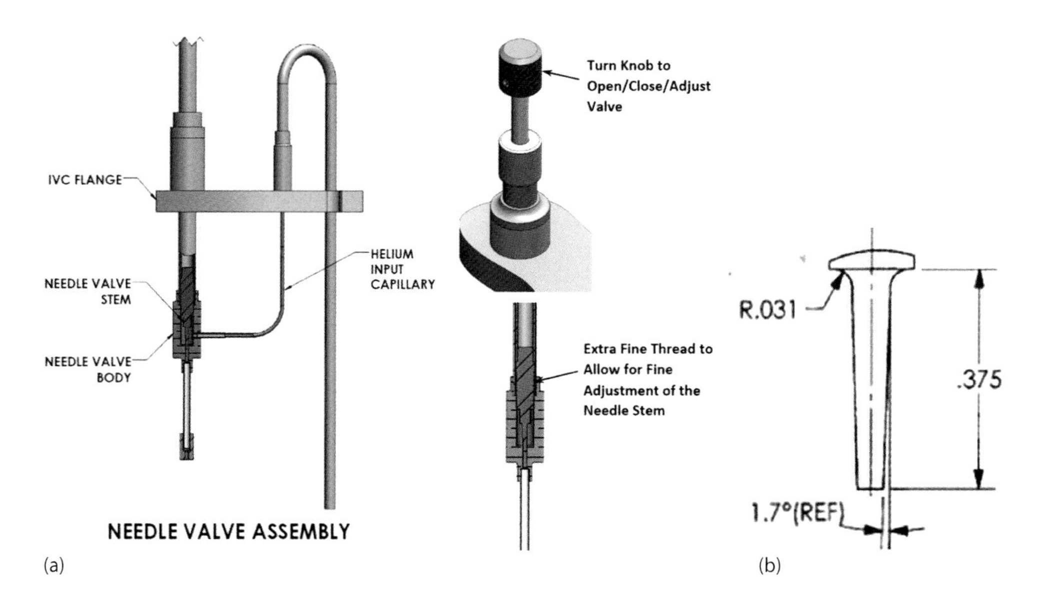

FIGURE 6.11 (a) Typical adjustable JT valve. (b) Needle.

An outlet orifice with typical diameter of 0.062″ is located at the bottom of the seat. The seat does not move, and the gap space between the needle plunger and the seat depends on the position of the needle plunger.

Operation mechanism: The operating shaft can be operated manually or by a step motor controlled by computer. The operator is able to adjust the gap space between the needle and the seat by regulating the shaft and adjusting the position of the needle plunger, thus achieving a fine control of flow of the fluid.

When the plunger is completely moved into the seat and "blocks" the inlet orifice, the fluid flow is significantly impeded, or completely shut off.

Tips:

1. This needle valve is an adjustable cryogenic impedance, but not necessarily a hermetic cryogenic valve. In other words, the needle valves are not designed for simple shutoff applications when hermetic sealing is required.
2. In order to achieve the fine control of the fluid flow, fine threads such as #1/4-40 thread or finer should be implemented for proper flow control.
3. The long and thin plunger with small taper is easy to get bent and damaged. The tip is often deformed by excessive turning force when an inexperienced user tries to "close" the needle valve. "Finger tight" pressure is proper if the needle valve needs to be "closed."
4. The needle is prone to get stuck in the seat due to the small taper, and then get damaged. In order to avoid this problem, the seat should be made of a different material from the needle. Brass is a commonly used material for non-UHV compatible systems. Gold-plated alloy 182 can be used for the seat in the UHV compatible system. Alloy 182 is a "general purpose" nickel-chromium-iron flux-coated electrode used for joining many dissimilar combinations of nickel-based alloys, of the nickel chromium type, to themselves or to stainless or mild steels. Readers can refer to [28] for more details. Copper is not a good candidate because it is too soft.
5. Further approaches can be taken to prevent the needle valve tip from being stuck by coating the plunger surface with a liquid Teflon-based lubricant [29].

   An alternative approach is to coat it with molybdenum disulfide (also called MoS_2 coatings, or Moly coating). This coating is a dry film lubricant, and it provides effective lubrication sacrificially by transferring lubricant between the two mating surfaces, which helps to reduce wear and coefficient of friction. The reader can refer to [30] for more details.
6. The needle valve assembly should be hermetically isolated from any vacuum space inside the cryostat.
7. It has become a common practice to operate the needle valve with a computer-controlled step motor. The position of the needle is crucial to the system performance. An alternative approach, such as limit switches or encoders, can be implemented with the proper program to position the needle valve at desired positions.

   In order to keep a stable temperature at the "1 K pot" (e.g., in the continuous flow 1.5 K system to be discussed in Section 8.2), a computer-controlled closed-loop PID (Proportional Integral Differential) feature can be integrated in to the system. The computer takes information such as the temperature and/or the vapor pressure inside the 1 K pot and feeds it back into the control software. The position of the needle will then be optimized from the feedback information.

6.2.6 Complete Joule–Thomson Expansion System

A typical complete closed Joule–Thomson expansion system is illustrated in Figure 6.12.

It includes the compressor, the recuperative heat exchanger, and the JT valve, which can be a fixed impedance or a throttling valve for expansion. The incoming gas is precooled by the cold out-going gas inside the heat exchanger to a lower temperatures. The compressor is used to maintain a high enough head pressure on the gas before expansion. The combined effect of the heat exchanger and the compressor ensures that the temperature and the pressure of the incoming gas are located within the inversion curve as shown in Figure 6.3. Details of the recuperative heat exchanger are described in Section 11.3.1.

If the incoming gas is precooled to a low enough temperature that the gas is located within the inversion curve below the atmospheric pressure, the compressor is no longer necessary. This is the case for most of the cryogen-free dilution refrigerator systems as described in Chapter 11.

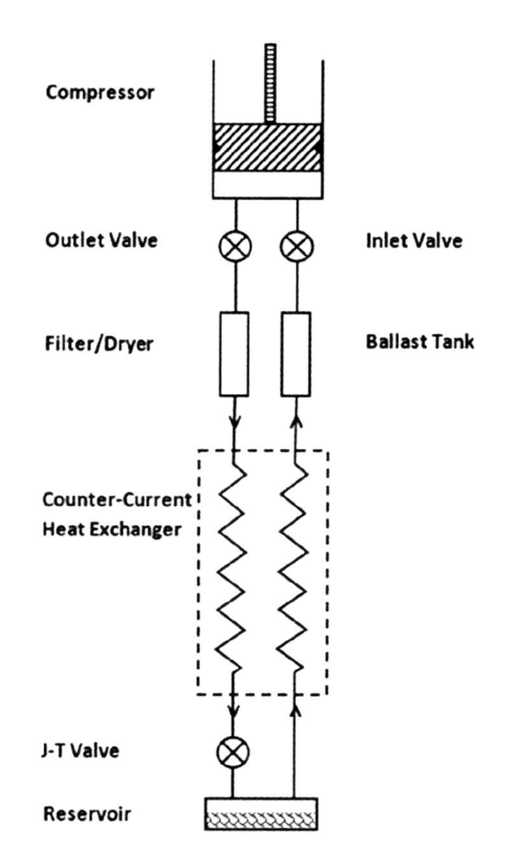

FIGURE 6.12 Complete Joule–Thomson expansion system.

6.3 Heat Switch

6.3.1 Heat Switch and Its Important Parameters

Heat switches (referred to as HS in future discussions), also known as thermal switches, are devices that would establish or cut off the thermal connections between items at different temperatures.

For an example, cryocoolers are used to cool devices to a few Kelvin (depending on the model of the cryocooler). A good thermal link is needed between the cryocooler and the devices to be cooled.

However, if the device needs to go to lower temperatures after it reaches the "precooled" temperature, the thermal link between the cryocooler and the devices needs be "cut off" before it is further cooled down by different approaches (pumping on liquid 3He to reach below 300 mK, with adiabatic demagnetization refrigerator (ADR) to reach below 50 mK, running a dilution refrigerator to reach below 10 mK, etc.).

A heat switch needs to be installed between the cryocooler and the device. It should be engaged when a thermal link between the device and the cryocooler is needed. It should be disengaged when the thermal link is not needed.

The following terminologies will be used in our future discussions:

Cooler and **Object**

Cooler refers to the cryocooler or other types of refrigerators.

Object refers to the items or devices that need to be cooled by the **Cooler** through the heat switch.

ON state and OFF state of heat switch

The heat switch is "engaged" (or in "ON" state) and functions as a good thermal conductor when the thermal conduction is established. The heat switch is "disengaged" (or in "OFF" state) and functions as a good thermal insulator when the thermal conduction is cut off.

ON/OFF Switching Ratio (or **Switching Ratio**)

Switching Ratio is defined as the ratio of the thermal conductance between the **Cooler** and the **Object** when the HS is in the ON state to that when the HS is in the **OFF** state.

Time Constant

It is defined as the time needed for the thermal conductance to change (either decreases when the HS is turned off or increases when the HS is turned on) by 90% of its full value. Shorter **Time Constant** is always desired.

Active heat switch and **Passive heat switch**

External energy is needed to actuate an **Active heat switch**, for example, mechanical heat switch, gas gap heat switch, piezoelectric heat switch, etc.

No external energy is needed to actuate a **Passive heat switch**, such as Paraffin heat switch, differential expansion heat switch, etc.

Mechanical heat switch (MHS) and gas gap heat switch (GGHS) are the two most frequently used types of heat switches. MHS are widely used in ground instruments, while GGHS are particularly important for space instruments.

This section will also give a brief introduction to the superconducting heat switch (SCHS) and heat pipe.

6.3.2 Mechanical Heat Switch (MHS)

Mechanical heat switches (MHS) control heat flow through the mechanical contact of two surfaces.

An MHS usually contains the following basic components as illustrated in Figure 6.13:

- Actuation mechanism: such as manipulator, step motor driven shaft, actuator, etc.
- Movable cold finger made of high-thermal-conductivity materials
- Flexible thermal link between the movable cold finger and **Cooler**

The simplest MHS is the so-called *Single-Stage MHS (SMHS),* as shown in Figure 6.14.

The movable **cold finger** is thermally linked to the **Cooler** with flexible copper braids.

It can move down by a manipulator and make mechanical contact to the **Object**, that is, the SMHS is engaged (ON state). The mechanical contact will be removed when the **cold finger** is lifted by manipulator, that is, the SMHS is disengaged (OFF state).

The tapered design of the contact interface is to generate a larger normal force and enhance the thermal conductance.

Some **Object** contains multiple stages that need to be precooled simultaneously. A good example is the dilution refrigerator stage that normally contains a Still, Intermediate Cold Plate (ICP) and Mixing Chamber (MC). (The reader should refer to Chapter 11.) A multiple-stage mechanical heat switch (*MMHS*) is needed.

Illustrated in Figure 6.15 is an example of MMHS with the following main components:

- Cylindrical rigid cold thermal link thermally connected to a low-temperature reservoir (e.g., cryocooler second stage) as the cooling source;
- Moving shaft with low thermal conductivity located inside the cylindrical rigid thermal link;
- "House" with multiple spring-loaded thermal contactors (cold "feet"); the "house" can "slide" along the cylindrical rigid cold thermal link, and the "feet"

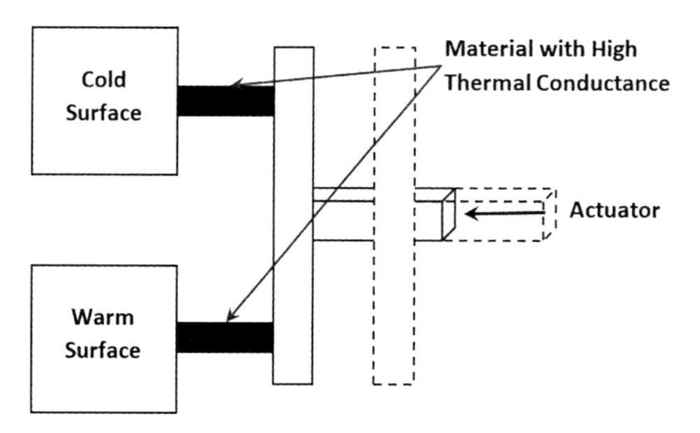

FIGURE 6.13 Concept of mechanical heat switch.

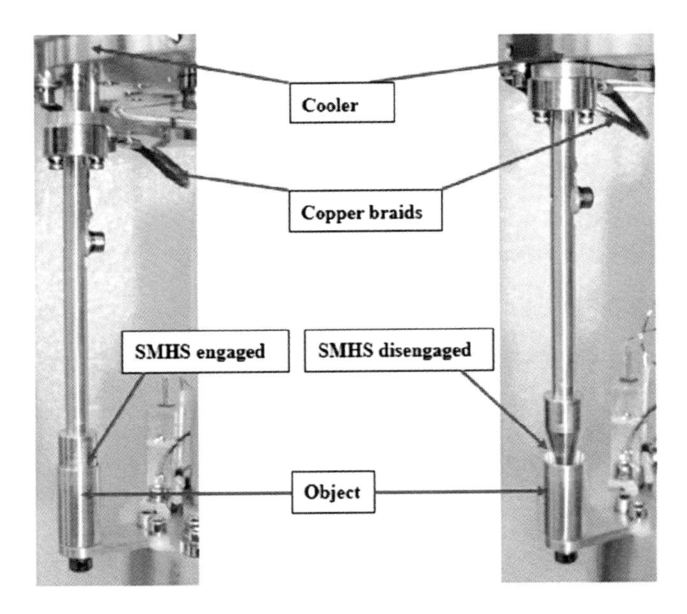

FIGURE 6.14 Typical single-stage mechanical heat switch. (Courtesy of Janis Research Company, Woburn, MA.)

FIGURE 6.15 Spring-loaded multiple-stage heat switch (MMHS). (Courtesy of Janis Research Company, Woburn, MA.)

are located a few millimeters ("gap") above each component to be cooled when the MMHS is in the OFF state;

A "key" (not shown in the figures) is attached to each "house" so that the "house" (and the "feet") will move down when the shaft pushes on the "keys"; and

- Flexible thermal link (copper braids) between the "feet" and the rigid thermal link.

The MMHS is engaged by moving the shaft down with a manipulator or actuator. The shaft in turn pushes the "keys" and moves the all the spring-loaded "feet" down until they make contact with the components to be cooled. The MMHS provides a thermal link to all stages simultaneously.

Figure 6.16 illustrated an alternative design of a vise-type mechanical heat switch (referred to HS-Vise) developed by High Precision Devices, Inc. [31].

The HS-Vise is mounted on a cold plate of cryocooler. An electric motor located at a higher temperature is connected to the vise via a rotating shaft. The motor can be placed either in the vacuum space or outside the cryostat in atmosphere. Conductive "fingers" are attached to the element being cooled by the vise, and these fingers are then situated so that they are suspended between the vise jaws.

When the vise is closed, that is, the heat switch is *engaged* as shown in Figure 6.16a, the jaws create a high-pressure contact between the fingers and the vise jaws, allowing heat transfer. When the vise is opened, that is, the heat switch is *disengaged* as shown in Figure 6.16b, a gap is created between the fingers and the vise jaws to eliminate the thermal conduction between the members.

Due to the highly leveraged mechanical linkage, this HS-Vise provides a high clamping force, yields good thermal conductance at the contact surface, and generates a very high **ON/OFF Switching Ratio**.

The HS-Vise is designed to be self-centering, that is, the vise will apply the same amount of force on each side of the conductors that are being clamped between the vise jaws. The undesired net sideways force on the conductors for delicate systems is avoided.

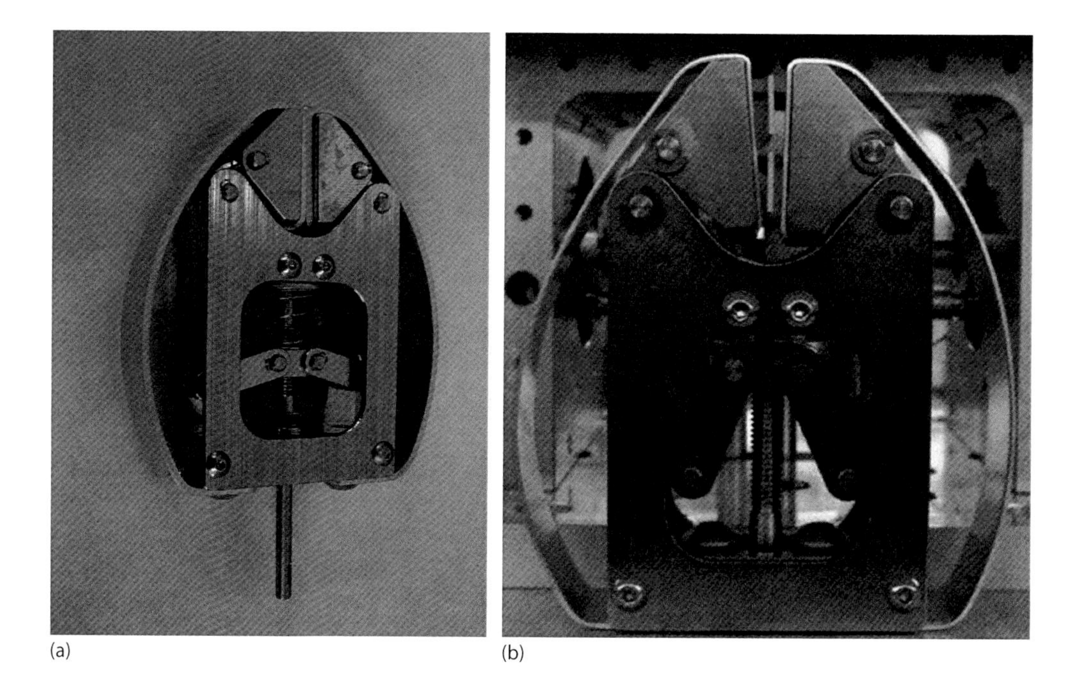

(a) (b)

FIGURE 6.16 (a) HPD heat switch vise assembly engaged. (b) HPD heat switch vise assembly disengaged.

The design also has the thermal conductor integral to the vise and eliminates any additional conductive members to transfer the heat from the object attached to the cold stages to the cooling source.

Advantages and disadvantages of the mechanical heat switch

Advantages:

- Mechanically simple for the *SMHS*
- Easy to operate
- Short time constant
- Large ON/OFF switching ratio
- Applicable at cryogenic temperatures
- Additional features can be added to the HS finger (shutter, basket, etc.)

Disadvantages:

- Mechanically complicated for the *MMHS*
- Thermal shrinkage of the HS or the system may reduce (or even remove) the contact thermal conductance. This problem can be improved with the following approaches:
    - Employ a spring-loaded shaft as shown in Figure 6.14
    - The thermal contact is achieved between the **cold finger** and a multi-contact shown in Figure 6.17 so that the thermal conduction is still secured when the **cold finger** moves.
- Larger forces are usually needed for effective MHS

Alternative mechanical heat switch

The MHS can be implemented with alternative forms, such as the piezoelectric cryogenic heat switch and differential thermal expansion coefficient heat switch.

The piezoelectric cryogenic heat switch is actuated by piezoelectric device(s)/positioned(s).

The positioner will move when voltage with proper polarity is applied to the piezoelectric device until mechanical (thermal) contact is established between the two components at different temperatures. The contact will be removed when voltage with opposite polarity is applied [32].

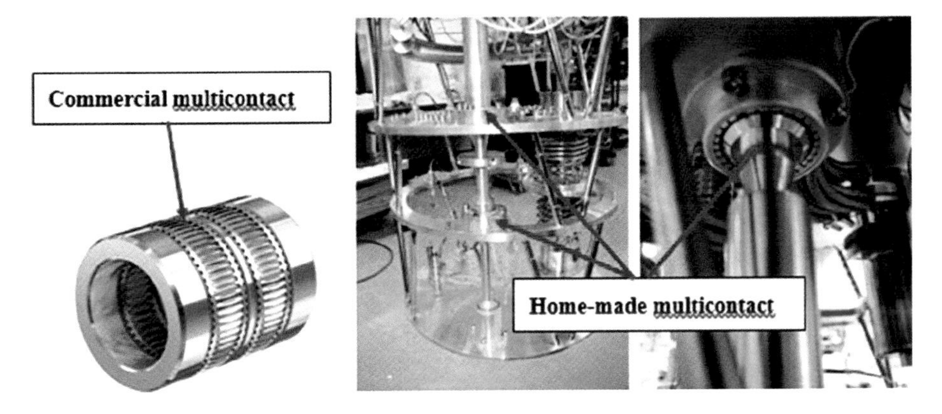

FIGURE 6.17 Multi-contact. (Courtesy of Janis Research Company, Woburn, MA.)

The differential thermal expansion coefficient heat switch makes use of large differential thermal expansion coefficients at different temperatures [33].

The heat switch material expands at higher temperature and makes contact between the **Cooler** and the **Object**. When the temperature drops, the heat switch material shrinks and removes the contact.

6.3.3 Gas Gap Heat Switch (GGHS)

6.3.3.1 What Is GGHS?

A gas gap heat switch (GGHS) typically consists of two cylindrical pieces separated by small gaps, which is filled with a conductive gas (e.g., helium gas) as shown in Figure 6.18.

A typical GGHS has the following structure and main components:

- Cooler plate with fins as a cooling source (CPF);
- Object plate with fins to be cooled (OPF) (typical gap between CPF and OPF can be 100 micrometers);
- Gas absorber with sorbent material (usually activated charcoal);
- Absorber heater and thermometer;
- Capillary connecting the GGHS body to the charcoal absorber;
- Typical gases used in GGHS at cryogenic temperatures are neon, hydrogen, and helium, depending on the applications; and
- Typical switching ratio: >1,000.

Advantages and disadvantages of GGHS

Advantages:

- No moving parts
- Vibration-free operation
- Simple to operate and control
- Various types of gases available for different applications at different temperature ranges [34]
- Due to its unique feature and operation approaches, GGHS has been widely used in space instrumentation.

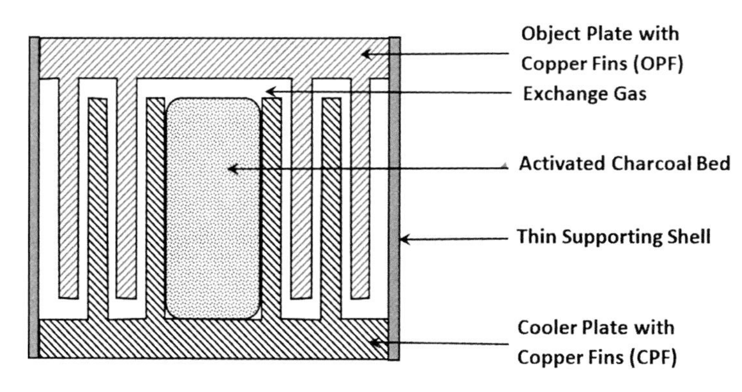

FIGURE 6.18 Conceptual sketch of typical gas gap heat switch.

Disadvantages:

- The HS chassis must be hermetic
- Higher fabrication cost due to the small gaps between CPF and OPF
- Higher radiation heat load to the OPF to be cooled when the GGHS is **OFF**
- Sensitive to vibration because of the very narrow gap
- The long and thin gas delivery capillary between the absorber and the GGHS makes the gas in the gap difficult to be evacuated, leading to longer switching times
- Reliability is always a concern for GGHS
- The sorbent material may have limited lifetime

6.3.3.2 How Does GGHS Work?

6.3.3.2.1 Mean-Free Path (MFP) of Gas Molecules

Mean-free path of a gas molecule is defined as the average distance this molecule travels between collisions with other moving gas molecules.

It can be expressed by Eq. 6.13 when gas in a system is treated as an ideal gas with the assumption that all gas molecules are identical hard spheres with the same radius.

$$\lambda = [RT]/\left[\sqrt{2}\pi d_a^{\,2} LP\right] \tag{6.13}$$

where:

R is the (gas constant) = 8.314 joule. K^{-1} $mole^{-1}$,
T is the temperature in K,
L is the (Avogadro constant) = 6.0221367×10^{23} joule.s^{-1},
d_a is the kinetic diameter of the gas molecule (Kinetic diameter is the size of the sphere of influence that can lead to a scattering event, or "collision"), and
p is the pressure in Newton/m^2.

The MFP of gases is a function of temperature and gas pressure.

Table 6.8 [35] has listed the MFPs of some selected gases at 273.15 K under 1.0 atm. The MFP of these gases at different temperatures under different pressures can be deducted with Eq. 6.13.

Table 6.8 Mean-Free Path of Selected Gases at 273.15 K under 1.0 Atmospheric Pressure

Gas	Mean-Free Path (nm)
Argon	62.6
Helium	173.6
Hydrogen	110.6
Neon	124.0
Nitrogen	58.8
Oxygen	63.8

The reader can also refer to a user-friendly formula of Eq. 6.14 [36] to estimate the MFP of helium gas at different temperatures and pressures:

$$\lambda = 2.87 \times 10^{-5} T^{1.147} / P \tag{6.14}$$

where λ is the MFP in meters, T is the temperature in Kelvin, and P is the pressure in Pascal.

This is a useful formula since helium is one of the most commonly used exchange gasses in GGHS because of its low condensation temperature.

6.3.3.2.2 Different Types of Gas Flow

Define Knudsen number as

$$K_n = \lambda / d$$

where:

λ is the MFP (m), and
d is the diameter of the flow channel (m).

The value of the Knudsen number characterizes the type of gas flow, where

Continuous (or viscous) flow when $K_n < 0.01$
Knudsen flow when $0.01 < K_n < 0.5$
(Free) molecular flow when $K_n > 0.5$

Profiles of the various types of flow regimes are shown in Figure 6.19.

6.3.3.2.3 Gas Viscous Flow

Viscous flow is also known as continuous flow when the gas molecules collide each other frequently. The MFP of the gas molecules is significantly shorter than the dimensions of the flow channel in viscous flow.

The gas thermal conductance in the viscous flow region is independent of the gas pressure.

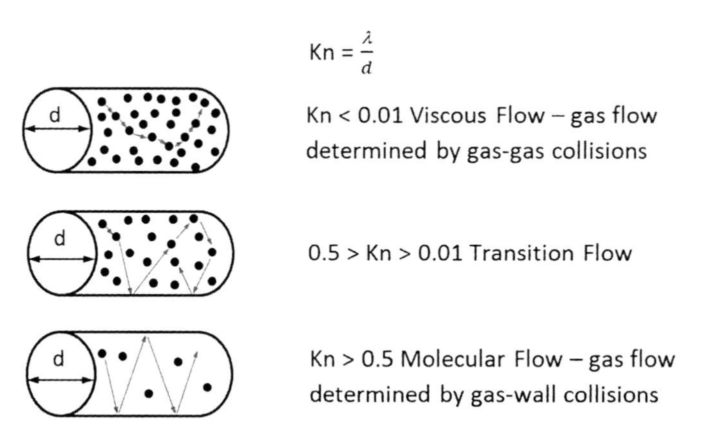

FIGURE 6.19 Different types of flow regimes.

6.3.3.2.4 Gas Free-Molecule Flow

Free-molecule flow is also called molecular flow.

This is the region where the MFP of the gas molecules is comparable or larger than the size of the flow channel.

The gas thermal conductance in the free-molecule region decrease linearly when the gas pressure decreases.

6.3.3.2.5 Transition between Viscous Flow and Molecular Flow

The MFP (and the Knudsen number) varies when the gas temperature and/or the pressure changes. A transition between **viscous flow** and the **molecular flow** takes place when the temperature and/or pressure changes and the Knudsen number reaches between 0.01 and 0.5.

6.3.3.2.6 Temperature Dependence of the Gas Thermal Conductivities

The gas thermal conductivity in the continuum (or viscous flow) region is a weak function of temperature as illustrated in Figure 6.20 [37].

Figure 6.20 depicts the gas thermal conductivities of some commonly used gases in GGHS in the continuum regime within the range of 0.1–1.0 mWatt/cm. K below 500 K. The gas thermal conductivities are not as high as many people expect!

Based on the preceding discussion, the GGHS performance involves four factors:

- Operation temperatures (T) determined by the application;
- Gas pressures (P) at ON state and OFF state, respectively;
- The gap (or separation) (*d*) between the CPF and OPF; and
- MFP of the gas under the operation conditions.

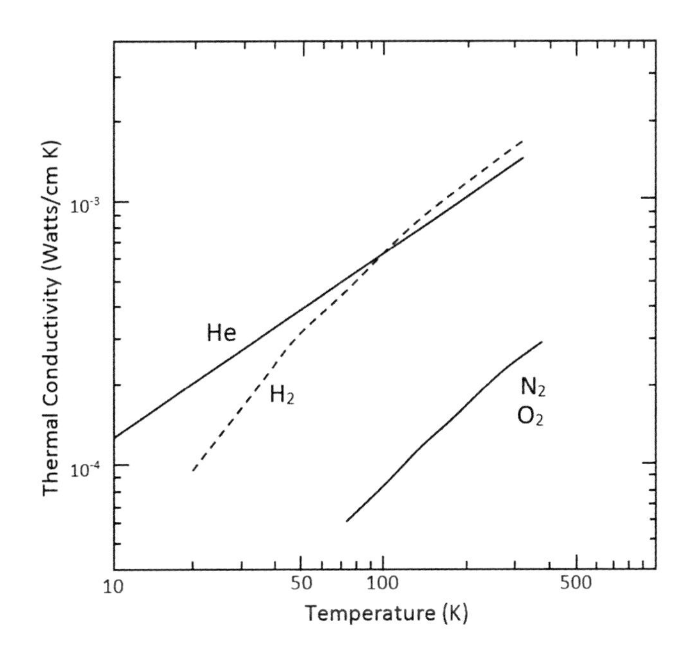

FIGURE 6.20 Thermal conductivity versus temperature for various gases in the continuum regime. (From Minter, C.C., Naval Research Lab Report: The experimental curve showing the effect of pressure thermal conductivity, 1963.)

where

The temperature, ***T***, is determined by the application;
The gap, ***d***, is predetermined by the design; and
T, *P*, and MFP are related to each other.

Once ***d*** and ***T*** are determined, the remaining key issue is to determine the right pressure for the GGHS operation.

The minimum **ON** state pressure for a fixed spacing and temperature takes place when the MFP is **equal** to the spacing, ***d***.

If we call this pressure$\boldsymbol{P}_{mfp=d}$, the maximum crossover pressure when transition between viscous flow and free-molecule flow takes place is approximately 1% of $\boldsymbol{P}_{mfp=d}$.

6.3.3.2.7 Helium Thermal Conductivity

In his book *Experimental Techniques for Low-Temperature Measurements* [39], Ekin gives an example of heat conduction through helium gas between two parallel copper plates spaced 1.0 cm apart as a function of pressure at various gas temperatures.

The heat conduction shows a linear dependence at lower gas pressures, and it remain constant at higher pressures as shown in Figure 6.21. The crossover pressure is inversely proportional to the plate separation, ***d***.

Ekin has also listed in Table 6.9 the crossover pressure between viscous and free-molecule behavior for helium at different temperatures when the distance between CPF and OPF is 1.0 cm.

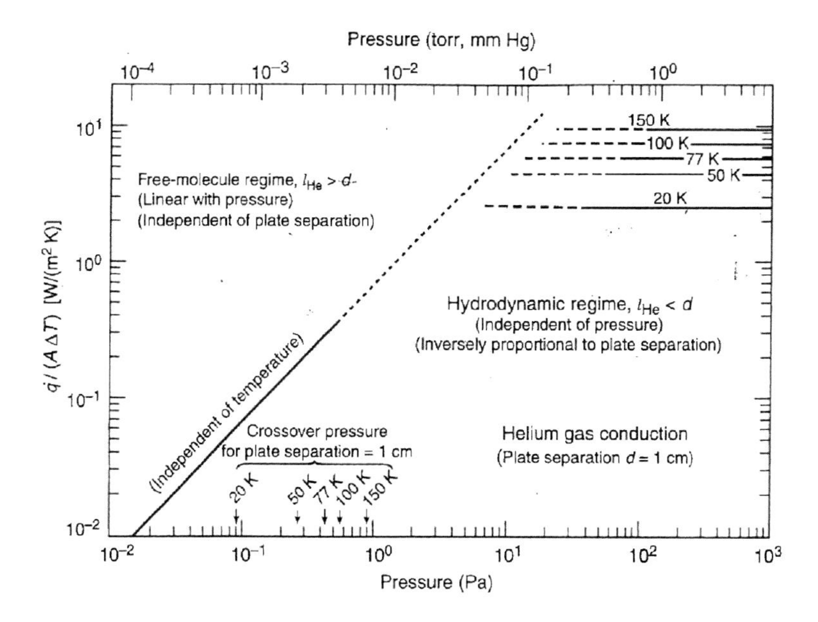

FIGURE 6.21 Heat conduction through helium gas between two parallel copper plates spaced 1.0 cm apart as a function of pressure at various gas temperatures. Note the crossover pressure as a function of pressures and temperatures under a fixed spacing. (From Ekin, J.W., *Experimental Techniques for Low-temperature Measurements*, Oxford University Press, Oxford, UK, 2006.)

Table 6.9 Crossover Pressure of Helium Gas between Hydrodynamic and Free-Molecule Behavior at Different Temperatures When the Distance between CPF and OPF Is 1.0 cm

Temperature	Pressure	
(K)	(Pascal)	(mmHg)
4.0	1.4×10^{-2}	1.1×10^{-4}
10	4.0×10^{-2}	3.0×10^{-4}
15	6.4×10^{-2}	4.8×10^{-4}
20	8.9×10^{-2}	6.7×10^{-4}
30	1.4×10^{-1}	1.1×10^{-3}
50	2.6×10^{-1}	1.9×10^{-3}
77	4.2×10^{-1}	3.1×10^{-3}
100	5.7×10^{-1}	4.2×10^{-3}
150	9.0×10^{-1}	6.8×10^{-3}
200	1.2×10^{0}	9.4×10^{-3}
250	1.6×10^{0}	1.2×10^{-2}
300	2.0×10^{0}	1.5×10^{-2}

Starting from this table, the reader can estimate the minimum **ON** state pressure of GGHS for helium at different temperatures (***T***) with different gap (***d***).

The crossover pressure for a separation (***d*** in cm) other than 1.0 cm can be determined by multiplying the pressure in Table 6.9 by a factor of [1.0 cm/d (cm)].

6.3.3.2.8 Examples of ON and OFF State of GGHS

We have learned the following from the above discussions:

- GGHS is under ON (thermally conducting) state when the gas is in its viscous flow regime, while it is under OFF (thermally non-conducting) state when the gas is in the molecular flow regime under very low pressures.
- The maximum crossover pressure for a fixed spacing and temperature is approximately 1% of that when the MFP is equal to the spacing.

Take the GGHS from [40] as an example. The GGHS has a gap of 100 micrometers and has been operated by both neon gas and hydrogen.

A sorption pump equipped with heater and thermometer contains of 30 g of active charcoal. It is connected to the GGHS by a stainless-steel capillary.

The temperature to actuate the switch with neon ranges from 17 K up to 40 K.

The temperature to actuate the switch with hydrogen ranges from 9.5 K up to 55 K.

This GGHS can also be actuated with helium gas, and it works at temperatures up to 15 K.

Measured values for the thermal ON conductance are 74 mW/K at 20 K for neon and 110 mW/K at 11 K for hydrogen, respectively.

The **ON/OFF** switching ratio is about 220 at 20 K for neon and is around 440 at 11 K for hydrogen.

6.3.3.3 GGHS Design [41]

A typical GGHS is illustrated in Figure 6.22.

The major challenge of a GGHS design is achieving satisfactory operation when large temperature differentials exist across the narrow gap between the **Cooler plate** (CPF) and **Object plate** (OPF). The GGHS design is a complicated process involving parameters of temperature (**T**), pressure (**P**), and dimensions (***d***).

The general requirements for a GGHS performance include:

- Right temperature range for the specific application,
- Proper selection of the gas,

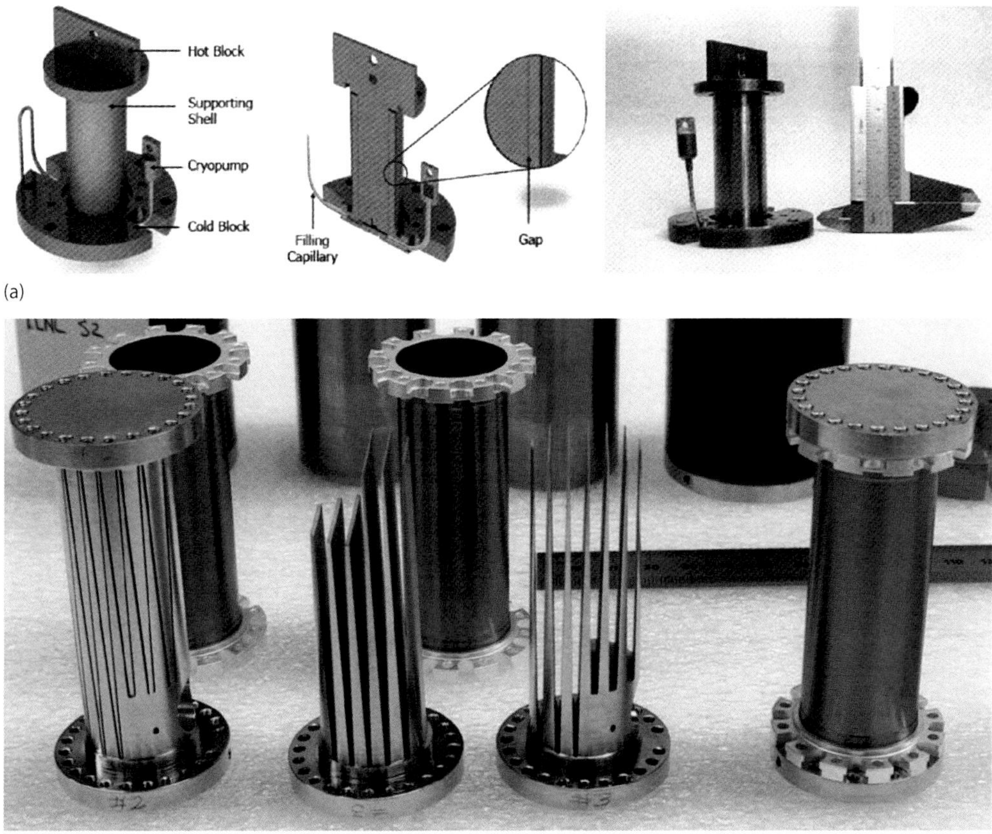

(b)

FIGURE 6.22 (a) Scheme of typical GGHS. (From Barreto, J. et al., Gas gap heat switch for a cryogen-free magnet system, CEC 2015 IOP Publishing, *IOP Conference Series: Materials Science and Engineering*, 101, 012144, 2015.) (b) Interior structure GGHS. (From Kimball, M.O. et al., *Passive Gas-Gap Heat Switches for Use in Low Temperature Cryogenic Systems*, CEC-ICMC, Madison, WI, July 2017.)

- High switching ratio: low radiation and conduction losses (good vacuum) across the switch in the **OFF** state and high thermal conductance in the **ON** state, and
- Short switching time.

This section is to give a brief introduction to the designing steps of GGHS.

Step #1: Understand the application requirements, such as the switching temperature range, the maximum allowable thermal resistance of the switch in the ON state, and the minimum allowable thermal conductance in the OFF state.

Step #2: Select the proper type of gas.

- The gaseous working fluid should not be condensable at the operating temperatures. Table 6.10 is a useful reference to start with.
- The selected gaseous working fluid has a reasonably high thermal conductivity in the GGHS **ON** state.
- The choice of specific type gases depends on their MFP at the operation temperature range and the application for systems with specific dimensions.

Helium, hydrogen, and neon are often good candidates because they have low critical temperatures and high thermal conductivity at cryogenic temperatures.

Thermal conductivity (Watt/cm/K) is a function of *T* and *P*, where the crossover pressures are an important design parameter.

Reference [39] gives the pressure crossover line for 4He gas in a GGHS with a given gap. The readers should be able to develop illustrations for other gases based on Figure 6.21.

Reference [40] gives the minimum temperature for the ON state and the maximum temperature of the OFF state as a function of pressure for a GGHS having a gap of 100 μm with Ne gas and H_2 gas.

4He gas can be used between 5K and 15K.

It is worth mentioning that although it is not commonly used in GGHS due to the high cost, 3He gas with its lower condensation temperature and higher vapor pressures at low temperatures is the best candidate for applications at very low temperatures.

Step #3: Predetermine the gap size

Start from typical number of 100 μm if you have the manufacturing capability—*this is a design and fabrication challenge!*

Typical gap parameters include gap width, gap shape, gap surface area, material of the fins (both CPF and OPF), etc.

Table 6.10 Critical Temperatures of Gas for GGHS

Gas	Critical Temperature (K)
4He	5.20
Hydrogen	33.14
Neon	44.4
3He	3.32

The choices for gap width, shape, and support structure are interrelated.

In general, the gap should be as narrow as possible. The expected operating loads will determine how narrow the gap should be, while the manufacturing and assembly limitations will determine how narrow the gap can be. As the gap width is reduced, the likelihood of unwanted gap contact increases. Small external mechanical load may cause deflection of the support structure and lead to "thermal shorting."

Once the gap width, shape, support structure, and working fluid are determined, the required **ON** conductance will then determine the necessary surface area between CPF and OPF.

Material for both CPF and OPF should have good thermal conductivities at low temperatures. They are usually made of gold-plated oxygen-free high-conductivity (OFHC) copper.

Step #4: Estimate the maximum gas pressure for the GGHS OFF state

Starting from the operation temperatures, **T**, and the predetermined gap width, ***d***, we can estimate pressure when the gas MFP equals the gap space, $P_{mfp=d}$, with Eq. 6.15,

$$d = \left[RT\right]/\left[\sqrt{2}\pi d_a^{\,2} L P_{mfp=d}\right] \tag{6.15}$$

Recall the maximum crossover pressure is approximately 1% of $P_{mfp=d}$; make sure you can reach pressures below this crossover pressure with the adsorber (refer to Step #5).

Optimize the gap spacing and crossover pressure in your design.

Step #5: Select the proper reversible adsorbent materials; optimize the design of the adsorber including the gas charging/venting system so that the required switching pressures can be achieved.

The adsorbent (also referred to sorption pump) should provide the proper pressure and vacuum required for **ON** and **OFF** operation, respectively.

Common materials include zeolite and activated charcoal, including the bituminous charcoal as mentioned in [42(b)].

Listed below are some semi-empirical parameters of the activated charcoal for the reader's reference:

- Typical surface area: 1,000 m^2 per gram [40]
- Typical density of charcoal is approximately 0.3~0.4 gram per cubic centimeter.
- Typical packing factor (the ratio between the actual volume of the charcoal and the space volume occupied by the charcoal after it is packed) of the charcoal is approximately 0.33.
- 1.0 gram can absorb 12 STP liter helium gas at 4.2 K [43]
- Adsorption isotherms of 3He on charcoal as a function of the helium gas pressure is as shown in Figure 6.23 [44], and it gives the information of **ON** and **OFF** temperature when 3He gas is used in GGHS. (Similar isotherm applies to He-4 gas.)

Readers can refer to Section 9.3.11 for more discussion.

Because the adsorption pump needs to be heated to provide gas for the **ON** state, it must be thermally well-isolated from the switch body. It is usually located distantly from both **Cooler** and **Object**, then connected to the GGHS body with a long, thin stainless-steel capillary. This is part of the reasons why the GGHS usually has a longer switching time.

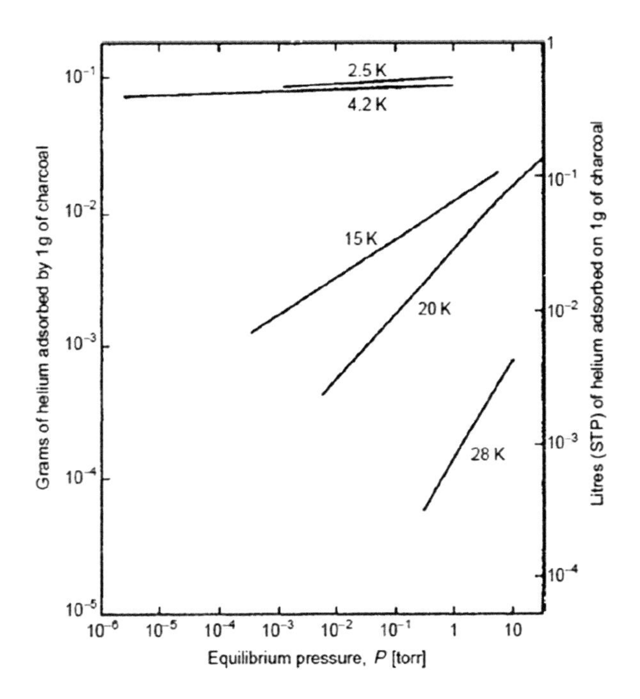

FIGURE 6.23 Adsorption isotherms of 3He on charcoal as a function of the helium gas pressure. (From Pobell, F., *Matter and Methods at Low Temperatures*, Third, Springer Berlin Heidelberg, New York, 2007.)

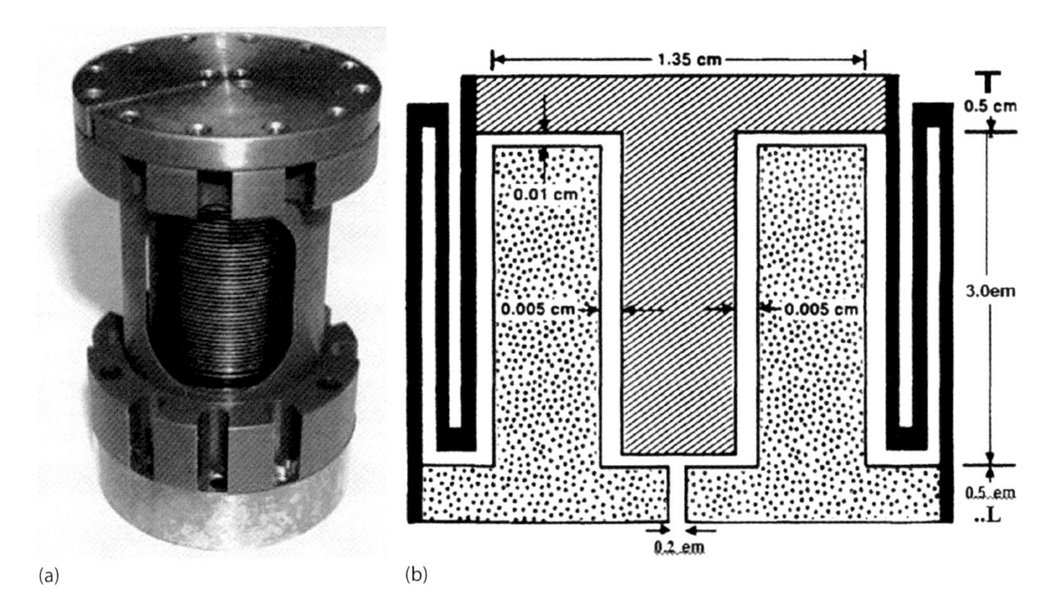

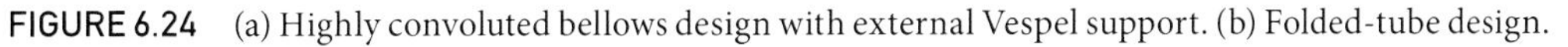

FIGURE 6.24 (a) Highly convoluted bellows design with external Vespel support. (b) Folded-tube design.

Step #6: Supporting structure (or enclosure)

Proper gap support structure design is crucial to maximize the thermal resistance when the GGHS is in the OFF state and optimizes the GGHS performance.

Typical examples include a highly convoluted bellow design [45] as shown in Figure 6.24a or folded-tube design [46] as shown in Figure 6.24b.

Both designs are made of thin-walled stainless steel, and they share the same idea of making the thermal path longer to minimize the thermal conduction between the CPF and OPF.

It is worth mentioning that GGHS has the unique contribution to space science, and NASA is undoubtedly holding the leading position for the development of GGHS. Readers are encouraged to visit references [42(b), 47] for more information.

6.3.4 Superconducting Heat Switch (SCHS)

A heat switch for cryogenic systems working at the mK or micro-Kelvin temperature range is challenging. A MHS is not a good candidate due to the extremely large contact resistance at that temperature range. GGHS will not be effective either, because their working temperature ranges are much higher than mK per our previous discussion.

Pure metals in general are good thermal conductors when they are at the normal state where their thermal conduction is dominated by the (free) electronic conduction. However, their thermal conductivity decreases dramatically when the materials become superconducting because of the depletion of the free electrons. The ratio of the electronic thermal conductivity in the normal state to the phonon thermal conductivity in the superconducting state can be larger than 10^5. An SCHS made of superconducting materials is the solution for (but not limited to) mK/micro-Kelvin systems.

For an example, aluminum is a good thermal conductor when it is in its normal state. Aluminum becomes an effective thermal insulator when it is in the superconducting state. Illustrated in Figure 6.25 are the thermal conductivities of pure aluminum at its normal state, $\boldsymbol{k}_n$, and at its superconducting state, $\boldsymbol{k}_s$, respectively.

Since superconductors can be forced to undergo a phase transition into its normal state when it is in a magnetic field higher than their critical fields, it is fairly easy to convert the material between its normal state and the superconducting state at low temperature, that is, to switch it between a thermal conductor state and a thermal insulator, by applying or removing a proper external magnetic field around the material. This property has been used in the SCHS design and fabrications.

We can conclude that the superconducting material is a good candidate for a highly efficient heat switch at the millikelvin (mK) temperature range.

6.3.4.1 Candidates of Superconducting Heat Switch Material

Ideal SCHS used in mK or sub-mK systems should have a superconducting transition temperature near or above 1 K, that is, orders of magnitude higher than the system base temperature(s). At the same time, the critical field of the SCHS material should be within the range that can be easily achieved.

Material and fabrication cost are also important factors for the proper candidate(s).

As shown in Table 6.11 and Appendix B, aluminum [50], tin (strictly speaking it is white tin, or β-tin) [50, 51], indium, lead [50], zinc [52], etc. are potential candidates, where the most commonly used materials are pure aluminum (Al) and pure (white) tin (Sn).

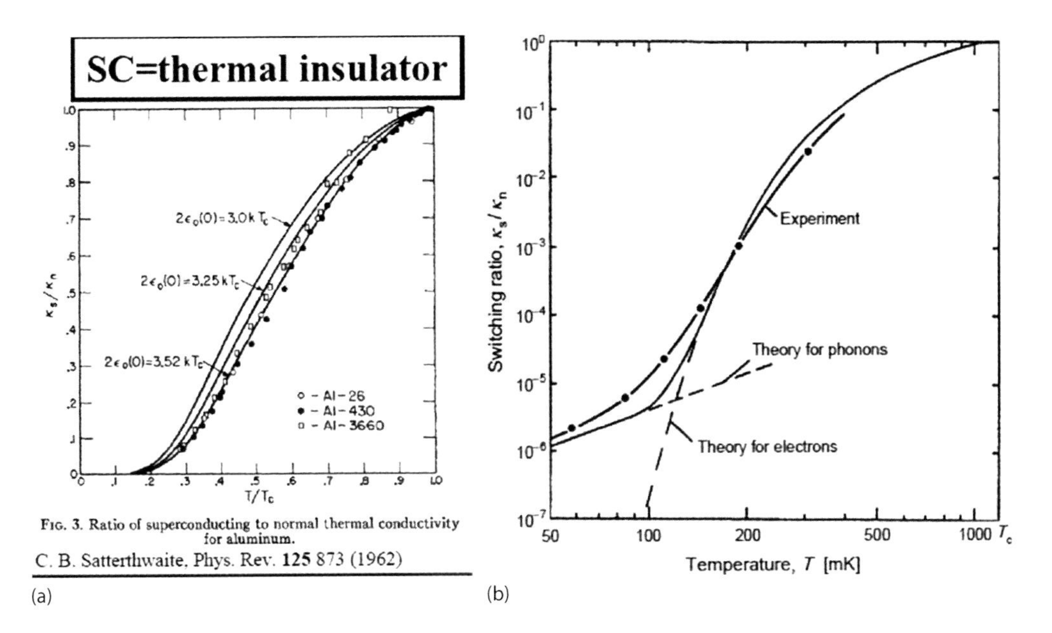

FIGURE 6.25 (a) Ratio of superconducting to normal thermal conductivity for aluminum. (From Satterthwaite, C.B., *Phys. Rev.*, 125, 873, 1962.) (b) Switching ratio, k_s/k_n, of the thermal conductivity of Al compared to theoretical predictions (note: Switching ratio is represented in k_n/k_s in most articles). (Reproduced with the permission from Mueller, R.M. et al., *Rev. Sci. Instrum.*, 49, 515, 1978. Copyright 1978 by the American Institute of Physics.)

Table 6.11 Critical Temperatures and Critical Fields of Some Candidate for SCHS

Material	SC Transition Temperature (K)	Critical Field (Gauss)	Melting Temperature (K)
Pb	7.19	803	601
Al	1.14	105	934
Sn	3.72	309	505
In	3.40	293	430
Zn	0.875	53	693

Source: Abe, S. and Matsumoto, K., *Cryogenics*, 62, 213, 2014.

6.3.4.2 Structure and Main Components

1. General requirements for SCHS
   - High switch ratio, $\boldsymbol{k_n/k_s} \sim \mathbf{T^{-2}}$ at very low temperatures ($T < 0.1T_c$)
   - Material selection for larger $\boldsymbol{k_n/k_s}$:
     - Low electrical resistivity at low temperatures to provide larger residual resistivity ratio (RRR)
     - High purity to increase $\boldsymbol{k_n}$
     - Large Debye temperature makes phonon conductivity ($\boldsymbol{k_{ph}}$) small at low temperature, that is, it decreases $\boldsymbol{k_s}$

- Short switching time
- Low heat capacity
- Low eddy current heating
- Minimum contact resistance between superconducting material and the heat switch body

  Sn, In, and Zn can be soldered onto copper or silver thermal link directly.

  Since aluminum usually has a firmly held oxidization layer on its surface, it needs to be gold-plated before soldering. An alternative approach is to hard press aluminum foil onto the heat switch body.
- Prevent the trapped flux in SCHS from becoming a thermal short

  The common approach is to make the direction of the thermal flow perpendicular to the field orientation of the superconducting solenoid (SCS) (refer to Section 6.3.4.3.2 Example 6.2).

2. Main components for a typical SCHS:
   - Superconductor as the SCHS switching material
   - Solenoid to generate a magnetic field and control the superconducting to normal state transition of the SCHS
   - Superconducting shield to screen the magnetic field generated by the heat switch solenoid from rest of the cryostat

6.3.4.3 Examples of Design and Fabrications of Superconducting Heat Switch

6.3.4.3.1 Conceptual Sketches

SCHS is exclusively used in temperature range of a few mK or lower, and it is the only proper heat switch for the Nuclear Adiabatic Demagnetization Refrigerator (NADR) system for 1.0 mK or lower [54].

Most of the NADR systems have a single-stage NADR stage as illustrated in Figure 6.25a. The SCHS is installed between the mixing chamber (MC) and the nuclear stage (NS).

The SCHS is at the normal state (ON-state) triggered by a solenoid surrounded (not shown in the sketch) during the precooling of the sample by the MC.

After the precooling, the SCHS is switched into superconducting (OFF-state) when the field from the solenoid is removed.

The sample will be further cooled down by the NADR stage.

In order to reach an even lower temperature, the double NADR stage system with two SCHS has been developed as depicted in Figure 6.26b. The first SCHS is installed between the mixing chamber and the first NS while a second SCHS is installed between the two NSs.

The first SCHS is turned to the OFF-state when the precooling is completed. Then the first NS stage cools down the second NS stage. The second SCHS is turned to the OFF-state when the first NS stage reached its base temperature. The sample is then further cooled down by the second NS stage.

Some details of SCHS are shown in schematic drawings of Figures 6.27 and 6.28.

6.3.4.3.2 Examples of SCHS with Different Superconducting Material

Example 6.1: Aluminum (Al) SCHS

Mueller et al. has measured the thermal conductivity of an aluminum foil made from a 5N aluminum rod in the normal and superconducting state down to 58 mK.

It gave a ratio for the thermal conductivities of $\boldsymbol{k_n/k_s}$ = 1600 T^{-2}. He has used that material and developed a SCHS. Listed below are some key design and fabrication step. Readers can refer to [54] for more details.

1. Proper preparation of aluminum is crucial since the aluminum surface can easily get oxidized and cause serious thermal contact resistance at the interface between the Al-SCHS and the thermal link. Zincate [56] is one of the most important steps. This is a mature technique and a typical preparing recipe used in [54] is described in Appendix C.

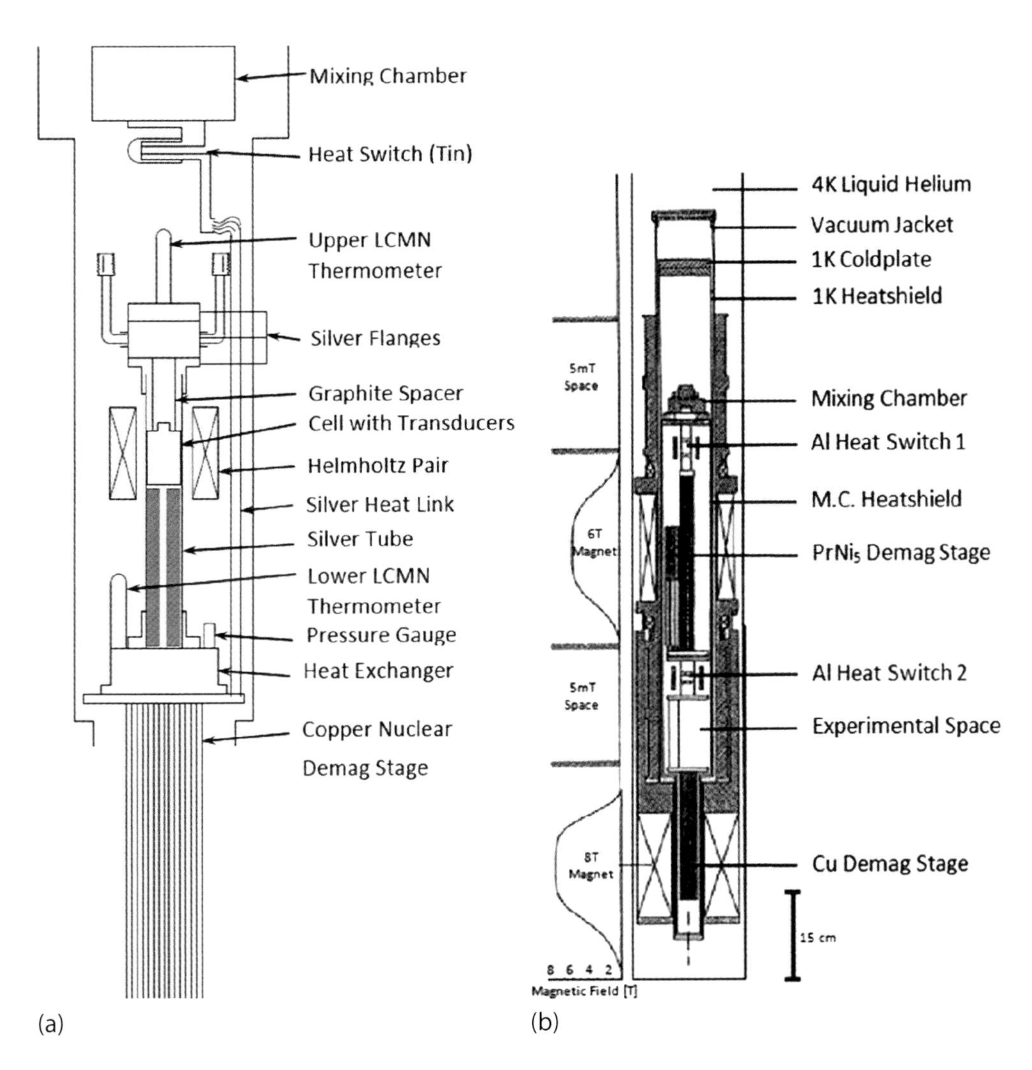

FIGURE 6.26 (a) Schematic conceptual design example of Tin SCHS. (From Zhao, Z., Study of nonconventional superfluid: Ultrasonic propagation in 3He-B and the microwave surface impedance of the heavy-fermion superconductor UPT_3, PhD Thesis, Department of Physics and Astronomy, Northwestern University, Evanston, IL, 1990.) (b and c) Schematic conceptual design example of Aluminum SCHS. (From Gloos, K. et al., *J. Low Temp. Phys.*, 73, 101, 1988; Yao, W. et al., *J. Low Temp. Phys.*, 120, 121–150, 2000.) *(Continued)*

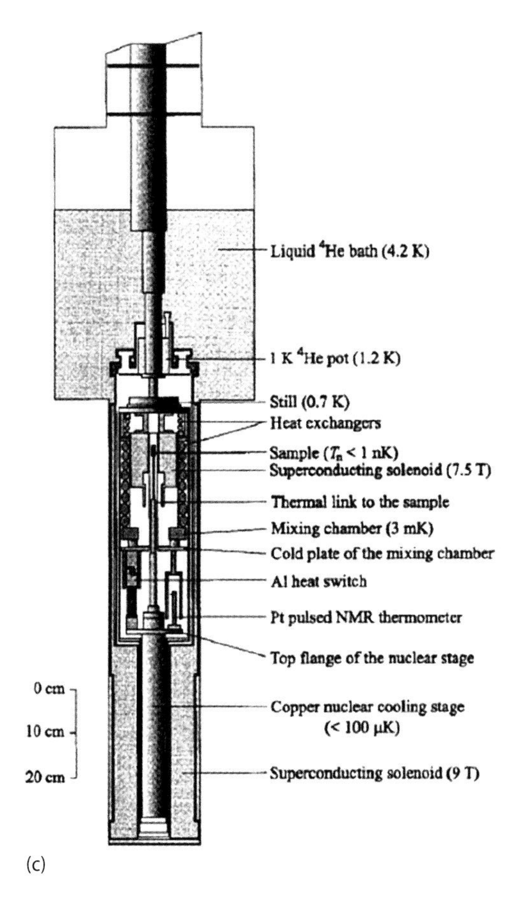

FIGURE 6.26 (Continued) (b and c) Schematic conceptual design example of Aluminum SCHS. (From Gloos, K. et al., *J. Low Temp. Phys.*, 73, 101, 1988; Yao, W. et al., *J. Low Temp. Phys.*, 120, 121–150, 2000.)

2. Press-contact joint between the SCHS material (aluminum) and the heat switch body (copper) is employed to achieve good electrical contact at low temperatures. Both parts were gold plated before the press.
3. An SCS is installed to complete the SCHS fabrication.
   This SCS will be used to generate a magnetic field above the critical field of aluminum and drive the SCHS to the normal state.
   An example of SCS is introduced in Example 6.2
4. A superconducting shield is installed around the SCS.
   An example of SCS is also introduced in Example 6.2

Example 6.2: Tin (Sn) SCHS

Wagner [51] has developed a tin SCHS for his PhD project.

Summarized below are some details of Tin-SCHS by Wagner and others [50,57].

1. The SCHS development started from four high-purity (6N) tin (Sn) foils. Each foil has the dimension of 2.0 × 6.0 × 0.8 mm, and they are soldered to massive copper fingers.

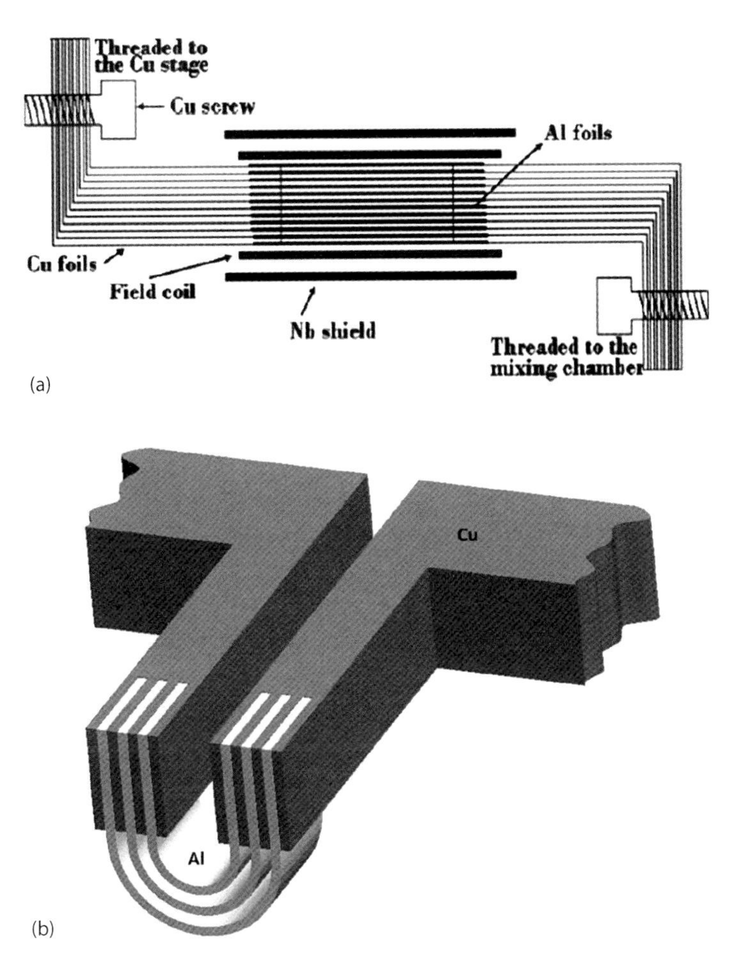

FIGURE 6.27 (a and b) Details of an aluminum SCHS. (From Gloos, K. et al., *J. Low Temp. Phys.*, 73, 101, 1988.)

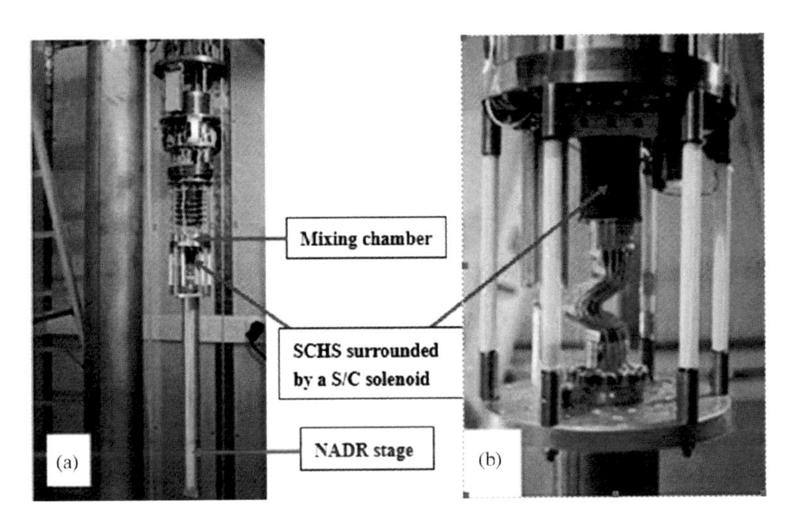

FIGURE 6.28 (a) SCHS set-up. (b) SCHS surrounded by an S/C solenoid. (Courtesy of Professor Rui-rui Du at Rice University, Houston, TX.)

This is a delicate soldering joint, and the thin tin foil may melt if it is overheated.

2. The heat flow directions along the SCHS should be perpendicular to the magnetic fields from the SCHS solenoid. Thus, if trapping flux lines are left in the material after the transition to the superconducting state, the normal "core" will not act as a short circuit for the thermal conduction through the switch. This may also have the advantages of making the heat switch solenoid capable of being installed after the heat switch is assembled.

A possible configuration is shown in Figures 6.29 and 6.30.

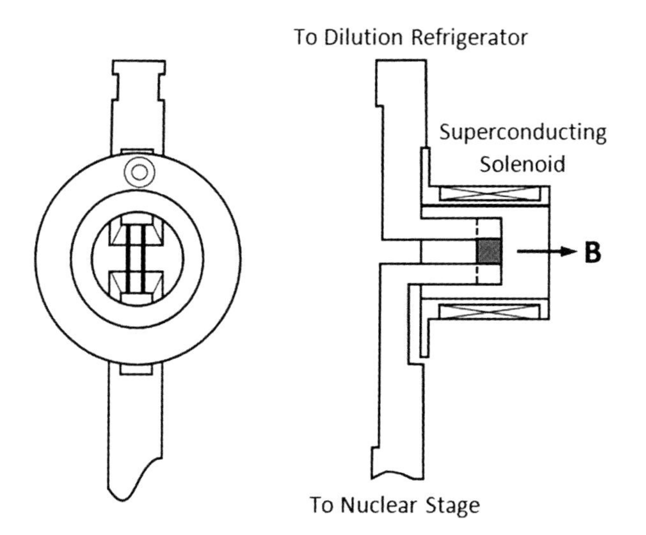

FIGURE 6.29 Schematic drawing of superconducting heat switch and solenoid to apply magnetic field. (From Abe, S. and Matsumoto, K., *Cryogenics*, 62, 213–220, 2014.)

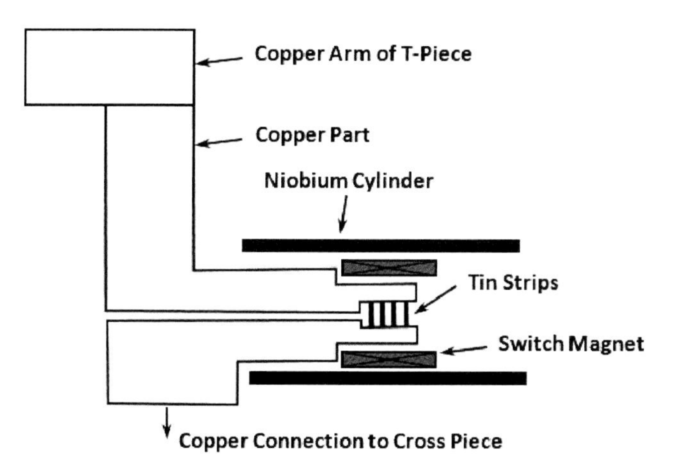

FIGURE 6.30 SCHS with SCS and SC shield. (From Abe, S. and Matsumoto, K., *Cryogenics*, 62, 213–220, 2014.)

3. Sketch of SCS by Wagner [51]
   Inner diameter: 11 cm
   Length: 11 cm
   Wire material: 0.1 mm NbTi/CuNi # of turns: 800
4. Nb-Ti superconducting shield by Raymond [51]
   Diameter: 14 mm
   Length: 40 mm
   When a superconducting shield is put around the heat switch solenoid, it must be located sufficiently far from the windings so that the screening current in the shield does not cancel out the field at the center of the solenoid.
5. White tin and gray tin
   Tin has two different forms, or allotropes: the white (or beta, β) tin, and gray (or alpha, α) tin.
   White tin has a body-centered tetragonal crystal structure, and it becomes superconductor at 3.72 K in zero magnetic fields. Its critical field is around 309 gausses.
   The gray tin has a face-centered cubic structure, and it is a "semiconductor" due to a small energy gap.
   A potential serious disadvantage of Sn SCHS is its possible transformation from white tin (β-tin) to the gray tin (α-tin).
   β-tin tends to transform spontaneously into the brittle, non-metallic α-tin during a rapid cooldown, a phenomenon known as "tin pest." The object will expand because the gray tin has smaller density and decomposes into powder during the transformation and finally disintegrates. Once the gray tin prevails, the SCHS will not work well anymore because the gray tin is not a good thermal conductor.
   Some impurities, such as Bi, Sb, and Pb, can slow down the tin pest [58]. An alternative example of Sn SCHS made of 15 × 3/8″ long, 0.020″ o.d. Sn wires is described in [59].

Example 6.3: Indium SCHS

Indium is another good candidate for SCHS. Pure (5N and even 6N) indium is commercially available, and it can be soldered onto copper and silver directly without any difficulties.

An Indium SCHS was reported by Abe and Matsumoto [53] with the following features:

1. Material: high purity (6N) indium plates with RRR > 10^4 without annealing
2. Area: 5.0 × 5.0 mm^2
3. Directly soldered onto the copper thermal link
4. Switching ratio $k_n / k_s \sim 46\ T^{-2}$
5. SCS was installed with its field orientation perpendicular to the heat flow.

Example 6.4: Zinc SCHS

Zinc has manifested itself with unique advantages in the family of SCHS candidates as listed in Table 6.11:

- Lowest superconducting transition temperature.
- Lowest critical field.
- Higher melting temperature compared to the other directly solderable material (Sn, In, etc.). So it can be easily soldered to a thermal link with Sn or In without any concerns. This feature not only simplifies the fabrications but also helps to make the joint more reliable with smaller joint resistance, which is particularly important for systems with larger heat flow.

A Zn SCHS has been reported by Krusius et al. [52].

The SCHS was made of 9 × 0.17-mm thick zinc foils that were directly soldered onto the copper thermal link with indium.

6.3.4.4 Advantages and Disadvantages of the Superconducting Heat Switch

Advantages:

- SCHS is the only heat switch that works in mK and sub-mK temperature range
- Having high switching ratio

Disadvantages:

- Complicated construction with higher cost

6.3.5 Heat Pipe

The heat pipe is a common passive, capillary-driven heat transfer device. It works in a similar manner to the so-called two-phase thermal switch [60].

It is usually an evacuated vessel with a shape of pipe that contains of a small quantity of working fluid.

The working fluid vaporizes at the hot interface of a heat pipe (or called "evaporator"). The vapor then travels along the heat pipe to the cold interface (or called "condenser"), releases the latent heat, and condenses back into a liquid. The working fluid is then returned to the evaporator either by capillary forces in the porous wick structure or by gravity, depending on the relative orientation of the evaporator and the condenser.

Each heat pipe with different working fluid applies to a specific temperature range. In other words, the heat pipe working fluid selection depends on the operating temperature range of each heat pipe. Table 6.12 lists some commonly used working fluids of cryogenic temperature heat pipes, including their melting and boiling points at atmospheric pressure, as well as their useful temperature ranges.

Illustrated in Figure 6.31 are heat pipes installed on a cryogen-free dilution refrigerator. It is employed together with a cryocooler to precool the dilution refrigerator stage and other massive items.

Alternative cryogenic heat switches (not necessarily working below 4 K) also exist, such as magnetic levitation suspension, shape memory alloys, cryogenic diode, and magnetoresistive. Interested readers can refer to the review article in [61] for details.

Table 6.12 Candidates of Working Fluids of Cryogenic Temperature Heat Pipes

Material	Melting Point (K)	Boiling Point (K)	Useful Temperature Range (K)
Helium	N/A	4.21	2–4
Hydrogen	13.8	20.38	14–31
Neon	24.4	27.09	27–37
Nitrogen	63.1	77.35	70–103
Argon	83.9	87.29	84–116
Oxygen	54.7	90.18	73–119
Methane	90.6	111.4	91–150
Ethane	89.9	184.6	150–240
Freon 22	113.1	232.2	193–297

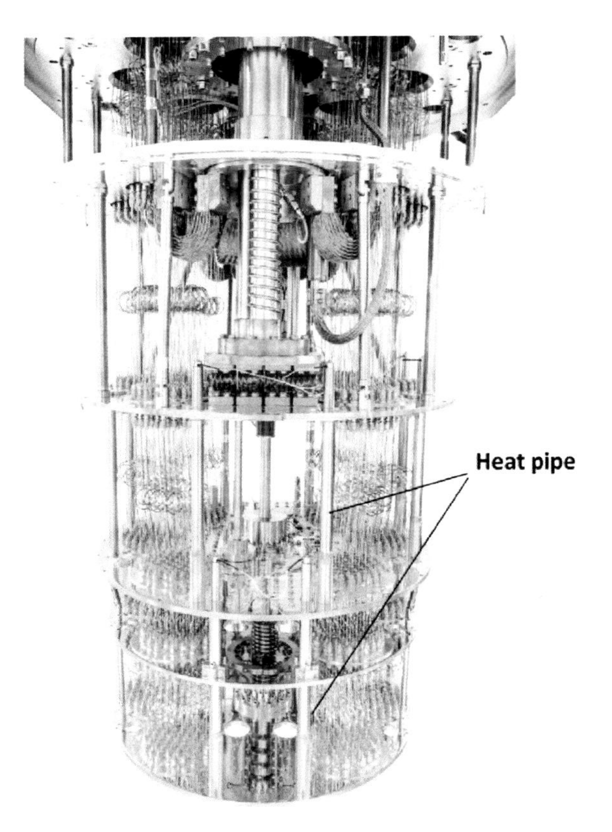

FIGURE 6.31 Heat pipe used to precool the cryogen-free dilution refrigerator. (Courtesy of BlueFors Cryogenics Oy, Helsinki, Finland.)

6.3.6 Flexible Thermal Links

Flexible thermal links are frequently used in the cryogen-free system. It usually consists of two rigid end mounting pieces connected by a flexible middle section with high thermal conduction, such as a bunch of fine metal wires or thin metal foils.

An ideal flexible thermal link should have the following features:

- Good thermal conductivities at the cryogenic temperatures
- Reliable and does not break during normal operation
- Flexible; does not cause stress or transmit vibration
- Easy to fabricate
- Low fabrication cost

Pure (4N or better) oxygen-free high-conductivity (OFHC) copper is the perfect candidate.

High-purity aluminum is another commonly used material when lighter weight is desirable.

The reader can refer to Figure 6.32b and compare the thermal conductivities of copper and aluminum to that of other selected classes of material.

6.3.6.1 Commercial Flexible Thermal Link Made of OFHC Copper

Commercial flexible thermal links are available. Two most commonly used thermal links offered by Technology Application, Inc. [63] are presented below.

1. **Thermal strap made of flexible, pure (4N or better) OFHC copper wires/ropes**
   This type of thermal strap is composed of copper braids made of 99.99%–99.998% pure OFHC copper wires as illustrated in Figure 6.33. It offers excellent flexibility-to-thermal-conductivity ratio.

   The copper strands are cold-pressed in slotted copper blocks (also called mounting lugs), which are also made of 99.99–99.998% pure OFHC copper. The pressing operation is carried out in open atmosphere by means of a mechanical press.

   Dhuley et al. [64] have measured the thermal conductance (*not thermal conductivity!*) of one of these types of thermal links (TAI's Model P5-502 Copper Thermal Straps) within a temperature range of 0.13–10 K.

   The strap he has used has the following dimensions:
   Rope cross section, A: 5E-6 m^2;
   Rope porosity, p 0.645;
   # ropes, n 36;
   Braid/cabling length, L 0.15 m, 0.1 m*
   Illustrated in Figure 6.33 is the thermal strap developed by Technology Applications, Inc.

   The results of Dhuley's measurement with power law fitting is depicted in Figure 6.34.

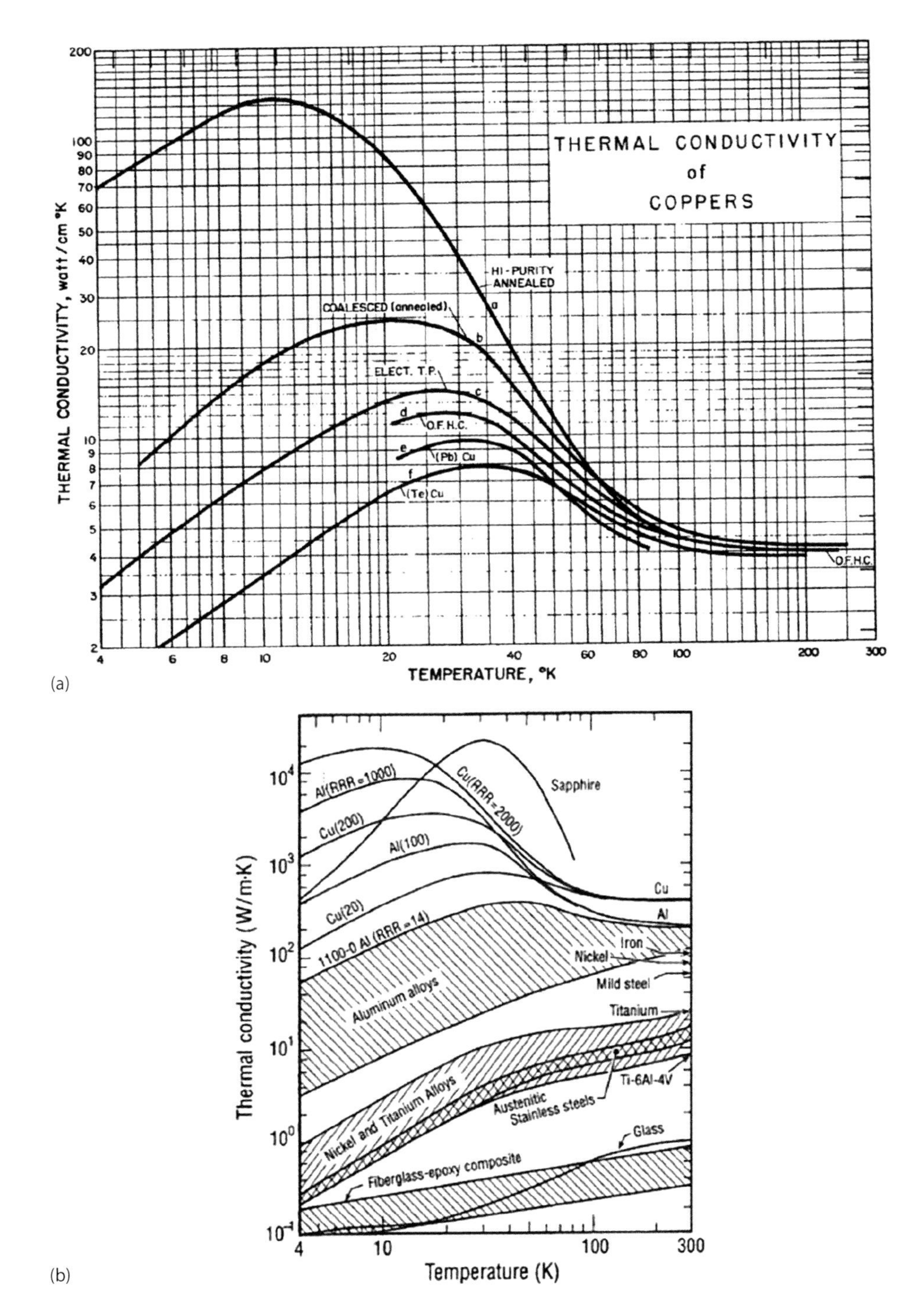

FIGURE 6.32 (a) Thermal conductivity of copper. (From Powel, R.I. and Blanpied, W., *Thermal Conductivity of Metals and Alloys at Low Temperatures,* NBS Circular 556, Blanpied, W.A.) (b) Thermal conductivity of selected classes of material. (From Ekin, J.W., *Experimental Techniques for Low-temperature Measurements*, Oxford University Press, Oxford, UK, 2006.)

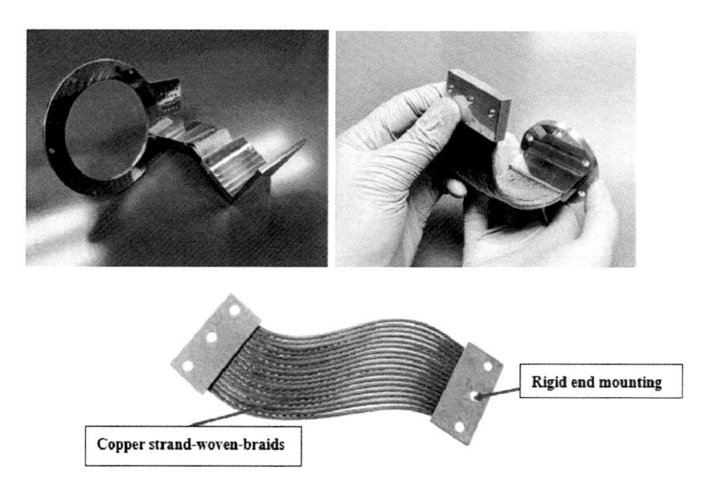

FIGURE 6.33 TAI's thermal strap made of flexible pure OFHC copper strand-woven rope braid and the end mounting lugs. (Courtesy of Technology Applications, Inc., Boulder, CO; Technology Applications, Inc. 5303 Spine Rd, Boulder, CO 80301, USA (P) 303.867.8145 (C) 720.917.4606 https://www.techapps.com/contact-us.)

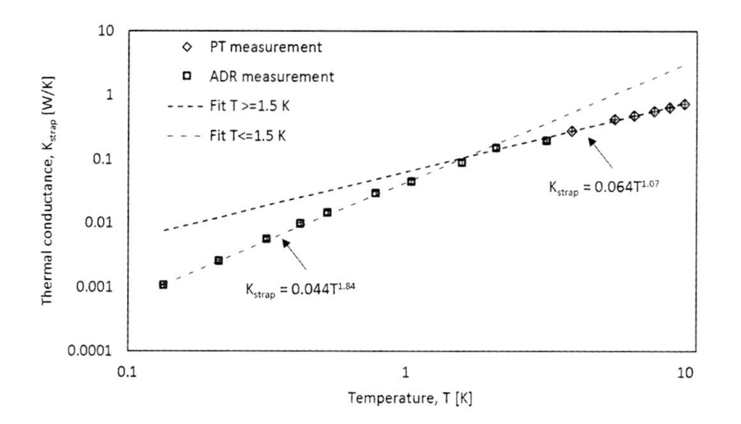

FIGURE 6.34 Measured thermal conductance of TAI's Model P5-502 Copper Thermal Straps as a function of temperature in the 0.13 ~ 10 K. (From Dhuley, R.C. et al., *Cryogenics*, 86, 17–21, 2017.)

As Dhuley et al. mentioned in their article [64], the measured thermal conductance is a combination of two contributions:

$$1/K_{\text{strap}} = 1/K_{\text{braid}} + 1/K_{\text{pressed}}$$

where

K_{strap} is the total thermal conductance of the strand,

K_{braid} is the thermal conductance of the copper rope, and

K_{pressed} is the thermal conductance at the pressed contact surface between the copper braid and the copper mounting lugs.

Since $1/K_{\text{strap}}$ is equal to the thermal resistance of the copper strap, an alternative statement should be that the total thermal resistance of the copper strap is a combination of thermal resistance of the copper rope and that of the pressed contact surface between the copper braid and the copper mounting lugs.

Between the two contributions, the contact conductance, $K_{pressed}$, is of great importance to the practical applications. Dhuley et al. made two important statements in his paper: (1) $K_{pressed}$ is independent of the macroscopic or apparent contact area, as the heat is carried across the joint through the contacting microscopic asperities; and (2) near-liquid helium temperatures the conductance follows a power law with temperature with the exponent between one and three. However, Dhuley et al.'s discussion did not address the force dependence of $K_{pressed}$ at the pressed contact surface between the copper braid and the copper mounting lugs. It is worth mentioning that $K_{pressed}$ near the helium temperature is, to the first order approximation, a linear function of the applied force normal to the contact surface.

2. **Thermal strap made of flexible pure (4N or better) OFHC copper foils**

   When the thermal link does not need three-axis flexibility it can be made of thin and flexible OFHC copper foils. Copper foil straps are attached to the copper mounting blocks with a cold press/swage process as shown in Figure 6.35. This type of thermal strap is more compact.

6.3.6.2 Homemade Flexible Thermal Link Made of OFHC Copper

New England Wire Products (NEW) [65] offers highly flexible copper "ropes." Each rope contains of approximately 10,374 #AWG 46 (0.00157″ diameter) annealed OFHC copper fine wires as shown in Figure 6.36. Users can use this type of copper rope and custom design thermal links to fit their special geometries and applications.

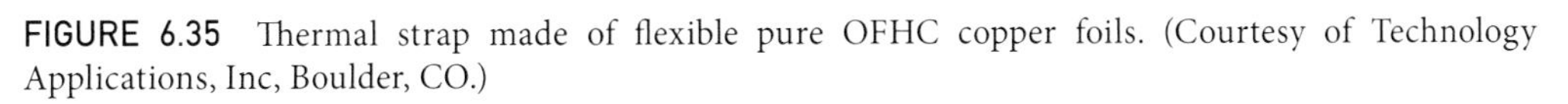

FIGURE 6.35 Thermal strap made of flexible pure OFHC copper foils. (Courtesy of Technology Applications, Inc, Boulder, CO.)

FIGURE 6.36 (a) Highly flexible copper rope from New England Wire Product; (b) custom designed thermal link using the highly flexible copper rope from New England Wire Product. (Courtesy of New England Wire Technologies (NEW), New England Wire Technologies, Lisbon, NH, https://www.newenglandwire.com, and Janis Research Company, Woburn, MA.)

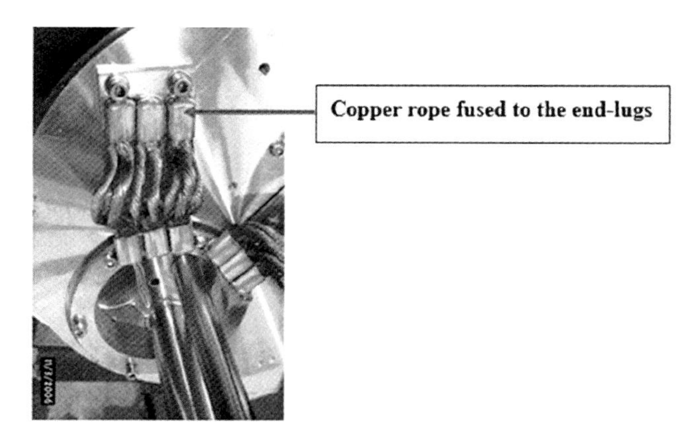

FIGURE 6.37 Copper rope fused to the end lugs.

6.3.6.3 Improvement of Thermal Conductance

The thermal conductance can be improved by the following approaches:

1. Use annealed copper wires/ropes: This approach greatly reduces the number of defects inside the copper wires and enhances the RRR.
2. Fuse the end of the copper: The pressed conductance can be improved by fusing the ropes to the end lugs as shown in Figure 6.37. The fusion converts the mechanically pressed contact into a solid copper connection and therefore eliminates the contact thermal resistance contribution at the pressed surface.
   Soldering the copper ropes to the mounting lugs with flux should be avoided because solder may "wick" into the fine copper wires and destroy the flexibility.
   In addition, copper will get oxidized when water is used to remove the flux.

6.4 Ultra-High Vacuum Technologies

6.4.1 Introduction

A chamber is under vacuum when the molecular density inside it is lower than that in our atmosphere, that is, the pressure inside the chamber is less than the atmospheric pressure. When the pressure inside the chamber in the range of 10^{-9} torr (some users demand 10^{-10} torr) at room temperature after it is baked properly (e.g., at 100°C for a few days), we generally say that the chamber is in ultra-high vacuum (UHV). UHV is required in many applications, including particle accelerators, gravitational wave detectors, microscopy techniques such as STM and AFM, surface analytic techniques such as X-ray photoelectron spectroscopy (XPS), Auger electron spectroscopy (AES), thin film growth and preparation techniques, and angle resolved photoemission spectroscopy (ARPES).

6.4.2 UHV Realization

The main obstacle of UHV realization is outgassing, which refers to the release of gas dissolved, trapped, or frozen within a solid or adsorbed on the surface.

The rate of outgassing is defined as the amount of gas (e.g., number of atoms) released from a unit area of surface in unit time, and the commonly used units are:

- Torr × liter per square centimeter per second (mostly in the USA)
- Millibar × liter per square centimeter per second (mostly in Europe)
- Pascal × cubic meter per square meter per second (the SI unit)

In his paper of PHYSICS OF OUTGASSING [66] J. L. de Segovia has described the physical bases of the gassing phenomena in UHV, including the gas permeation through solids in vacuum, which is one of the forms of outgassing, how to estimate the outgassing rate, the phenomena of adsorption, desorption and absorption, etc.

The key issue to achieve UHV is to minimize the outgassing [67]. It needs proper baking and pumping, right material selection, professional design and fabrication, and adequate surface treatment.

6.4.2.1 Material Selection

Using the right material is the first step to realizing UHV.

All materials, including metals, "evaporate", and reach its Saturated Vapor Pressure (usually just called Vapor Pressure in most discussions) in a closed system, which is defined as the pressure exerted by a vapor in thermodynamic equilibrium with its environment (we only discuss solid state here) at a given temperature. The equilibrium vapor pressure is an indication of the material's evaporation rate [68,69].

As an example, the vapor pressure of many pure elements at room (or higher) temperature(s) can be estimated by the Antoine (or the extended Antoine) equation [70]:

$$\log\left(p / Pa\right) = 5.006 + A + BT^{-1} + C\log T + DT^{-3} \tag{6.16a}$$

p is the pressure in Pascal
T is the temperature in K

or

$$\log\left(p/\text{atm}\right) = A + BT^{-1} + C\log T + DT^{-3} \tag{6.16b}$$

p is the pressure in atmospheres:
T is the temperature in K

where A, B, C, and D are numerical parameters, which depend on the units used for calculations.

This is a semi-empirical formula. Attention should be paid to its valid pressure range with the right units when this formula is being used.

Table 6.13 gives coefficients in an equation for the vapor pressure of some selected metals in the pressure range of 10^{-10}–10^{2} Pa (10^{-15} to 10^{-3} atm).

Table 6.13 Coefficients in an Equation for the Vapor Pressure of a Few Selected Metals

Element	A	B	C	D	T range (K)
Ag	9.127	−14999	−0.7845	0	298-melting point
Al	9.459	−17342	−0.7927	0	298-melting point
Au	9.152	−19343	−0.7479	0	298-melting point
Cu	9.123	−17748	−0.7317	0	298-melting point
In	5.991	−12548	0	0	298-melting point
Pb	5.643	−10143	0	0	298-melting point
Sn	6.036	−15710	0	0	298-melting point
Ti	11.925	−24991	−1.3376	0	298-melting point
Zn	6.102	−6776	0	0	298-melting point

This equation reproduces the observed vapor pressures to an accuracy of ±5% or better.

The reader can refer to "Karlsruher Nuklidkarte" [71(a)] for values of vapor pressures of different metals at different temperatures. The reader should pay attention to the temperatures when these materials reach 10^{-8} Torr range (which is generally not rated as UHV level), as it would help to tell whether they are adequate for UHV systems or not. Take zinc as an example; multiple measurements and calculations [71(b)] showed that the vapor pressure of zinc would reach above 10^{-8} Torr at around 400 K (123°C), which is a typical baking temperature for UHV systems. So zinc is not a good candidate for UHV systems.

Low outgassing materials (i.e., with low saturated vapor pressure) at high (minimum at baking) temperature are needed for both the "system" and its "environment" (usually the vacuum chamber). It is the order of the magnitude rather than the accurate number of the saturated vapor pressure that is important for our material selections.

The most commonly used metals include stainless steel, copper, aluminum, titanium, etc.

Indium can be used in UHV systems, but the baking temperature must be kept much lower than its melting temperature of 156.6°C. This restriction of baking temperature can be compensated by more powerful pumps and longer baking time.

Brass is a copper alloy with zinc. It should be avoided in UHV systems because zinc has a higher vapor pressure at higher temperatures as discussed previously. In addition, brass rods sometimes have cracks along the axial direction that would generate leaks. The crack can also form a "pocket" that would cause a virtual leak (refer to more detailed discussion in **5. Virtual leak—UHV system design and fabrications**).

Commonly used non-metallic UHV compatible materials include Teflon, PEEK, Kapton, type SP-1 Vespel, Ceramic (Alumina, Sapphire, Macor, Shapal-M), etc.

Vespel is the trademark of a range of durable high-performance polyimide-based plastics manufactured by DuPont. It is available in multiple grades with difference components, among which SP-1 Vespel should be used in UHV systems.

Teflon (i.e., PTFE or FEP) is generally (not for all users) categorized as UHV-compatible material, but some customers prefer PTFE to FEP.

Kapton (Polyimide) is categorized as UHV-compatible material, including Kapton tape with silicone adhesive (not acrylic adhesive).

Peek is generally categorized as UHV-compatible material, but glass-filled Peek is usually avoided in UHV-compatible systems.

Cautions should be taken for the above-mentioned materials when they are used in thick bulk pieces since porosities in non-metallic material will trap air inside that may become the source of a virtual leak.

Low outgas epoxies for UHV applications do exist (EPO-TEK H77, EPO-TEK E4110, EPO-TEK H70, etc.).

The following materials should be avoided in UHV system: plastics, rubber O-ring seals, G-10, glues, greases (although Apiezon-L grease has a vapor pressure of 1.0E-11 torr at 20°C, it can hardly be baked since it melts at 50°C), 60/40 soft solder, zinc, brass, and cadmium due to their high vapor pressures during system baking.

Multilayer superinsulation (MLI) [72] is often used in cryogenic systems. As far as the material is concerned, certain MLI can be used in UHV systems, for example, an MLI sheet made of gold- or aluminum-coated Kapton as shown in Figure 6.38.

However, an MLI blanket usually ends up with very large surface area when it is installed in a system. MLI is also highly adsorbent to water vapor, and it is extremely difficult to pump the system down to UHV level even the MLI blanket is baked.

Some users make the compromise by installing just a few (say, three to five) layers to reduce the total surface area of the MLI so that the system is much easier to pump down to the UHV after it is baked.

Components such as wires, coaxial cables, feedthrough, connectors, adhesives (epoxies), and solders have to be selected carefully. Readers can refer to Appendix D for more detailed information. Since few venders are willing to claim (or guarantee) UHV compatibilities for their products, the user needs to make sure that the material and the components selected are able to reach the level of UHV.

Coaxial and triaxial cables are different from the wires for DC measurements. They have an inner conductor, dielectrics, and outer conductors. The dielectrics tightly "wrap" the inner conductor, and in turn the outer conductor tightly "wraps" the dielectrics (in semi-rigid coaxial cables). Water vapor can be trapped in between the tight space,

FIGURE 6.38 Gold-coated Kapton used in space science.

and it usually takes a much longer baking time. The final acceptance of any component depends on if the system can reach UHV after proper baking or not.

It is worth mentioning that more and more contemporary scientists need non-magnetic or low magnetic UHV systems for their researches, which is more demanding to the material selection. For example, 316 or 321 series (including 316LN) instead of 304 series stainless steel should be used for low magnetic UHV systems.

6.4.2.2 Surface Finish

Surface finish is another crucial step to UHV realization.

1. **Electro-polish (EP)**
   Electro-polishing is an electrochemical polishing or electrolytic polishing process that removes surface contaminants and improves the corrosion resistance.

   All stainless steel made items in UHV system should be electro-polished, except commercial items, such as the stainless-steel VCR fittings and the stainless-steel CF flanges.

   Typically, the work piece is immersed in a temperature-controlled bath of electrolytes and serves as the anode, as it is connected to the positive terminal of a DC power supply while the negative terminal is attached to the electrolytes as the cathode.

   A current passes from the anode, where metal on the surface is oxidized and dissolved in the electrolyte, to the cathode. At the cathode, a reduction reaction occurs, which normally produces hydrogen. Electrolytes used for EP are most often concentrated acid solutions, such as mixtures of sulfuric acid and phosphoric acid.

   Other EP electrolytes include mixtures of perchlorates with acetic anhydride and methanolic solutions of sulfuric acid.

   The EP process also deburrs as it polishes the surface. The improved microfinish reduces product adhesion and contamination buildup. Users should be very careful with the temperature and timing of the EP process if a conflat flange (CF) is to be electro-polished because the process may "deburr" the knife edge.
2. **Gold plating**
   In order to avoid oxidization, all copper-made items in UHV system should be gold plated except for the copper gaskets for the CFs. Various types of gold plating for different applications are as listed in Table 6.14 for readers' reference.

   Nickel undercoating on the work piece before the gold plating is a common practice. It will act as a **diffusion barrier** to prevent the copper from migrating into the gold layer over time. It also provides a good adhesion simultaneously.

   However, the existence of nickel is sometimes highly undesirable because of its higher level of magnetism. In this case, gold will have to be plated onto the copper work piece directly.

   An alternative is to use low-magnetic material such as **palladium** as the diffusion barrier.

   It is not easy to have a nice gold-plated surface on the solder joints, which is usually a silver alloy. The solution is to apply "copper strike" on the solder joint in advance, that is, to plate a thin layer of copper on the solder joint before the gold plating.

Table 6.14 Types of Gold Plating for Reference

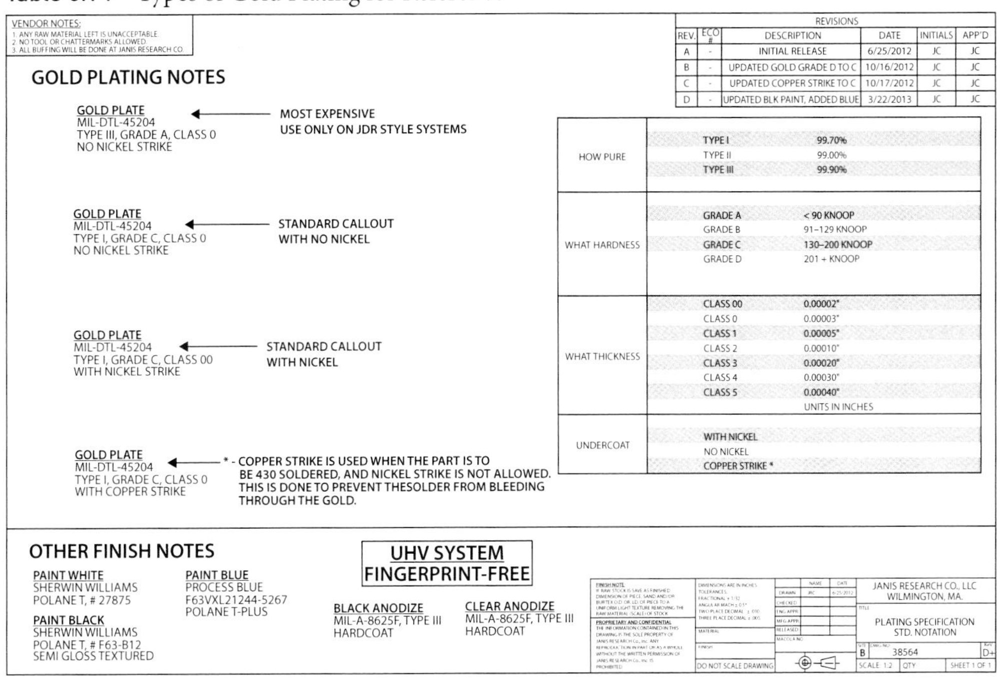

VENDOR NOTES:
1. ANY RAW MATERIAL LEFT IS UNACCEPTABLE.
2. NO TOOL OR CHATTERMARKS ALLOWED.
3. ALL BUFFING WILL BE DONE AT JANIS RESEARCH CO.

REV.	ECO #	DESCRIPTION	DATE	INITIALS	APP'D
A	-	INITIAL RELEASE	6/25/2012	JC	JC
B	-	UPDATED GOLD GRADE D TO C	10/16/2012	JC	JC
C	-	UPDATED COPPER STRIKE TO C	10/17/2012	JC	JC
D	-	UPDATED BLK PAINT, ADDED BLUE	3/22/2013	JC	JC

GOLD PLATING NOTES

GOLD PLATE
MIL-DTL-45204
TYPE III, GRADE A, CLASS 0
NO NICKEL STRIKE
← MOST EXPENSIVE
USE ONLY ON JDR STYLE SYSTEMS

GOLD PLATE
MIL-DTL-45204
TYPE I, GRADE C, CLASS 0
NO NICKEL STRIKE
← STANDARD CALLOUT
WITH NO NICKEL

GOLD PLATE
MIL-DTL-45204
TYPE I, GRADE C, CLASS 00
WITH NICKEL STRIKE
← STANDARD CALLOUT
WITH NICKEL

GOLD PLATE
MIL-DTL-45204
TYPE I, GRADE C, CLASS 0
WITH COPPER STRIKE
← * - COPPER STRIKE IS USED WHEN THE PART IS TO BE 430 SOLDERED, AND NICKEL STRIKE IS NOT ALLOWED. THIS IS DONE TO PREVENT THESOLDER FROM BLEEDING THROUGH THE GOLD.

HOW PURE	TYPE I	99.70%
	TYPE II	99.00%
	TYPE III	99.90%
WHAT HARDNESS	GRADE A	< 90 KNOOP
	GRADE B	91–129 KNOOP
	GRADE C	130–200 KNOOP
	GRADE D	201 + KNOOP
WHAT THICKNESS	CLASS 00	0.00002"
	CLASS 0	0.00003"
	CLASS 1	0.00005"
	CLASS 2	0.00010"
	CLASS 3	0.00020"
	CLASS 4	0.00030"
	CLASS 5	0.00040"
		UNITS IN INCHES
UNDERCOAT	WITH NICKEL	
	NO NICKEL	
	COPPER STRIKE *	

OTHER FINISH NOTES

PAINT WHITE
SHERWIN WILLIAMS
POLANE T, # 27875

PAINT BLACK
SHERWIN WILLIAMS
POLANE T, # F63-B12
SEMI GLOSS TEXTURED

PAINT BLUE
PROCESS BLUE
F63VXL21244-5267
POLANE T-PLUS

UHV SYSTEM
FINGERPRINT-FREE

BLACK ANODIZE
MIL-A-8625F, TYPE III
HARDCOAT

CLEAR ANODIZE
MIL-A-8625F, TYPE III
HARDCOAT

JANIS RESEARCH CO., LLC
WILMINGTON, MA.
TITLE: PLATING SPECIFICATION STD. NOTATION
SIZE: B DWG NO: 38564 REV: D+
SCALE 1:2 QTY SHEET 1 OF 1
DO NOT SCALE DRAWING

Source: Courtesy of Janis Research Company, Woburn, MA.

An often asked question is: "Can or should the copper pieces in UHV systems be buffed before they are gold plated?" The answer is *NO* to start with.

UHV systems normally have two important requirements: vacuum level and cleanliness.

The vacuum level mainly depends on the degas level from the system materials and their surface finish, while cleanliness means *freedom from impurities*.

A buffing compound is a source of impurity, and it is too naïve to believe that a thin gold plating will prevent the buffing compound from escaping into the vacuum space during baking and pumping, resulting in sample contamination. Copper parts can be buffed in order to provide a reflective finish for cosmetic reasons only if the application is not especially sensitive to contamination. A high vacuum level can be reached after gold plating, but there is no guarantee on being free of contaminations after buffing.

3. **Special cleaning process**

   Proper cleaning after surface treatment is often needed. General guidance includes:

   - Using ultrasonic cleaning in a dedicated ultrasonic cleaner.
   - All personnel should wear gloves (latex or nitrile) to work on the UHV systems after items are cleaned.

   1. *Large items cleaning* (too big and heavy to put it in the ultrasonic cleaner)
      - Remove welding burn or discoloration with soft sand pad, then wipe with acetone.
      - Wipe all surfaces with acetone thoroughly.

- Wipe with 2-Propanol (= isopropyl alcohol = isopropanol).
- Cover the system with a clean plastic bag or UHV-compatible aluminum foil.

2. *Small items that can fit the ultrasonic cleaner*
   - Remove welding burns or discoloration with soft pad, then wipe with acetone.
   - Soak the items in 60°C DI water in ultrasonic cleaner for minimum 10 minutes.
   - Clean with acetone in ultrasonic cleaner for minimum 10 minutes.
   - Clean with propanol in ultrasonic cleaner for minimum 10 minutes.
   - Cover the parts with clean plastic bags or UHV-compatible aluminum foils.

4. Baking and pumping

A proper baking and pumping process is necessary to remove the substances adsorbed on the material surfaces.

Water is a significant (quite often is the main) source of outgassing because a thin layer of water vapor will be rapidly adsorbed to all surfaces whenever they are exposed to the air. The adsorbed water presents a continuous level of background contamination after the system is closed, even when the system is being pumped to vacuum.

The solution is to bake and pump the system simultaneously. Removal of water and similar adsorbed gases generally requires baking the system up to 100°C (373.15 K) or higher temperatures, depending on the material. A proper high-vacuum pumping station is necessary. High-power turbo molecular pump or cryopump is normally employed to pump out the system during baking.

Baking can be implemented by an "External Baking Process" as illustrated in Figure 6.39a and b or/and an "Internal Baking Process" as illustrated in Figure 6.39c.

Illustrated in Figure 6.39a is a system wrapped with a high-power heating blanket for the baking process. This approach usually applies to items that cannot be removed from the complete experimental setup. Heating tapes can also be used for baking as shown in Figure 6.39b. This approaches applies to both portable systems as well as fixed items.

An alternative approach is the "*Internal Baking*" as shown in Figure 6.39c, where a so-called baking assembly with heaters and thermometer is installed inside the system. It is not uncommon to install two heaters in parallel for higher heating power. A thermometer is used to set up the baking temperature needed. This type of baking assembly can be mounted to any local place inside the system where heating (baking) is needed.

The baking temperature depends on the individual system and application, and it should always be controlled and kept within the proper ranges.

5. Virtual leak—UHV system design and fabrications

One of the key issues of the correct UHV system design and fabrication is to avoid virtual leaks.

A virtual leak is not a real leak but refers to the source of gas that is physically trapped inside a pocket within the vacuum chamber with only a small, very low conductance or path for evacuation. The virtual leak constantly releases the trapped gas and makes it highly difficult (or even impossible) to pump the system down to UHV. The situation is worse if trapped water vapor exists.

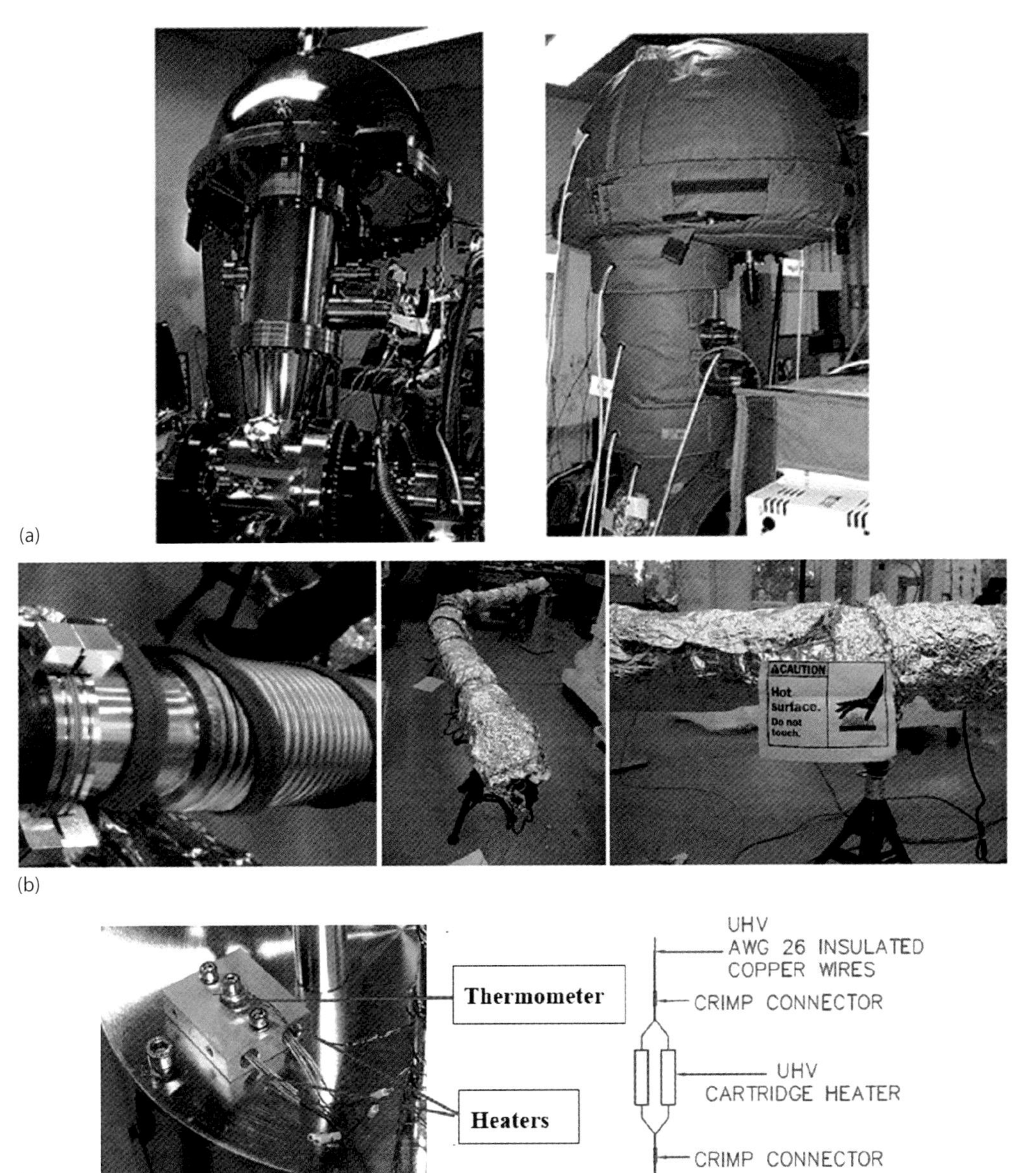

FIGURE 6.39 (a) Bake the system with heating blanket. (From http://www.hemiheating.se/en/concepts/uhv-urbakning/.) (b) Wrap the system with heating tape. (c) Bake the system with baking heaters installed inside the system. ([b,c] Courtesy of Janis Research Company, Woburn, MA.)

Typical "pockets" include:

- Gaps (e.g., have the welding joint at the wrong place),
- Cracks in material,
- Space in between the contact surfaces of two surfaces, and
- Trapped pockets such as blind-tapped holes (one of the most common causes of virtual leak).

In addition, hydrogen and carbon monoxide often diffuse out from the grain boundaries in stainless steel (the most commonly used material in UHV systems) and forms one of the common background gases in a UHV system. It is also referred to as a "virtual leak."

Solutions to avoid the virtual leak from an internal source

- Locate the welding joint at the right position as shown in Figure 6.40.
- Use vented screws/bolts for blind-tapped holes or add vent channel as shown in Figure 6.41a.
- Use lock washers as shown in Figure 6.41b.
- Make sure that there is no crack or porosity on any surface; also minimize the surface area.

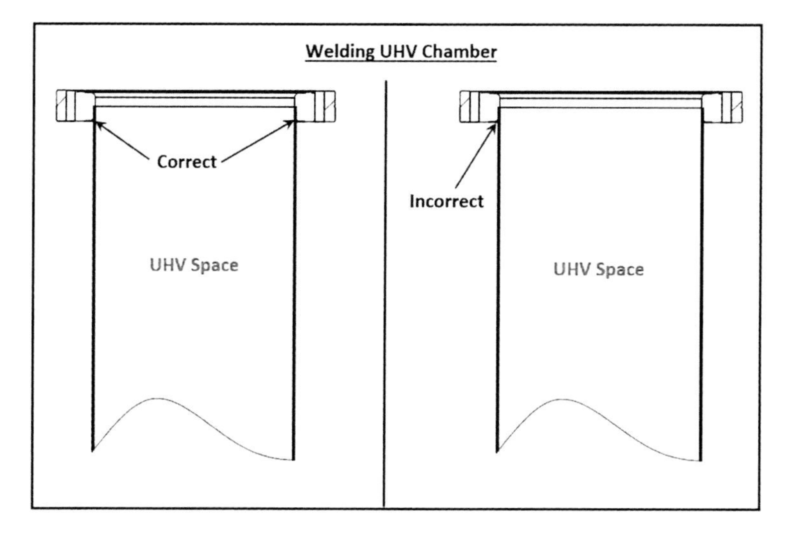

FIGURE 6.40 Correct location for welding joint.

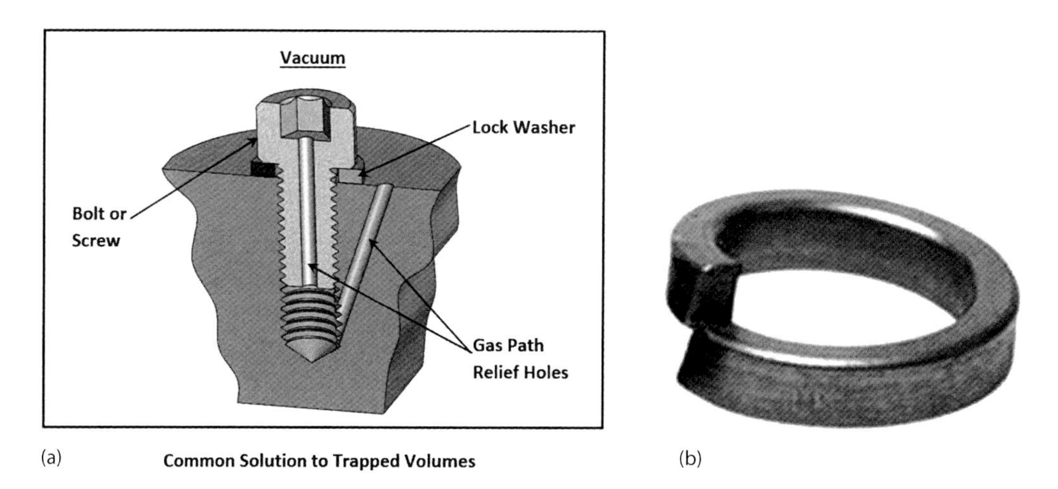

FIGURE 6.41 (a) Vented screws and vent channel. (b) Lock washer.

6. **Clean machinery tools**

   Clean machining tools without impurity particles should be used for UHV part fabrications. It is highly desirable that the machine shop has a separate set of clean cutting tools, which are to be used for machining UHV parts only. Proper coolants and lubricants are needed for UHV system fabrications as listed in Appendix D.

7. **Permeation**

   Permeation in our discussion refers to a diffusive process when gas penetrates the wall of the vacuum chamber from the surrounding environment. It is different from a virtual leak.

   It is worth mentioning that our previous discussion on the material selection for the UHV system was based on the outgassing rate of materials. However, gases can permeate through some UHV-compatible material as itemized in [66] (usually at room or higher temperatures).

   For an example, Kapton has very low outgassing rate, and it is widely used in UHV systems. However, the helium atom can permeate through Kapton due to its small size [74].

   Although it is that unlikely any UHV chamber be made of Kapton, a Kapton sample cell (or one with Kapton windows) containing a helium sample does exist. A Kapton seal against helium has also been used. Users should pay special attention for possible helium permeations through Kapton.

   Similar concerns also go to glass [75].

   Stainless steel is the most popular material to make a UHV-compatible vacuum chamber and UHV-compatible parts. Helium permeation through stainless steel is not a concern in normal cases, but the user needs to make sure that the material used is "perfect" without cracks or porosity.

   Hydrogen does permeate through stainless steel at higher temperatures as described in [76]. But it is usually not a concern because almost all UHV vacuum chambers are simply installed in the atmospheric environment at the room temperature, where the concentration of hydrogen is extremely low (approximately 0.00005%). Its permeation rate through stainless steel is pragmatically negligible.

6.4.3 UHV-Compatible Cryogen-Free Cryostat

UHV systems need to be baked above the room temperatures, including the cryogen-free systems with cryocoolers. However, the cold head of the cryocooler (in particular the regenerators) cannot be heated above 325 K. Approaches are needed to protect the cold head from being damaged during the baking process.

First of all, the exposed surfaces of the cold head should be made of UHV-compatible materials like stainless steel and copper. The joints should be implemented by brazing or welding.

A CF with a metal gasket should be employed for the vacuum seals as shown in Figure 6.42.

A key approach is to remove the thermal link between the cryostat (to be baked) and the cryocooler cold head during the baking process. An approach is to remove the cryocooler cold head from the system during the baking process.

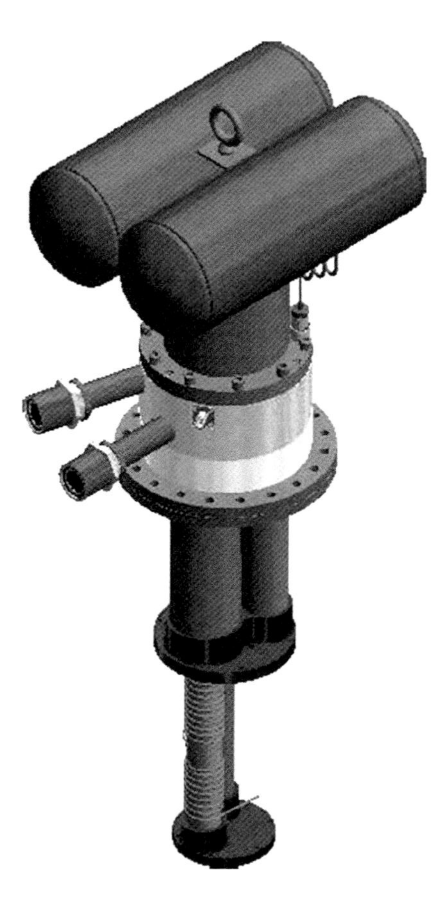

FIGURE 6.42 A two-stage pulse-tube cryocooler cold head with CF flange. (Courtesy of Cryomech Inc, Syracuse, NY.)

We present two alternative designs where the cryocooler cold head stays on the system during the baking process.

Design 1: Seat the cryocooler cold head on a linear motion stage as shown in Figure 6.43.

This manipulator can be lowered and both stages of the cold head pressed against the proper spots on the system for cooling. An approximately 160-lb force can be generated by the linear motion stage, which minimizes the temperature gradient between the system and the cryocooler. The contact between the cold head and the system can be lifted by the linear motion stage during the baking process.

This approach was implemented at Janis Research Company in a UHV-compatible 3He-ARPES with a "UHV-compatible" model PT-415 pulse-tube cryocooler per Figure 6.43 and it works well.

The cryocooler needs to be turned on and the cold head kept at a temperature below 325 K when the system is being baked. A cooling water loop can be installed on both stages of the cryocooler so that water cooling can be applied when necessary.

Design 2: Install a heat switch as the thermal link between the cryostat and the cryocooler cold head as shown in Figure 6.44.

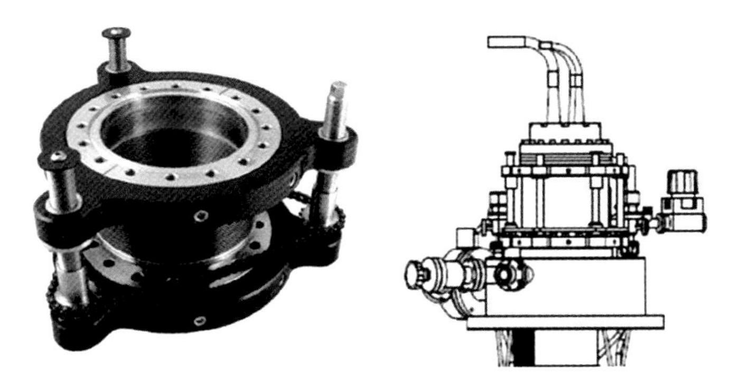

FIGURE 6.43 UHV-compatible cryogen-free helium-free 3He cryostat with pulse-tube cryocooler cold head seated on a McAllister linear motion manipulator. (Courtesy of Janis Research Company, Woburn, MA.)

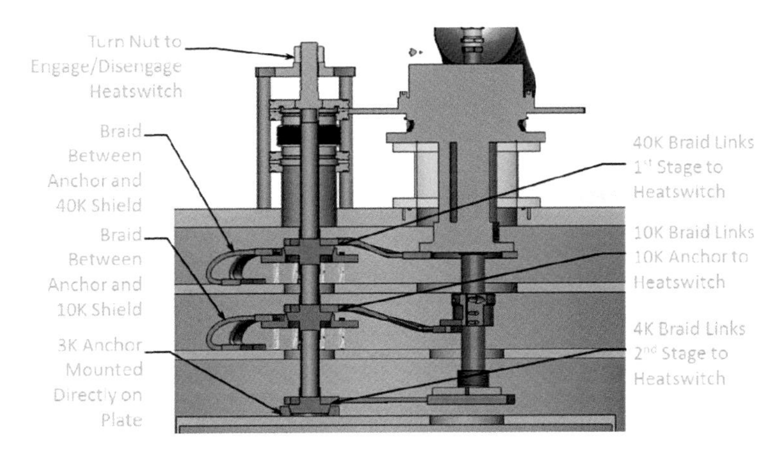

FIGURE 6.44 Install a heat switch as the thermal link between the cryostat and the cryocooler cold head.

The heat switch is engaged when the cryostat is cooled by the cryocooler. The heat switch is disengaged during the baking process.

Same as in Design #1, the cryocooler should be turned on during the baking process. A cooling water loop can be installed on the cryocooler.

Acknowledgments

The contributor would like to send his special thanks to Chao Wang for his review and editing work that made this chapter more professional.

Sincere thanks go to contributor's colleagues at Janis Research Company:

Thanks to Kerry Faulkner for his technical review of this section. Thanks to Chuong Diep for making all new figures. Thanks to Jeonghoon Ha and Sam Gonzales for their final proofreading. Last but not the least, thanks to Ann Carroll for her constant help in solving software difficulties, reference searches, copyright permission requests, etc.

References

1. Perry, R.H. and Green D.W. *Perry's Chemical Engineers' Handbook*, McGraw-Hill, New York (1984).
2. Roy, B.N. *Fundamentals of Classical and Statistical Thermodynamics*, John Wiley & Sons, Chichester, UK (2002).
3. Edmister, W.C. and Lee, B.I. *Applied Hydrocarbon Thermodynamics* Vol. 1 (2nd ed.), Gulf Publishing, Houston, TX (1984).
4. Reif, F. *Simple Applications of Macroscopic Thermodynamics, Fundamentals of Statistical and Thermal Physics*. McGraw-Hill, New York (1965).
5. Gans, P.J. Physical Chemistry I, V25.0651, *Joule-Thomson Expansion* (1992).
6. Klotz, I.M. and Rosenberg, R.M. *Chemical Thermodynamics, Basic Concepts and Methods*, 7th ed., John Wiley & Sons, New York (2008).
7. Hasan, S. and Şişman, A. Joule–Thomson coefficients of quantum ideal-gases, *Applied Energy* 70(1), 1–102, (2001).
8. McCarty, R.D. Thermodynamic properties Helium 4 from 2 to 1500 K at pressure to 10^8 Pa, *Journal of Physical and Chemical Reference Data*, 2(4), 923–1042 (1972).
9. http://www.phys.ufl.edu/courses/phy4550-6555c/spring11/expansion%20cooling.pdf.
10. https://www.coursehero.com/sitemap/schools/552-University-of-Florida/courses/718116-PHY4550.
11. Hendricks, R.C., Peller, I.C. and Baron A.K., *Joule-Thomson Inversion Curves and Related Coefficients for Several Simple Fluids*, NASA Technical Note, NASA TN D-6807 (July 1972), Lewis Research Center, Cleveland, OH, National Aeronautics and Space Administration, Washington, DC.
12. Johnston, D.C. *Advances in Thermodynamics of the van der Waals Fluid,* Morgan & Claypool, San Rafael, CA, September 2014.
13. Encyclopedia Britannica Jean Louis Marie Poiseuille (1799–1869).
14. Symon, K.R., *Mechanics* (3rd ed.). Addison-Wesley Publishing Company (1971).
15. Maxwell, J.C. On the viscosity or internal friction of air and other gases, *Philosophical Transactions of the Royal Society of London* 156, 249–268 (1866).
16. Maisel, J.E., Webeler, Rayjor W. H. and Grimes, H.H. *Use of a Torsional Pendulum as a High-Pressure Gage and Determination of Viscosity of Helium Gas at High Pressures*, NASA Technical Note, TN D-7241 May, 1973.
17. Reid, R.C., Pransuitz, J.M. and Poling, B.E. *The Properties of Gases and Liquids*, 5th ed., McGraw-Hill, New York (2004).
18. Elert, G. *The Physics Hypertextbook-Viscosity*, Physics.info. Retrieved November 14, 2010.
19. Schroeder, D.V. *An Introduction of Thermal Physics*, Addison-Wesley, New York (2000).
20. Sutherland, W. The viscosity of gases and molecular force, *Philosophical Magazine*, S. 5, 36, 507–531 (1893).
21. Bich, E., Millat, J. and Vogel, E. The viscosity and thermal conductivity of pure monatomic gases from their normal boiling point up to 5,000 K in the limit of zero density and at 0.101325 MPa, *Journal of Physical and Chemical Reference Data* 19, 1289 (1990).
22. Hurly, J.J. and Moldover, M.R. Ab initio values of the thermophysical properties of Helium as standards, *Journal of Research of the National Institute of Standards and Technology*, 105(5), 667 (2000).
23. Zhang, P., Huang, Y.H. and Chen, G.B., Applications of Helium-4 and Helium-3, National Defense Industrial Publisher (April 2006), National Defense Industry Press, Beijing, China.
24. Engineering Toolbox, http://www.engineeringtoolbox.com/gases-absolute-dynamic-viscosity-d_1888.html (April 1, 2019).
25. Angus, S., de Reuck, K.M. and McCarty, R.D., *International Thermodynamic Tables of the Fluid State Helium-4*, Elsevier, New York (2016).
26. Vollhardt, D. and Wolfle, P., *The Superfluid Phases of Helium* 3, Courier Corporation, Chelmsford, MA (2003).
27. Temperature dependence of liquid viscosity, https://en.wikipedia.org/wiki/Temperature_dependence_of_liquid_viscosity
28. Special metals, http://www.specialmetals.com/assets/smw_docs/we182.pdf
29. For an example: VAC Goop® Thread Lubricant, Fluorosilicone-Based, 1 oz. (29.5 cm^3) Tube Part No. MS-TL-VGT.

30. Molybdenum disulfide (MoS2) coating lubrication review, http://industrialcoatingsworld.com/low-friction-coatings/mos2-low-friction-coatings;http://www.engineersedge.com/lubrication/molybdenum_disulfide_characteristics.htm
31. High Precision Devices, Inc. 4601 Nautilus Court South, Boulder, CO 80301.
32. Ukrainczyk, N., Kurajica, S. and Sipusic, J. Thermophysical comparison of five commercial paraffin waxes as latent heat storage materials, *Chemical and Biochemical Engineering Quarterly*, 24(2), 129–137 (2010).
33. Guo, L., Zhang, X., Huang, Y., Hu, R. and Liu, C. Thermal characterization of a new differential thermal expansion heat switch for space optical remote sensor, *Applied Thermal Engineering* 113, 1242–1249 (2017).
34. Noble gas (data page), https://en.wikipedia.org/wiki/Noble_gas_(data_page)
35. Hirschfelder, J.O., Curtiss, C.F. and Bird R.B. *Molecular Theory of Gases and Liquids*, John Wiley & Sons, New York (1954).
36. Frost, W., *Heat Transfer at Low Temperature*, Plenum Press, New York (1975).
37. Brookhaven National Laboratory Selected Cryogenic Data Book, Compiled and Edited by Jensen, J.E., Tuttle, W.A., Stewart, R.B., Brechna, H., Prodell, A.G., revised August 1990. Brookhaven National Laboratory, Associated Universities, Inc., under contract NO. DE-AC02-76CH00016 with the United States Department of Energy.
38. Minter, C.C., Naval Research Lab Report: Effect of pressure on the thermal conductivity of gas (February 20, 1963), U.S. Naval Research Laboratory, Washington, DC.
39. Ekin, J.W., *Experimental Techniques for Low-Temperature Measurements*, Oxford University Press, Oxford, UK (2006).
40. Catarino, I., Afonso, J., Martins, D., Duband, L. and Bonfait, G. Gas gap thermal switches using neon or hydrogen and sorption pump, *Vacuum*, 83, 1270–1273 (2009).
41. Park, I., Yoo, D., Park, J. and Jeong, S., Development of a passive helium heat switch for fast cool down by two-stage cryocooler, Cryocoolers 18, edited by Miller, S.D. and Ross, Jr., R.G., *International Cryocooler Conference, Inc.*, Boulder, CO (2014).
42. (a) Barreto, J., Borges de Sousa, P., Martins, D., Kar, S., Bonfait G. and Catarino, I. Gas gap heat switch for a cryogen-free magnet system, CEC 2015 IOP Publishing, *IOP Conference Series: Materials Science and Engineering* 101 (2015) 012144; (b) Kimball, M.O., Shirron, P.J., Canavan, E.R., Tuttle, J.G., Jahromi, A.E., Dipirro, M.J., James, B.L., Sampson, M.A., and Letmate, R.V., *Passive Gas-Gap Heat Switches for Use in Low Temperature Cryogenic Systems*, CEC-ICMC, Madison, WI, July (2017).
43. Mate, C.F., Harris-Lowe, R., Davis, W.L. and Daunt, J.G., 3He cryostat with adsorption pumping, *The Review of Scientific Instruments*, 36(3), 369 (1965).
44. Pobell, F., *Matter and Methods at Low Temperatures*, 3rd ed., Springer Berlin, Germany (2007).
45. DiPirro, M. and Shirron, P., Heat switches for ADRs, *Cryogenics*, 62, 172–176 (2014).
46. Chan, C.K. and Ross, Jr., R.G. Design and application of gas-gap het switch, Final Report of Phase-II, JPL Publication (1990).
47. Shirron, P., Canavan, E.R., DiPirro, M.J., Jackson, M., Panek, J. and Tuttl, J.G., Passive gas-gap heat switches for use in adiabatic demagnetization refrigerators, *Advances in Cryogenic Engineering*, 613, 1175 (2002); Dipirro, M. and Shirron, P., Design and test of passively operated heat switches for 0.2 to 15 K, *Advances in Cryogenic Engineering*, 49B, 436 (2003).
48. Satterthwaite, C.B., Thermal conductivity of normal and superconducting aluminum, *Physical Review*, 125, 873 (1962).
49. Mueller, R.M., Buchal, C., Oversluizen, T. and Pobell, F. Superconducting aluminum heat switch and plated press-contacts for use at ultralow temperatures, *Review of Scientific Instruments*, 49, 515 (1978).
50. Schuberth, E., Superconducting heat switch of simple design, *Review of Scientific Instruments*, 55(9), 1486 (1984).
51. Raymond, W. The observations of 3He crystals at temperatures down to 1 mK using a cooled CCD camera, PhD dissertation 1996, Kamerlingh Onnes Laboratorium, Leiden University, Leiden, the Netherlands.
52. Krusius, M., Paulson, D.N. and Wheatley, J.C. Superconducting switch for large heat flow below 50 mK, *Review of Science Instruments* 49(3), 396–398 (1978).
53. Abe, S. and Matsumoto, K. Nuclear demagnetization for ultra-low temperatures, *Cryogenics*, 62, 213–220 (2014).

54. Mueller, R.M., Buchal, C., Folle, H.R., Kubota, M. and Pobell, F. A double-stage nuclear demagnetization refrigerator, *Cryogenics*, 20, 395 (1980).
55. (a) Zhao, Z. Study of non-conventional superfluid: Ultrasonic propagation in 3He-B and the microwave surface impedance of the heavy-fermion superconductor UPT_3, PhD Thesis (1990), Department of Physics and Astronomy, Northwestern University, Evanston, IL. (b) Gloos, K., Smeibidl, P., Kennedy, C., Singsaas, A., Sekowski, P., Mueller, R.M. and Pobell, F. The Bayreuth nuclear demagnetization refrigerator, *Journal of Low Temperature Physics*, 73, 101 (1988). (c) Yao, W., Knuuttila, T.A., Nummila, K.K., Martikainen, J.E., Oja, A.S. and Lounasmaa, O.V. A versatile nuclear demagnetization cryostat for ultralow temperature research, *Journal of Low Temperature Physics*, 120, 121–150 (2000).
56. ASTM B253, *Standard Guide for Preparation of Aluminum Alloys for Electroplating*, Standard by ASTM International (2011). http://www.techstreet.com/standards/astm-b253-11?product_id=1823197
57. Richardson, R.C. *Low Temperature Lectures*, (1981) (private communications).
58. Richardson, R.C. and Smith, E.N. *Experimental Techniques in Condensed Matter Physics at Low Temperatures*, Addison-Wesley Publishing Company, Inc, New York, (1988).
59. Mast, D. Ultrasonic investigation of the two order-parameter collective modes in superfluid 3He-B, PhD Thesis, (1982), Department of Physics and Astronomy, Northwestern University, Evanston, IL.
60. Velson, N.V., Tarau, C. and Anderson, W.G. Two-phase thermal switch for spacecraft passive thermal management, Advanced Cooling Technologies, Inc., Lancaster, PA, 17601, *45th International Conference on Environmental Systems ICES-2015-51* July 12–16, 2015, Bellevue, Washington.
61. Shu, Q.S., Demko, J.A. and Fesmire, J.E. Heat switch technology for cryogenic thermal management, *Materials Science and Engineering*, 278, 012133 (2017).
62. Powel, R.I. and Blanpied, W. *Thermal Conductivity of Metals and Alloys at Low Temperatures,* NBS Circular 556, Blanpied, WA.
63. Technology Applications, Inc. 5303 Spine Rd, Boulder, CO 80301, USA (P) 303.867.8145 (C) 720.917.4606 https://www.techapps.com/contact-us.
64. Dhuley, R.C., Ruschman,.M., Link, J.T. and Eyre, J., Thermal conductance characterization of a pressed copper rope strap between 0.13 K and 10 K, *Cryogenics* 86, 17–21 (2017).
65. New England Wire Technologies, 130 North Main Street, Lisbon, NH 03585, USA, Tel. 603-838-7077 https://www.newenglandwire.com.
66. Segovia, J.L. de. *Physics of Outgassing, Instituto de Física Aplicada, CETEF* L. Torres Quevedo, CSIC, Madrid, Spain. CAS—CERN Accelerator School: Vacuum Technology, Snekersten, Denmark (1999).
67. NASA Outgassing Database http://outgassing.nasa.gov (April 1, 2019).
68. Alcock, C.B., Itkin, V.P. and Horrigan, M.K., Vapor pressure of the metallic elements, *Canadian Metallurgical Quarterly*, 23, 309 (1984).
69. Safarian, J. and Engh, T.A., Vacuum evaporation of pure metals, *Metallurgical and Materials Transactions A*, 44(2), 747–753 (2012).
70. Thomson, G., The Antoine equation for vapor-pressure data. *Chemical Review*, 38(1), 1–39 (1946).
71. (a) Magill, J., Dreher, R. and Sóti, Z.S. *Karlsruher Nuklidkarte* (2018); (b) Kash Lab, Case Western Reserve University, *Calculation of Zinc Pressure* (2014).
72. Savage, Chris J. Thermal control of spacecraft. In *Spacecraft Systems Engineering* (3rd ed.), Peter W. Fortescue, J. Stark, G. Swinerd. John Wiley & Sons, Southampton, UK, pp. 378–379 (2003).
73. http://www.hemiheating.se/en/concepts/uhv-urbakning/ (April 1, 2019).
74. Schowalter, S.J., Connolly, C.B. and Doyle J.M., Permeability of noble gases through Kapton, butyl, nylon, and SilverShield, *Nuclear Instruments and Methods in Physics Research A* 615, 267–271 (2010).
75. Kawasaki, K. and Senzaki, K., *Permeation of Helium Gas through Glass,* Copyright (c) 1962 The Japan Society of Applied Physics, Tokyo, Japan.
76. Tezuka, M., Mizuno, T. and Sato, S. *Study of Hydrogen Permeation Through the Stainless Steel,* Graduate School of Engineering, Division of Quantum Energy Engineering, Hokkaido University, North 13 West 8, Sapporo, Japan.

7. Cryogen-Free Superconducting Magnets

Adam Berryhill, Michael Coffey, and Brian Pollard

7.1 Introduction to Superconducting Magnets

Superconductivity was discovered in 1911 by H. Kamerlingh Onnes [1] during his attempt to liquefy helium. He observed that wires made from lead, tin, and mercury lost all resistance below a certain temperature. Generally, the initial superconductors were somewhat disappointing in that they could not maintain superconductivity in a field larger than a few hundred gauss. By the 1930s, superconductors were being divided into what we now call Type I, soft, and Type II, hard, superconductors based on their critical field. Compounds were discovered in the 50s and 60s, which allowed for higher fields to be generated in small solenoid magnets. In the 60s, superconducting solenoids were being developed by Kunzler et al. [2], showing an increase in field up to 15 kilogauss. It was also during this time when niobium titanium (NbTi) and Nb_3Sn (pronounced niobium three tin) were discovered. These two superconductors are the modern-day workhorses of superconducting magnets. These are called *low-temperature superconductors*, or LTS, as they must operate near liquid helium temperature to carry sufficient current and produce large fields. This is in contrast to modern-day *high temperature superconductors*, or HTS, which were discovered in 1986 and have extended temperatures from liquid hydrogen temperatures up to near liquid nitrogen temperatures. Limitations on these new conductors still exist, which keep them from widespread use today, the main one being cost. HTS presently used in magnet fabrication consist of magnesium diboride (MgB_2), bismuth strontium calcium copper oxide (BSCCO), and rare earth barium copper oxide (ReBCO) alloys.

Three properties define the usefulness of a superconductor, its *critical temperature*, its *critical field*, and its *critical current density*. The word *critical* refers to the point above when it loses its superconductivity and transitions back to a resistive state. A superconducting magnet must be designed so it operates below the surface of these three parameters. The farther below this surface, the higher the stability, or "margin," in the superconductor. Increasing stability or operating margin requires more superconductor and thus increases cost.

Applications for superconducting magnets are wide ranging and include high-energy physics, medical, industrial, military, and basic materials research applications. The most familiar application to the public is that of magnetic resonance imaging (MRI). MRIs use the strong magnetic field produced by a superconducting magnet to align the nuclear spins of hydrogen atoms in water and fat while radio-frequency (RF) pulses are used to generate signals, which can be turned into images of the human body. Preclinical MRIs are used in the development of new drugs for cancer research by imaging rodents that have been given cancer while the effects of the drugs on the size and number of tumors are studied (Figure 7.1).

Another medical application of superconducting magnets is their use in charged particle (usually proton) therapy systems for the treatment of cancer. The use of superconducting magnets in these systems can allow for the reduction of weight of a three-story rotating gantry from as much as 65 tons for the non-superconducting version to 15 tons for the superconducting version. This leads to a large reduction in cost of the facility needed to house these machines.

Industrial applications include furnace annealing of silicon wafers in a magnetic field, magnetic separation of ferromagnetic materials from liquid or solid solutions, and plasma heating for fusion reactors.

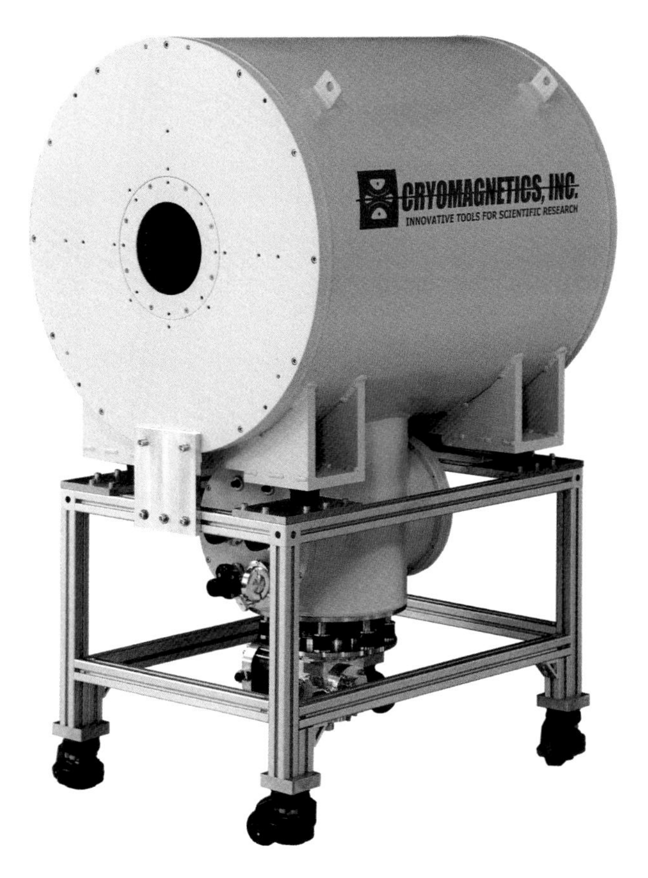

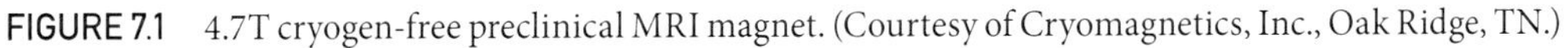
FIGURE 7.1 4.7T cryogen-free preclinical MRI magnet. (Courtesy of Cryomagnetics, Inc., Oak Ridge, TN.)

Military applications include non-lethal weapons systems such as pain rays, which use microwave frequency energy to excite the nerve endings in skin, superconducting motors for ships for size and weight reduction, and millimeter wave radar systems for the tracking of space debris.

Research applications can vary widely and range from small magnets for material science studies, such as a vibrating sample magnetometer to large dipole and quadrupole magnets for high-energy physics.

7.2 Superconducting Magnet Design

Three parameters can generally be said to define the design of a superconducting magnet, *the required operating field intensity*, *the required bore access to the field*, and *the shape or uniformity of the field*. In specialty magnets, there may be other requirements such as field gradient, field integral, harmonics, or field profile. Magnet configurations can range from simple solenoids, to split pairs (Helmholtz configurations), to dipoles, higher-order multipole magnets, or toroids. Once these parameters are known, the coil geometry can be designed, which will meet those requirements. Since mostly we are dealing with electromagnets with current carrying wire, most field calculation codes refer to the Biot-Savart Law to calculate the field from a current element at a point and integrate this.

$$\vec{B} = \frac{\mu_0}{4\pi} I \int \frac{d\vec{l} \times \hat{r}}{r^2} \tag{7.1}$$

Superconducting magnets capable of generating fields less than or equal to 9T generally are made using NbTi due to its low cost and robust mechanical and electrical properties. To generate fields higher than 9T at 4.2 K Nb_3Sn or HTS is used, typically still in combination with NbTi due to cost considerations. Magnets using Nb_3Sn have achieved central magnetic fields over 21T and using HTS in excess of 32T in steady-state operation.

High-field solenoid and split pair magnets typically take the form of an insert (smaller coil) made of the more expensive (Nb_3Sn or HTS) material inside an approximately 9T NbTi "outsert" coil. Magnets made entirely from Nb_3Sn or HTS are prohibitively expensive unless there are other motivating factors (e.g., if the entire magnet needs to operate at an elevated temperature).

Prior to the discovery and development of HTS, to achieve DC fields higher than 21T required the addition of resistive inserts or water-cooled Bitter magnets inside the bore of a superconducting magnet. HTS inserts are now being used to increase magnetic fields in place of the resistive inserts.

The simplest superconducting magnet is a straight solenoid in which a wire made of superconducting material is wound around a coil form. Various configurations of split pair and multi-axis designs are possible through the use of multiple solenoids in series or operated independently to affect the magnetic field over a given sample region. Careful design is used to find a fine balance between wire composition, diameter, and distribution along the axis of the coil form.

As part of the design process, many variables are considered both with respect to the general field profile but also the manner in which the magnet will be used. Proper design assures a robust winding while avoiding excessive cooling losses due to excessive charging current or inadequate homogeneity.

Superconducting magnets must operate below both the critical temperature and the critical field of the material from which they are constructed. Table 7.1 illustrates critical temperatures and fields of the most common superconductive materials used to fabricate magnets.

Note that the numbers in Table 7.1 show the critical temperature when there is zero applied magnetic field. This is reduced considerably once field, and its related effects starts being applied. Most magnets are run in liquid helium (4.2 K), in superfluid liquid Helium (<2.2 K), or with a mechanical refrigerator in the case of cryogen-free magnets.

Figure 7.2 shows how the capabilities of NbTi vary as a function of both temperature and field. The critical current of the superconductor will vary according to wire size and copper to the superconductor ratio. For a superconducting magnet manufactured

Table 7.1 Properties of Common LTS Superconductors

Material	T_c (K)	H_c (kG)
NbTi	9.8	120 @ 4.2 K 148 @ 1.2 K
Nb3Sn	18.05	221 @ 4.2 K

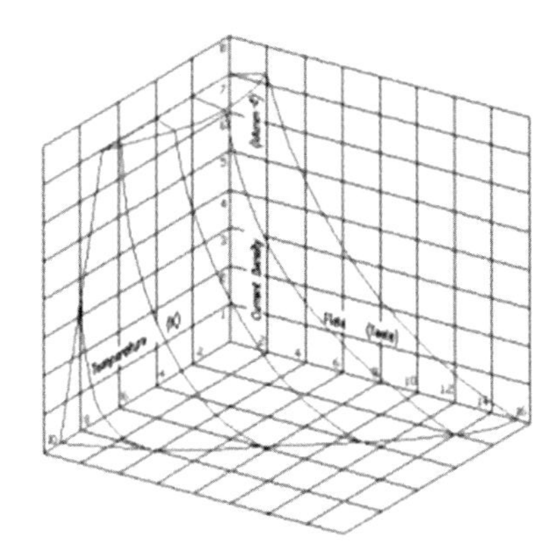

FIGURE 7.2 NbTi critical current surface. (Courtesy of Cryomagnetics, Inc., Oak Ridge, TN.)

using NbTi, operation in the superconducting state is only possible below the surface indicated in this three-dimensional graph.

For a superconducting magnet operating beneath the surface shown in Figure 7.2, that is, below the critical temperature, current and field of the superconductor used to make it, the superconducting state is maintained and the magnet operates properly. If, however, one tries to operate the magnet above this surface, a transition to resistive state or "quench" will occur.

For instance, if one is operating a magnet at the constant temperature of 4.2 K, the surface in Figure 7.2 reduces to a single two-dimensional curve of current density vs. field. If one changes the "current density" scale to simply "current" by considering a particular wire size and type, the curve in Figure 7.2 can be represented by something similar to that shown in Figure 7.3. A curve such as Figure 7.3 is frequently referred to as a

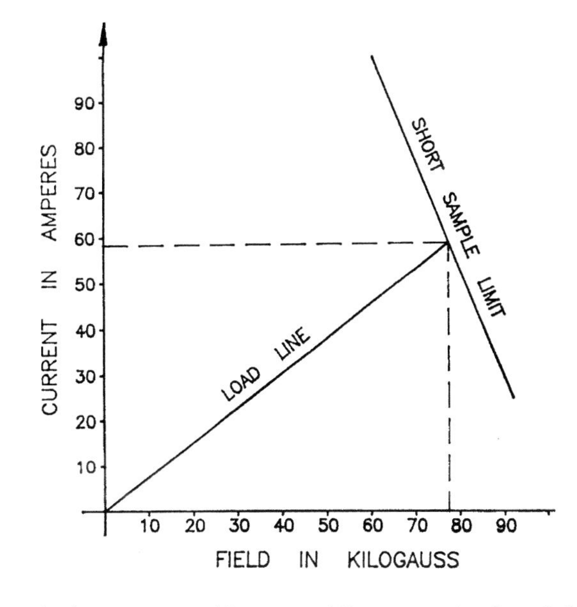

FIGURE 7.3 Short sample characteristics. (Courtesy of Cryomagnetics, Inc., Oak Ridge, TN.)

"short-sample I-H curve" since it represents the behavior of a short piece of a particular wire. For an LTS round wire superconductor, this short sample curve is independent of orientation of the superconductor; however, we will see later that this is not necessarily the case for HTS flat tape superconductors.

If the superconducting magnet is operating from a single power supply with all windings in series, then superposition implies that the operating point of the magnet can be represented by a "load line" showing the current in the magnet vs. the peak magnetic field on the windings. Such a load line is shown in Figure 7.3. In this case, as long as the magnet is operated below 58 amps (which corresponds to a peak field of 7.6 Tesla in the windings), superconductivity is maintained. If one attempts to drive the magnet past the short sample limit of the wire, the point where the peak field exists will undergo a transition back to the normal (resistive) state. This resistive region will quickly heat up due to the high current through it, and the resistive region (called a "normal zone") will propagate until (a) all of the energy stored in the magnet is dissipated, (b) the entire magnet becomes resistive, or (c) sufficient cooling is provided to stop the propagation. Usually either (a) or (b) occurs, resulting in a complete quench of the magnet.

7.2.1 Forces within Magnets

When a superconducting magnet is charged, there are many forces that begin acting upon it—both internal and external. The basic physics formula

$$\vec{F} = \vec{I} \times \vec{B} \tag{7.2}$$

where F = force, I = current, and B = field. This follows the familiar "right-hand rule" where F, I, and B are all vector quantities. Since there is usually considerable current in a superconducting magnet's windings and it is generating high magnetic fields, it follows that the forces within the windings can become extreme. The magnetic field can be thought of as exerting a pressure proportional to $B^2/2m_o$ (where m_o is the magnetic permeability of free space) on the surfaces perpendicular to the field lines. This force for a 10T magnet is 4×10^7 N m^{-2}, which is about the yield strength of annealed copper.

The forces on and in an axisymmetric magnet system can usually be broken down into axial and radial components. Axial components consist of forces between coil sections, such as the attraction between two coils in a split-pair Helmholtz design, or the repulsion between two coils forming a magnetic field gradient. Even simple solenoids with only one coil have axial forces trying to compress the windings, so long coils must take this into account.

Radial forces, sometimes called "hoop forces," are those forces that are trying to make the magnet expand in diameter. In magnets with small inner and outer diameters, hoop forces are usually not difficult to restrain since the windings themselves are put in tension by these forces and can easily support the stresses. These hoop forces give rise to strain ($\Delta L/L$) within the superconductor that must be maintained below appropriate levels or else the current carrying capacity of the superconductor will be reduced. Typically, this is less than 0.1% for NbTi and 0.2% for Nb_3Sn.

Magnets with large bores, however, can have problems with restraining radial forces and sometimes need additional force restraints other than the windings themselves.

Generally, this is done by winding aluminum or stainless wire over top of the magnet windings or by sleeving the coil with bands of aluminum or stainless. Since the coils shrink as they are cooled, one can take advantage of the higher shrinkage rates of differing materials to provide preload to the coil and support to outward hoop forces during charging.

The forces described above try to force the wires within a superconducting magnet to move. This movement must be completely restrained because even the smallest wire movement will generate heat due to friction and can quench the magnet. To restrain wire movements, most superconducting magnets are potted using epoxy, ceramic, or some other material. These materials are usually proprietary to magnet manufacturers since they are crucial to the magnet's performance. The materials have special characteristics in terms of their thermal and mechanical specifications.

7.2.2 Quench Protection

A superconducting magnet will quench when driven beyond its superconducting limits. This happens by trying to operate beyond the wire's critical limit of field and/or current. This is a materials limitation, and it can also occur whenever one exceeds the critical temperature surface in Figure 7.2.

Raising of the temperature of a superconducting wire can occur due to:

- A small wire movement creating frictional heating within the winding pack,
- Excessive heat generated from eddy currents or hysteresis loss in the superconductor during charge or discharge of the magnet,
- Magnet warming due to a loss of system vacuum, or
- Heating due to loss of cooling power—for example, when power is lost to the cryorefrigerator compressor or when a cryocooler failure occurs.

Typically, a cryogen-free magnet has only tens of seconds to a few minutes of time before a quench occurs due to loss of power.

In low-temperature superconductors, the quench energy spreads out within a few seconds across the entire conductor volume and releases all energy stored in the magnet. Because the heat capacity of materials is generally very low at low temperature, the quench warms the entire coil very quickly. The energy stored in the magnet is given by the equation below.

$$E = \frac{1}{2} LI^2 \tag{7.3}$$

where E is the stored energy, L is the magnet inductance, and I is the operating current. In HTS materials, a quench is much more difficult to deal with since the higher temperature operation allows for higher heat capacity, which does not spread out the quench energy as quickly.

Most small laboratory magnets have maximum stored energies on the order of a few thousand to a few hundred thousand joules (kilojoules). Larger magnets frequently are capable of storing several million joules (mega-joules).

Within wet superconducting magnet systems, liquid helium surrounding the superconducting magnet helps dissipate heat generated during a quench. Cryogen-free superconducting magnets are typically installed in a vacuum environment, which complicates how to remove heat from the magnet during a quench.

This sudden release of energy can lead to hot spot temperatures in the superconducting wire and high voltages inside the coil, which will damage the coil and its electrical insulation. When designing a superconducting magnet, limiting the peak temperature and voltage present in the coil during a quench is of utmost importance. There are many ways to approach this, but the two most common methods are either through *active* or *passive* quench protection.

Active quench protection refers to protection of a magnet using active (usually externally driven) means. There are generally two types of active quench protection: (1) detect and dump and (2) forced quench propagation. Either method is based on detection of the onset of a quench and taking action to manage the release of magnet energy to prevent damage. Sometimes the two techniques are combined.

As the name implies, "detect and dump" protection detects the onset of a quench, and to prevent damage to the magnet, current is dumped by rapid discharge of the coil. This technique often uses a resistor in parallel with a coil or coils to dissipate the energy once a quench is detected.

Accurately detecting the quench in the first place is essential, as the time scales involved typically allow at most tens of milliseconds for detection before starting to take action. Voltage taps are usually distributed at strategic locations in the coil, and voltages are measured at several points. Voltages generated during charging are monitored, subtracting out the inductive component in order to detect the onset of a resistive section in the coil. Any voltage above a certain threshold (usually also for a threshold amount of time, for noise immunity) triggers a relay to switch in the dump resistor to the circuit and dissipate the current. The power supply must also stop supplying current to the magnet during this time. A typical setup of a magnet with voltage taps, relay, and dump resistor is shown in Figure 7.4.

A superconducting magnet can be thought of as a pure inductive load. Figure 7.4 shows a magnet, represented by a series of inductors, L1 through L4, on the left and a power supply, V1, with a dump resistor, R1, on the right. Voltage taps, labeled V1_out to V5_out, are used to measure the voltage throughout the inductor and open the switch shown below the power supply. In most cases, the voltage of one-half of the coil is subtracted from the other half to zero the inductive voltage during ramping of the magnet. This is so that a quench during ramping may also be detected. Typically, one wants to keep the voltage across the dump resistor to a reasonably safe value and therefore sizes it based on the current through the magnet and maximum voltage tolerable. A fast decay time of the current is preferred to reduce the maximum temperature in the coil but must be balanced against the high voltage produced across the resistor.

"Forced quench propagation" is another form of active quench protection. It is sometimes used in connection with "detect and dump." In this quench protection scheme, quench heaters are used to help distribute the heat generated in a quench over the entire magnet's windings. A high thermal conductive epoxy with which the windings are impregnated also helps speed the normal zone's propagation. Rapid normal zone

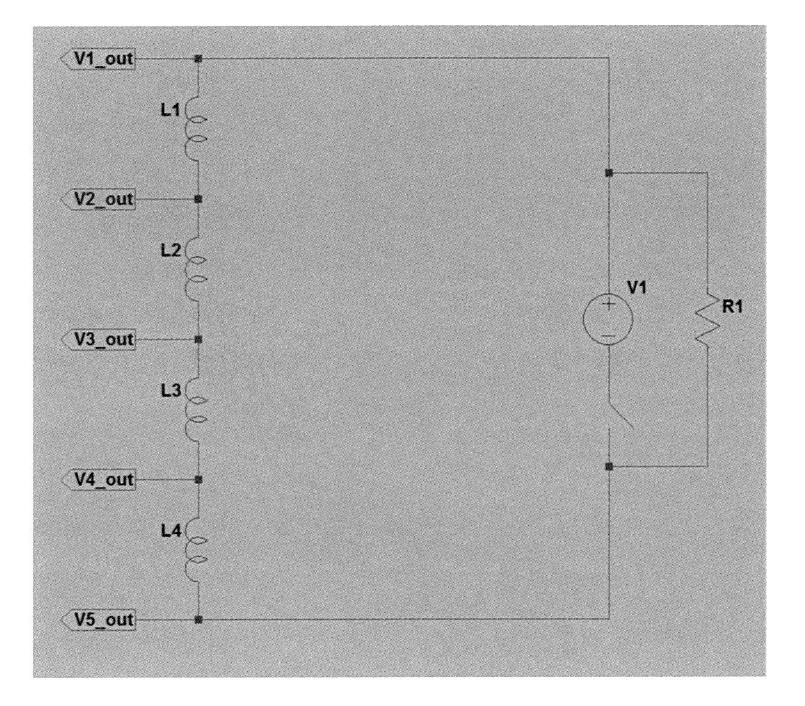

FIGURE 7.4 Active quench protection diagram. (Courtesy of Cryomagnetics, Inc., Oak Ridge, TN.)

propagation is usually the key to keeping the heat generated from becoming localized and damaging the superconductor.

Passive quench detection relies on the magnet protecting itself during a quench. This is done through proper choice in superconductor, specifically copper content of the wire, and through subdivision of the coil, that is, dividing the coil into smaller and smaller inductive sections through the use of bypass diode and resistors combinations to manage voltages that can occur within the winding pack. The maximum voltage allowed in any one section during the quench is clamped to prevent overheating or arcing. This helps spread the quench out through the coil. Most small-scale laboratory magnets rely on this method of quench protection. A typical passive quench protections system using subdivision with only diodes is shown in Figure 7.5.

In this figure, the inductor, or magnet, is divided into smaller sections by diodes. During a quench, one can think of one inductor section being replaced by an inductor in series with a resistor whose value is increasing as the quench propagates. Voltage is produced across the resistor, and this voltage grows until it eventually turns on the diode, allowing the current to bypass the resistive section. Because the coil sections are in close thermal contact with each other, what follows is a thermal avalanche where subsequent sections quench, helping to distribute the energy. The power supply must recognize this effect, by either a voltage or current change measurement, and shut itself off immediately. This is so as to not continue to supply current to the diodes and magnet possibly damaging one or the other. Occasionally, series resistors, of a few hundred milliohms, can be added to the diodes to dissipate energy as well.

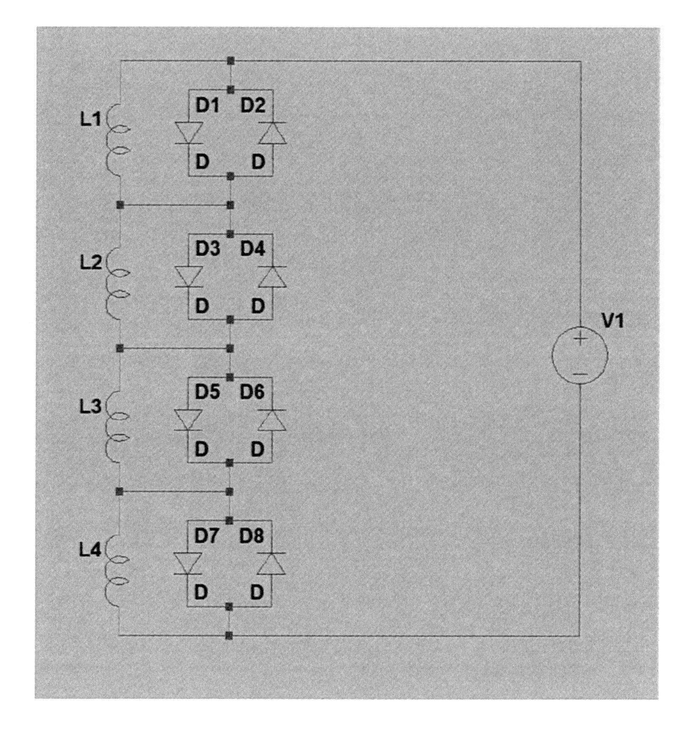

FIGURE 7.5 Passive quench protection diagram. (Courtesy of Cryomagnetics, Inc., Oak Ridge, TN.)

Finding appropriate diodes that function reliably at high current, at cryogenic temperatures, and in high magnetic fields can often be difficult. Testing must be done on the diodes to determine their suitability, turn on voltage at operating temperatures, and ability to handle rapid thermal changes. The diode's ability to withstand surge currents (which frequently will exceed the operating current of the magnet) must also be characterized. If resistors are used in series with the diodes, then their suitability must also be determined. Typically, metal film resistors show suitable characteristics for use at cryogenic temperatures.

It should be noted that quench protecting *high-temperature superconducting magnets* is still an area of active research. When these magnets operate at high temperature, the heat capacity of the wire is much higher than that of typical LTS. This brings with it a new dilemma in that the quench energy cannot be propagated fast enough with normal passive methods. As a result, the hot spot temperature and voltage can easily damage the superconductor. Most HTS magnets to date therefore use a form of active quench protection.

Because most cryogen-free magnets utilize HTS current leads, there is also some inherent risk in these systems of a quench in the HTS lead pair, generally due to a loss of cooling power or excessive magnetic field on the HTS material. HTS leads are fabricated using low thermal conductivity HTS wire and can only propagate the heat along its length. Quench protection of HTS current leads is very challenging, and one usually plans to quickly ramp down or quench the magnet in order to protect the leads in the case of a loss of cooling power.

7.2.3 Training and Stability

The objective in any useful superconducting magnet is to have it generate the desired magnetic field reliably and without the interruption of excessive quenching. *Training* is a phenomenon that typically may occur during the first time or two a magnet is charged. Training is a premature quench (i.e., a quench before the short sample limit of the magnet is reached in terms of field and current) due to wire movement. Training in a well-designed and constructed solenoid is extremely rare; however, in multiple coil systems where there is a wider variety of forces acting in many different directions, training is not so surprising. Usually, once a magnet has been trained to the short sample limit of the wire, training will no longer occur. This is because wires have moved into place and are unlikely to return to their initial position without some additional disturbance.

Training can occur when a wire in the bulk of the coil is allowed to move relative to the coil pack. This movement can generate heat via friction, which raises the temperature of the wire locally above the critical temperature. This local "hot spot" then becomes resistive and begins generating heat. A run-away condition ensues and the magnet quenches.

Since the heat capacity of metals is very small at 4 K, the amount of heat needed to raise the conductor above the critical temperature is also very small. Since there is no helium present in cryogen-free magnets, this can be treated as an adiabatic case, and the heat needed is only a few millijoules per cm^3 of material. This correlates to movements on the orders of micrometers.

Another possible type of training occurs due to flux jumps in the wire itself. Flux jumps are a result of the decay of screening currents in the superconductor and their resulting flux motion [3]. This motion produces heat that reduces the current density of the superconductor, creating a positive feedback effect. This was a problem during the early days of superconducting wire development and magnet design but has since been rectified through better design of multifilamentary wire. In a multifilamentary wire design, it is possible to allow several filaments to momentarily become resistive as flux penetrates the wire while other filaments share the current. This prevents localized heating that could cause a quench.

In cryogen-free magnets, guarding against such events often requires a higher critical current margin in the superconductor, or a lower I_{op}/I_c value (operating current over critical current), since the magnet is cooled by conduction only and liquid helium does not contact the windings. Superconductor margins near 50% I_{op}/I_c begin to fall into the range of "pseudo-cryostable" magnets and are much less susceptible to training. The term "cryostability" was coined in the mid-60s [4] and refers to having enough copper in the magnet wire for the joule heating to be balanced by the surface cooling of the copper.

In a magnet cooled with liquid helium, a parameter can be defined, the Stekly parameter, a, and is given below. For values less than one a magnet is completely cryostable.

$$\alpha = \frac{\rho I_c^2}{hpA(T_b - T_c)} \tag{7.4}$$

Chapter 7

where ρ is the resistivity of the copper; I_c is the maximum current a superconductor can carry at temperature, T_b, in a magnetic field, H; A is the area of the superconductor; h is the heat transfer coefficient (typically 2000 W/m^2); p is the wetted surface of the coil; and T_b and T_c are the helium bath temperature and critical temperature, respectively. If one replaces the bath temperature, T_b, with the temperature of the surface of the magnet, then a similar convention holds for cryogen-free magnets.

This concept gives rise to a "minimum propagation zone" needed for heat generated to exceed the cooling power of the conductor and a quench to occur. If pure NbTi wire were to be used, then we would have a minimum propagation length of:

$$l_{mpz} = \sqrt{\frac{2\lambda\left(T_c - T_0\right)}{\rho J_c^2}} \tag{7.5}$$

where λ is the thermal conductivity of the conductor and all other parameters are as before. Using values for pure NbTi gives one a length of approximately 1 µm, which is typically less than the filament diameter, and this is why copper must be used to help stabilize the conductor and raise this number to 10–100s of micrometers.

7.2.4 Losses and Heat

Superconducting materials generate heat during charging or discharging. What are the sources of this heating, or losses, in a superconducting magnet? The losses in question are the result of time-varying fields or currents within the superconductor, within the matrix surrounding the superconducting filaments and in other electrically conductive materials nearby. The losses are mainly due to *hysteresis*, *coupling*, and *eddy current loss*. These are also referred to as AC losses.

Changing the magnetic field on any electrical conductor will cause a current to flow in the conductor. In superconducting systems, this can be particularly prominent since the electrical conductivity of both the superconductor itself and other nearby materials (parts of the cryostat, magnet bobbin, even the non-superconducting parts of the wire itself) are so high.

The electrical resistance of copper is typically 100x lower at 4 K than at 300 K. This means that when it is placed in a changing magnetic field, such as in a superconducting magnet being energized, an eddy current is induced in it. This current generates heat or *eddy current losses*.

To calculate this power loss in the conductor, one must know something about the resistivity of the metal (normal conductor(s) in the wire itself as well as any other metals in the changing field) as well as the current density flowing through the conductor. This would involve a complex numerical solution that is quite difficult to accurately calculate for real-world situations. Even when software is used, one must often compare results with real-world examples and modify the parameters to obtain realistic results.

In cryogen-free magnets, one must be careful to minimize the size of any closed-current loops since these result in additional eddy current heat generated during charging or discharging of the magnet, reducing the amount of cryocooler power available for handling hysteresis heating and thermal conduction and radiation. Often this is done by trying to ensure that a minimum of conducting loops are present around the magnet or in its bore when possible.

Hysteresis losses are heating effects due to the magnetization of the superconductor. When a superconductor goes through a full cycle of being charged and discharged the work done is equivalent to the area of the m-H curve. Calculating hysteresis loss is difficult and imprecise due to the complex nature of the magnet. Details on calculating hysteresis loss using Bean's slab model can be found in Wilson's [5] and Iwasa's [6] books.

Generally, the method to reducing these hysteresis losses involves using very fine filaments and high-count filament superconducting wire. Superconductors with 500–10,000 filaments exist but are often much more expensive than so-called standard superconductors, which may have at most 100 filaments. Unless very fast ramp rates are required, or an AC operation of the coil is envisioned, most superconducting magnets use "standard" superconducting wire types.

Coupling loss is heating due to the circulation of currents between the superconducting filaments. These currents do decay once the ramping process stops. Again, both Wilson's and Iwasa's book cover this process.

7.3 Cryogen-Free Magnets

7.3.1 Historical Development

Prior to commercialization of cryogen-free magnet systems, there was considerable development in conduction-cooled magnets. One of the more popular applications for conduction-cooled magnets was adiabatic demagnetization refrigerators (ADRs). See Chapter 10 for details on ADR operation.

Since these conduction-cooled ADR magnets were not submerged in a bath of liquid helium, many of the same issues existed for these as future cryogen-free magnets:

- Proper magnet design concerning safety margins
- How to effectively cool all the magnet windings
- Cooling interfaces
- Quench protection
- Current input lead cooling

Experience gained in manufacturing conduction-cooled magnets combined with enabling technologies, such as 4.2 K cryocoolers and HTS materials, allowed for the successful development of cryogen-free magnet systems.

What is believed to be the first commercially available cryogen-free superconducting magnet system cooled by a 4.2 K pulse-tube cryocooler was successfully manufactured and tested in 1999 (NIH SBIR Grant # 1 R43 GM59564-01) (Figure 7.6).

At the time, the Cryomech PT-405 4.2 K pulse-tube cryocooler was newly developed. This cryocooler was capable of 0.5 watts at 4.2 K, 25 watts at 65 K.

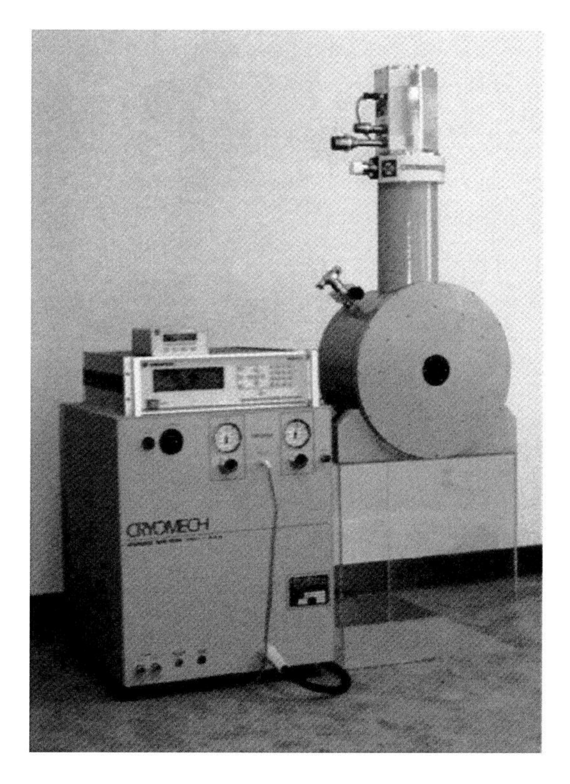

FIGURE 7.6 Early cryogen-free system. (Courtesy of Cryomagnetics, Inc., Oak Ridge, TN.)

7.3.2 Motivation

There are many benefits to operating superconducting magnets in a cryogen-free environment.

7.3.2.1 Ease of Use

Cryogen-free magnets are among the easiest to use superconducting magnets. One simply turns it on almost as easily as a light switch. The magnet only needs a reasonable vacuum in the cryostat in addition to power and cooling water for the cryocooler compressor.

However, since cryogen-free magnets rely on the cryocooler alone to cool the magnet and provide thermal stability, there are a few differences one must consider when using the cryogen-free magnet as opposed to the wet or liquid-cooled magnet.

7.3.2.2 Safety

Cryogen-free magnets are safer to operate than liquid helium magnets, as there is no handling of liquid cryogens involved. These magnets are also not subject to the possibility of explosion if a vent port is closed off or iced up and the magnet was to quench or suddenly lose vacuum. When dealing with liquid helium magnets, one must analyze the room in which the magnet sits to ensure that if a quench were to occur, no oxygen deficiency hazard would be present. Cryogen-free magnets do not have this concern since there are no liquid cryogens.

7.3.2.3 Operating Cost

Without the need to replenish expensive liquid helium and other cryogens, there are clear benefits to operating costs of cryogen-free superconducting magnet systems. Depleting reserves of helium continue to drive costs up and limit availability of helium (see Chapter 2 for more details about helium). Operating without liquid helium requires a trade-off though. Refrigeration using a cryocooler does not come without increased electrical costs, and maintenance of the cryocooler and compressor must be considered.

As an example, in the 2018 Tennessee area and at the time of this writing, electricity costs were $0.09/kW-hr and water was approximately $3.20 per 100 cubic feet. For one year of continuous operation, this gave a power bill of approximately $6,300 and a water bill of approximately $4,500 for an 8-kW compressor (assuming tap water was used to cool the compressor). This means operation of a single cryocooler costs approximately $10,800/yr. in utilities. Some research facilities have water-chilling capability in the building, which can further reduce cryocooler operational costs.

If one assumes a liquid helium consumption of 50 L per week (the equivalent boil-off rate of a typical 1W cryocooler) and $12/L for helium cost (in many cases this may be higher), the cost of operation is about $31,000/yr.

Considering a ~ $40,000.00, cryocooler cost, the break-even time would be two years with ~ $20,000.00, gained every year afterward (not including cryocooler maintenance cost).

7.3.2.4 Cooldown Time

The cost of time is another operational consideration. The cooldown of a cryogen-free magnet is slower than can be achieved when using liquid nitrogen precooling and liquid helium cooling, so if the system will be thermally cycled (warmed to room temperature and recooled to operating temperature), this needs to be considered. A typical 9T laboratory magnet that can be cooled down in the lab within a few hours using liquid cryogens often takes 12 hours or more to cool using a cryocooler alone. For large magnets of 250 to 1,000 kg, it can take one to two weeks of cooling for a 1W cryocooler, depending on the details of the design.

The ramp rates used in energizing cryogen-free magnets are also often slower than those of an equivalent wet magnet. This is due to the difficulty of relying solely on the surface area in contact with cooling sources to pull out the heat and thermal resistances of the path. This heat is easier to extract when all surfaces are submerged in liquid. To achieve higher ramp rates, one must employ high-filament-count superconducting wires to minimize the hysteresis, coupling, and eddy current losses mentioned before.

Persistent switches, when installed in cryogen-free magnets, operate much like the persistent switches in wet magnets. However, less power may be required to warm the wire above the superconducting transition temperatures. Additionally, the resistance of the persistent switch at room temperature is often higher. This is also done to reduce the power applied when cold. Even with these considerations, a persistent switch-equipped cryogen-free magnet must sweep slower than a cryogen-free magnet without a switch.

If a cryogen-free magnet quenches, the recovery time can be much longer than the time it takes to refill a liquid helium dewar. Cooldown time of the magnet before it can be reenergized can be 4 to 10 times as long as recooling with liquid cryogens. Additionally, one must keep an eye out for other related issues that can happen in a cryogen-free system during a quench, such as the outgassing of the magnet or a shield due to a higher temperature. Such outgassing has been known to overpower a single cryocooler and not let the system recool. Reconnecting a turbo pump to the system can quickly remedy the issue but only once it has been identified.

7.3.2.5 Utility Reliability

Loss of utilities can be an issue for a cryogen-free system. If power is lost, then one typically has tens of seconds to a few minutes before the magnet may quench if charged. If quench initiates in the HTS leads, this can be catastrophic since HTS leads are not normally quench protected. This is normally only a concern in magnets, which are low field and high inductance, as the margin in the superconductor can be quite large and the critical temperature is very high. Some power supplies will also begin to ramp down any non-persistent magnets, although this ramp rate is typically slow. Options to protect against such possibilities often rely on force quenching the magnet from an embedded heater in the windings or through the use of backup power for the cryocooler and control system. Since the cryocoolers have power requirements typically higher than 8 kW, this often means a backup three-phase generator is required.

4 K cryocoolers also require water cooling as well; therefore, if the water pressure or flow is lost, then the compressor can overheat and will shut down. This too can cause a quench in the magnet. Water flow issues are one of the most common causes of a cryogen-free magnet shutting down or quenching. Air-cooled compressors are available but tend to not be popular for many laboratory applications.

Generally, cryogen-free magnet systems will cryopump themselves, assuming the cryostat has a reasonably good initial vacuum. It is recommended to pump the vacuum space down when a cryogen-free magnet system is first received or after a long shutdown period of over a few weeks, in order to keep the cooldown time to a minimum. This can be done with just a roughing pump capable of less than 1×10^{-2} mbar base pressure and of sufficient size. Most cryogen-free magnet systems require a pump of around 5 m^3/hr or larger. Smaller pumps are not recommended due to the high outgassing rates of superconducting magnets and the superinsulation.

When using a rough pump, the system should be pumped below 50 mbar before starting the cryocooler. Once the cryocooler has been started, it is generally best to valve out the rough pump, as the magnet system can become a better pump than the rough pump quickly. If a turbo pump is used, this is not the case and can be left on the system while the cooldown process occurs, although it generally will not be needed once the magnet is below liquid nitrogen temperatures.

7.3.3 Enabling Technologies

Two main components have enabled cryogen-free magnets to develop and become common in today's laboratories: *cryocoolers* and *HTS materials*. While the cryocooler and its development has been covered in previous chapters, it is worth noting that without the development of 4.2 K cryocoolers in the 1990s cryogen-free superconducting magnets would not have been practical. Today's cryocoolers, which provide one to two watts of cooling power at 4 K, are ideal for small-scale superconducting magnet systems. Multiple cryocoolers are often installed on larger magnet systems or recondensing methods are employed.

7.3.3.1 Cryocoolers

The most commonly used cryocoolers for superconducting magnets are 4 K cryocoolers. These are divided into two types: Gifford-McMahon (GM) and pulse tube (PT). See Chapter 3 for a detailed discussion on the cryocoolers. In general, these are two-stage devices that provide approximately 40W of refrigeration at 40 K at the first stage, and 1–2 Watts at 4.2 K at the second stage.

The 4 K GM's are the older of the two technologies. These refrigerators rely on a moving piston inside the expander to produce the cold temperatures. This brings with it a major disadvantage. Since there are moving parts in a cold space, the wear on the seals is quite high and these systems often need servicing—typically every 10,000 hours.

Conversely, the only moving or wearing part in the PT cryocooler is the rotary valve. Since the rotary valve is at room temperature, it can generally last > 35,000 hours. This is much longer than displacer seals (for further details, see Chapter 4). Both devices generally use a scroll-type compressor producing a high-pressure supply and low-pressure return and use high-purity (99.999% minimum) helium gas as a refrigerant.

Typically, the first stage of the cryocooler is connected to a radiation shield around the superconducting magnet and to the initial thermal intercept point of the HTS current leads. This is done to reduce the radiation heat loads to a level well below the heat lift of the second stage. The radiation shield is supported internally by supports made from stainless, fiberglass, G-10, or other low-thermal-conductivity materials. Additionally, the radiation shield must block all possible line-of-sight from high-temperature surfaces, including from samples in the bore of the magnet.

The second stage of the cryocooler is thermally connected to the magnet, either directly or through flexible thermal links. All cooling is done by conduction through high-conductivity copper or high-purity aluminum supports. The magnet is suspended from the shield with low thermal conductivity supports. The magnets themselves are specially designed with conductive cooling paths to the coils so that any heat generated during charging may be rapidly removed.

If more heat than what a single cryocooler can reject is present, then additional cryocoolers may be needed. In some cases, a high heat load on the first stage (such as from high current leads) may require a single-stage cooler just for the current leads or the shield of the magnet.

When using cryocoolers with superconducting magnets a few additional things must be considered.

- If the application requires the cold head of the cryocooler to be tilted more than 30° from vertical, a GM cryocooler must be used. PT cryocoolers must remain vertical or nearly vertical to operate properly. GM cryocoolers may be tilted or even inverted with only a minor penalty in cooling performance.
- If vibration is of concern, a PT cryocooler should be used. Its vibration at the cold surfaces can be an order of magnitude or greater lower than a GM cryocooler.
- The motor of any cryocooler typically must reside outside of the 500-gauss surface of a superconducting magnet. Sometimes localized magnetic shielding is used around the motor in magnets with high stray field.
- It is also advisable to keep the regenerator material (which can be magnetic) outside the 1,000-gauss stray magnetic field surface.

7.3.3.2 HTS Current Leads

The discovery and development of HTS materials also made cryogen-free superconducting magnets possible. Liquid magnet systems often rely on using the enthalpy of the escaping helium gas to cool the current leads and minimize the heat load for the superconducting magnet. Cryogen-free magnets use HTS current leads to minimize the heat load to the second stage of the cryocooler and thus the magnet. Were these to be built using only metal wires, the heat load to the second stage of the cryocooler would be too high for most superconducting magnets.

The basis behind current lead design is an energy balance equation of the incoming heat to the 4 K surface. In wet magnet systems, this is given by the one-dimensional equation below:

$$\frac{d}{dx}\left(Ak(T)\frac{dT}{dx}\right)-\dot{m}_I\,c_p(T)\frac{dT}{dx}+\frac{\rho(T)I^2}{A}=0 \tag{7.6}$$

Where dx is the differential element of length, A is the cross-sectional area, $k(T)$ is the thermal conductivity as a function of temperature (T), $\dot{m}_I$ the helium mass flow rate, $c_p(T)$ the heat capacity as a function of temperature, current in the leads I, and $\rho(T)$ the electrical resistivity as a function of temperature. The first term represents the heat conducted down the lead, the second term the heat removed out of the lead by the gas flow, and the last term the joule heating produced. The total heat is zero when the mass flow out of the system equals the conducted and joule heating into the system.

For cryogen-free magnet systems, the middle term can be dropped since there is no helium flow out of the system and the equation becomes:

$$\frac{d}{dx}\left(Ak(T)\frac{dT}{dx}\right)=-\frac{\rho(T)I^2}{A} \tag{7.7}$$

For an optimum copper current lead pair, the minimum heat into the liquid (or 4.2 K surface) is about 84 mW/A [7]. This is calculated by minimizing the heat conducted down the lead along with the joule heating produced. Thus, for a typical 100 A laboratory magnet system, this would generate 8.4 W at 4.2 K. Even if one intercepts the heat at 77 K and optimized the leads for the 77–4 K region, this would still require 18 mW/A for a pair of leads, or 1.8W/pair.

In wet magnets, this is solved by using the enthalpy of the gas to carry away the heat. In cryogen-free magnets with HTS leads, the current lead is divided into two parts, from room temperature to the first stage of the cryocooler (or shield), and from there to the second stage (or 4 K). The room temperature to first-stage section of the current lead uses resistive metals, typically brass or copper. The heat introduced to the first stage or shield region is around 66 mW/A per pair of leads.

The first-stage to second-stage section uses HTS material, either in bulk or tape format. Since there is virtually no joule heating, the only heat produced is via conduction and can be on the order of tens of milliwatts. Special grades of HTS tape are produced with very low thermal conductivities to minimize conducted heat. One item that the designer must be aware of, however, is the critical field of the HTS materials. Designers

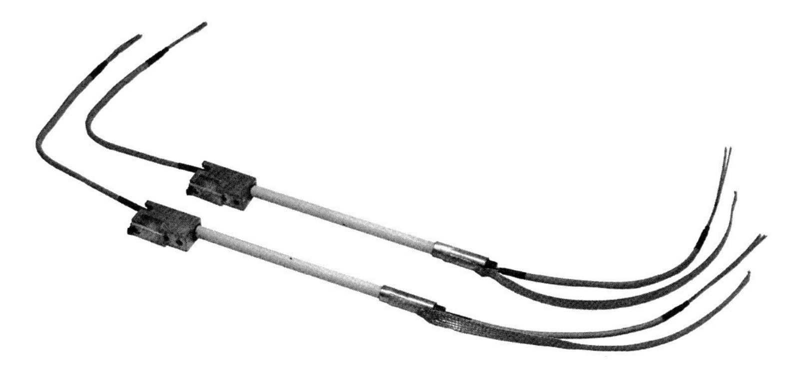

FIGURE 7.7 Typical 100 amp HTS leads. (Courtesy of Cryomagnetics, Inc., Oak Ridge, TN.)

using tape-based HTS materials must also be aware that the perpendicular critical field is often much lower than the parallel critical field. This can be the rate-limiting step for HTS leads, and background fields are often limited to a few thousand gauss. The current carrying capacity is also greatly increased when operating below 77 K. In most cases of cryogen-free systems, temperatures of the HTS components should remain below 65 K and 1,000 gauss during operation. From the second stage or 4 K area of the current lead, a high copper content LTS superconductor is often used as a buss wire for connection to the magnet. A photo of HTS leads with both copper to room-temperature wiring and LTS buss wire is shown in Figure 7.7.

7.3.4 Design Considerations

As discussed in Section 7.2 above, superconducting magnet designers must consider parameters such as critical field/current/temperature of the conductor and issues relating to forces, quench protection, and operating losses. Specific considerations apply to each of these when the magnet is to operate in a cryogen-free environment.

7.3.4.1 Conduction Cooling

Maintaining the operating temperature of a superconducting magnet is crucial to its performance. The magnet and overall system must have a reasonable cooldown time, and thermal stresses must be managed well. The thermal path from the cryocooler to the magnet must be considered, minimizing thermal resistance wherever possible.

At the same time, some components of the magnet may need to operate at different temperatures. For example, if the cryogen-free magnet has a persistent switch, it must be possible to raise the temperature of the switch above its critical temperature while keeping the magnet below its critical temperature. Yet the switch and magnet are necessarily linked electrically and thermally. The heat balance must be carefully considered.

In magnets cooled with liquid or gaseous helium, cooling through direct contact all around the coils takes place. Conduction-cooled magnets require all heat generated in the magnet during operation, in the event of a quench, or even through radiation, must be efficiently removed through, must be efficiently removed through thermal contact with the cryocooler.

7.3.4.2 Eddy Current and Hysteresis Heat

It is more important in cryogen-free magnets than liquid helium-cooled magnets to minimize heat sources of all types. This includes heat generated in the superconductor and coil former while charging, discharging, or while the magnet is in operation. Eddy current heat is minimized through proper choice of coil-former materials (including interfaces with the cryocooler) and any other cold electrically conductive materials. Choice of the superconductor used in the magnet is important too, since filament size, twist pitch, residue resistivity ratio (RRR) of the matrix of the wire, and other factors are important.

7.3.4.3 Quench Protection

It must be remembered that conduction-cooled magnets and their quench protection circuits operate in a vacuum environment. Any active components that will generate heat must be accounted for. Even wiring between the magnet and quench protection circuits must be thermally anchored and have sufficient normal conducting materials present to prevent burnout (think of all wiring like it is a light bulb).

7.3.4.4 Magnet Supports

It is important in conduction cooled superconducting magnets that thermal gradients are minimized. Heat conducting into the magnet through mechanical supports must be minimized and intercepted to avoid any hot spots on the superconducting coil(s).

7.3.4.5 Current Leads

Power leads to the magnet are a significant source of conducted heat that must be properly designed. Heat should be intercepted before it reaches the LTS superconductor. It is also important to insure heat is intercepted efficiently at the warm end of the HTS leads, and to design the resistive section of the leads and thermal interfaces properly considering both conducted and joule heating.

7.3.4.6 Power Failure

Conduction-cooled magnets are directly dependent upon the refrigeration they receive. If the cryocooler experiences a power failure or loss of water cooling, the magnet will begin to warm quickly—especially if the magnet begins to discharge (if it does not have a persistent switch). Risk of a quench can be reduced by having a source of backup power for the cryocooler compressor, but it must react quickly. Since cryorefrigerators require considerable power, usually backup power must come from a generator rather than a faster UPS/battery source. Another alternative is to incorporate a small amount of liquefied helium inside the system (making it not truly "cryogen-free"), but this also has its challenges. The system designer needs to properly account for the possibility of power or other cooling failure, and also for routine maintenance of the cryocooler.

7.3.5 High-Temperature Superconductor Magnets

While it is possible to design and build cryogen-free magnets made with HTS materials, this is not widespread due to the high cost of the materials and due to additional design challenges in HTS magnets. Today's HTS materials are available in the form of tapes for rare earth bismuth copper oxides (REBCOs) and bismuth strontium calcium copper oxides (BSCCO-2223s), as well as some round wires, namely magnesium diboride (MgB_2) and Bismuth-2212. Each has its challenges for use with superconducting magnets, and especially in cryogen-free magnets.

7.3.5.1 Wire Performance

As discussed in Section 7.2, when designing superconducting magnets, one must consider the current carrying capacity of the wire (I_c), its operating temperature, and the magnetic field on the wire. HTS materials have clear advantages over LTS in that they can operate at elevated temperature, but they have other challenges with regard to performance.

REBCO (usually YBCO) HTS material is currently only available in tape form. Tape is not only more difficult to wind into a coil than round or rectangular wire, but the magnet designer must take into account that the critical current performance of HTS conductor depends upon the direction of magnetic field on it. HTS tape is, in general, more sensitive to the magnetic field perpendicular to the plane of the tape than it is to the field parallel to the conductor plane. This means that at the ends of a solenoid, where a magnet typically experiences a higher perpendicular field component (since the magnetic flux lines are wrapping around the coil), REBCO tape with its lower I_c becomes a limiting factor in magnet design. This will limit the maximum field of the magnet.

For the magnet designer, smaller-width tapes are desirable, but the act of cutting the tapes to the required width introduces damage, reducing its current carrying capacity. Also, tapes are more difficult to wind. Considerable progress has been made in novel wire configurations, such as tape twisted into a helical shape to form a round cable-type wire (e.g., Advanced Conductor Technologies development of CORC HTS wires).

In general, tape HTS magnets are made using double pancake designs. These magnets are challenging due to the need to make multiple splices on the inside and outside of the magnet, as shown in Figure 7.8.

FIGURE 7.8 A 3.5T @ 40 K ReBCO (YBCO) cryogen-free double pancake magnet. (Courtesy of Cryomagnetics, Inc., Oak Ridge, TN.)

MgB_2 and Bi-2212 HTS wires are round, and therefore practical magnet application is easier. But these superconductors require heat treatment after winding, making fabrication more difficult and costly. These magnets use *wind and react* processes where the magnet is first wound into the desired coil shape, then heated to high temperature in an inert or partially inert atmosphere to form the desired alloy, and finally vacuum impregnated with epoxy. This is a very similar process to how the low-temperature superconductor Nb_3Sn is produced.

Bi-2212 has an additional challenge in that for better performance, it requires the heat treatment be done at high pressures of 50 atm or more due to its powder-in-tube (PIT) architecture. The fact that both of these wire types require high-temperature heat treatments makes the cooling of them in a cryogen-free magnet application more difficult (but not impossible) due to the fact that the bobbins are often not of a high thermal conductivity material.

7.3.5.2 Quench Protection

A major challenge when designing HTS magnets is how to handle quench. This is even more of a challenge in a cryogen-free magnet design. Due to the high heat capacity of the superconductor at temperatures greater than 20 K, the quench propagation velocity is extremely slow. If a quench begins and the energy is not removed from the magnet quickly, hot spots and overvoltages will form that will damage the magnet. Most methods for dealing with quench protection of HTS coils are therefore active rather than passive protection, though other techniques are also being studied, such as No-Insulation magnets that rely on electrical contact between turns and layers throughout a coil.

Even considering all its challenges, the largest impediment to HTS cryogen-free magnets today is economics. The cost of HTS wire is simply too high to compete with traditional LTS wires. While the cryogenics may be simpler and reduced in terms of cost, the wire cost more than makes up for this. This results in a much higher cost for an HTS magnet compared to an LTS magnet of comparable specification.

7.4 Current Status and Examples

As cryogen-free magnet technology has matured, many magnet configurations previously developed for liquid helium-based magnet systems have become available in their cryogen-free equivalent. However, due to the complexities of integrating superconducting magnets, quench protection, cryorefrigeration, and current lead/instrumentation interfaces, cryogen-free superconducting magnets are typically sold as integrated packages, where in the past it was possible to buy components and assemble a wet system piece by piece.

Numerous configurations are available, depending upon field shape and intensity needed and how the user needs to access the magnetic field. Some of the most common products include room temperature bore access.

These systems are available with a variety of superconducting magnet configurations including solenoids, split-pairs, and multi-axis. Within these categories, there are many options including high field, high homogeneity, compensated, and actively shielded.

Vertical room temperature bore access allows for easy integration of commonly available variable temperature, 3He, and dilution refrigerator inserts.

Users often require both axial and perpendicular access to the magnetic field. This can be accomplished in split-pair and multi-axis magnet systems.

A major advantage of cryogen-free magnet systems is their compact size. Because no liquid helium volume is required, it is possible to have table-top systems where before a large support cryostat was required. Figure 7.9 is an example of such a compact system, having a 9T central field and an overall width of only 10 in.

This short distance from room temperature flange to magnetic center can be very important for many applications such as microscopy and other experiments.

The system in Figure 7.10 is a compact 5T split-pair system that can be installed in either the horizontal or vertical position. For such applications, a GM cryocooler must be used.

FIGURE 7.9 Cryogen-free compact system. (Courtesy of Cryomagnetics, Inc., Oak Ridge, TN.)

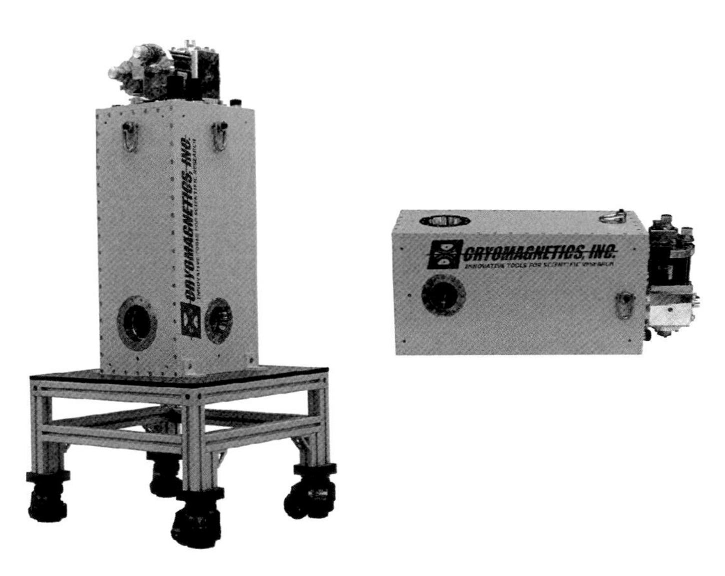

FIGURE 7.10 Cryogen-free multiposition magnet system. (Courtesy of Cryomagnetics, Inc., Oak Ridge, TN.)

7.4.1 Cryogen-Free Magnetic Research Platforms

The availability of scalable platforms using a single 4.2 K cryocooler has brought a new level of operating ease, convenience, and efficiency to the research community. These systems can include an integrated variable temperature insert, allowing for a versatile research instrument. Optical access configurations are also available. Typical sample temperature range is <1.6–300 K.

To increase efficiency, modules are available for specific areas of research, such as resistivity, Hall effect, thermal transport, and VSM. Modular upgrades are available, such as a 3He insert to achieve sample temperatures below 1 K. Frequently, these options can be added later as the customer's budget allows or their research needs change.

Although cryogen-free superconducting magnet systems add convenience and can reduce operating costs, they also can introduce unwanted vibration to the sample.

The source of this vibration can come from mechanical motors and pistons found in some 4.2 K cryocooler designs. It can also come from movement of the cold stages as high-pressure gas pulses during the refrigeration cycle.

Research applications such as scanning probe microscopy (SPM) require extremely low sample vibration. Figure 7.11 shows a 7T/1T/1T 3-axis, low-vibration cryogen-free system featuring a 4.2 K PT cryocooler. The sample is decoupled from the cryostat by using a heavily weighted triangular platform and flexible edge-welded bellows as seen on top of the system. This platform is supported by vibration-isolating pneumatic legs. Bedding on the compressor and vacuum lines helps further reduce vibration to the sample environment (Figure 7.11).

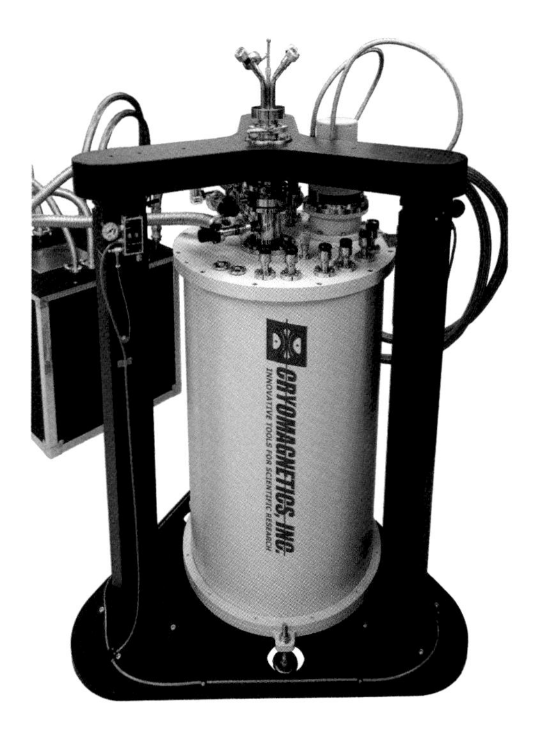

FIGURE 7.11 A cryogen-free, low-vibration, variable temperature/variable field materials research system. (Courtesy of Cryomagnetics, Inc., Oak Ridge, TN.)

7.5 Future Expectations

Not all superconducting magnet systems are compatible with cryogen-free operation. Large high-energy physics magnets are an example of this. Cooling complex superconducting magnets by conduction can make the magnet unreasonably complicated and, in many cases, impossible. But many magnet applications that currently use liquid helium for cooling can be designed to operate cryogen-free.

For applications where completely cryogen-free operation is impractical, one or more 4.2 K cryocoolers can be used to recondense, in many cases stopping the boil-off of liquid helium entirely. Next-generation superconducting undulators and wigglers are examples of high-energy physics magnets that operate by conduction cooling, but at the same time have helium recondensing to eliminate boil-off.

Some of the main areas that will benefit from cryogen-free magnet technology advances are applications where safety and magnet mobility are a concern. Cryogen-free superconducting magnet systems eliminate the safety hazards of liquid cryogens, making them safer for personnel working around the system. They also are more compact, making it more user-friendly to work around.

In applications where the superconducting magnet may be mobile—for example, mounted in a vehicle, or on the rotating gantry of a cancer treatment system—cryogen-free magnets greatly improve operation. Magnets with GM cryocoolers do not care if they are operated vertical, horizontal, upside down, or anything in between. But liquid cryogen systems are considerably more challenging when trying to operate in a mobile setting.

Technology continues to improve in many areas of cryogen-free superconducting magnet systems. Cryorefrigerator power at 4 K continues to increase. Reliability of cryocooler cold heads continues to improve, with maintenance intervals extending. Advanced materials are being developed that improve thermal conductivity at low temperature, providing better heat management in cryostat construction and more efficient cooling paths.

As cryogen-free magnet design technology matures, some of the biggest advances are coming from improvements in superconducting wire/cable/tape. New HTS materials are allowing for better operating margins and improved stability. Magnets can be operated at higher temperatures making refrigeration easier and more efficient.

A challenging area continues to be quench protection in advanced HTS magnets, particularly when these are conduction cooled. Heat management and quench propagation within a vacuum environment is more difficult without a liquid or gas environment around the magnet to help with energy dissipation. Much work still needs to be done.

A common theme for almost every area of research, medical, and industrial use of superconducting magnets is a desire for higher magnetic field. High-temperature superconductors are currently enabling a wave of significant advances in higher field magnets. Eventually, some of these advances will make it into cryogen-free systems.

An active area of research currently is in No-Insulation (NI) REBCO HTS magnets. This appears to be a promising way to manage quench protection issues and may eventually be applicable to cryogen-free magnet systems.

Research programs are underway to develop commercial, cryogen-free, 25–30T high-field laboratory-scale superconducting magnets using Bi-2212. While there are many obstacles to overcome, the goal of very high fields (25T+) in a cryogen-free, magnetic research platform as pictured in the previous section is in the near future.

References

1. Onnes, H. K. (1911). Leiden Comm. 120b, 122b, 124c.
2. Kunzler, E. A. (1962). Production of magnetic fields exceeding 15 Kilogauss by a superconducting solenoid. *Journal of Applied Physics* 32, 325.
3. Bean, C. P. (1962). Magnetization of hard superconductors. *Physical Review Letters* 8, 250.
4. Stekly, Z. J., & Zar, J. L. (1965). Stable superconducting coils. *IEEE Transactions on Nuclear Science* 12, 367.
5. Wilson, M. N. (1983). *Superconducting Magnets*. New York: Oxford University Press.
6. Iwasa, Y. (1994). *Case Studies in Superconducting Magnets*. New York: Plenum Press.
7. Ekin, J. W. (2006). *Experimental Techniques for Low-Temperature Measurements*. New York: Oxford University Press.

8. Cryogen-Free 4 K and 1.5 K Systems

Zuyu Zhao

8.1 4 K Cryogen-Free Systems

Base temperatures of 3–4 K can easily be reached using commercially available Gifford–McMahon (GM) or pulse-tube (PT) cryocoolers. Examples are shown in Figure 8.1. The readers should refer to **Chapter 3 ~ 4 K Regenerative Cryocooler** for cryocooler details.

A typical example of a cryostat including a two-stage cryocooler located in a vacuum shroud is illustrated in Figure 8.1.

The temperature of the first stage without an applied heat load is usually around 40 K depending on the cold-head model. A first-stage cold plate as well as a radiation shield made of material with high thermal conductivity and low emissivity, such as highly polished aluminum or copper is attached to the first stage. The radiation shield and radiation shield anchor thermally shield the second stage from the room temperature infrared radiation heat load from the vacuum shroud.

The second stage provides a cold platform for the user's experiments.

Multilayer super insulation (MLI) may be wrapped on the radiation shield in order to minimize the room temperature radiation heat load. This technology may be used in all non-ultra-high-vacuum (UHV) compatible cryogen-free cryostats and can play an important role in the performance of the cryostat.

Starting from a basic 4 K cryogen-free cryostat as illustrated in Figure 8.2, various types of systems for different applications have been developed as shown in the following examples. Several examples of cryogen-free systems operated near 4 K (usually referred to as 4 K cryogen-free cryostats) are presented below. A more detailed description is given

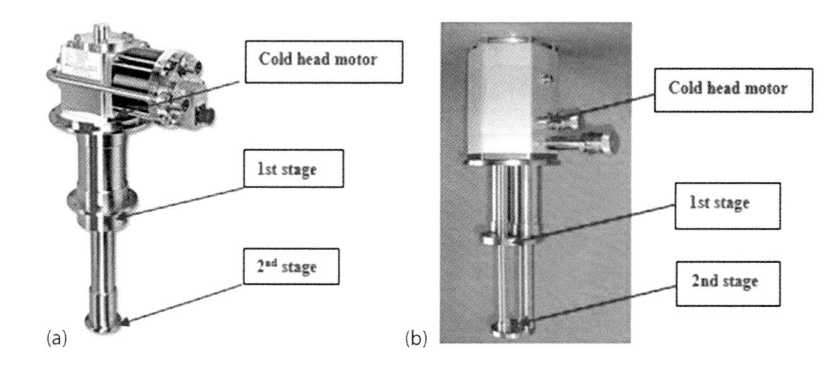

FIGURE 8.1 (a) GM cryocooler (Courtesy of Sumitomo Heavy Industries, Ltd., http://www.shicryogenics.com/products/4k-cryocoolers.) (b) Pulse tube cryocooler (Courtesy of Cryomech Inc.)

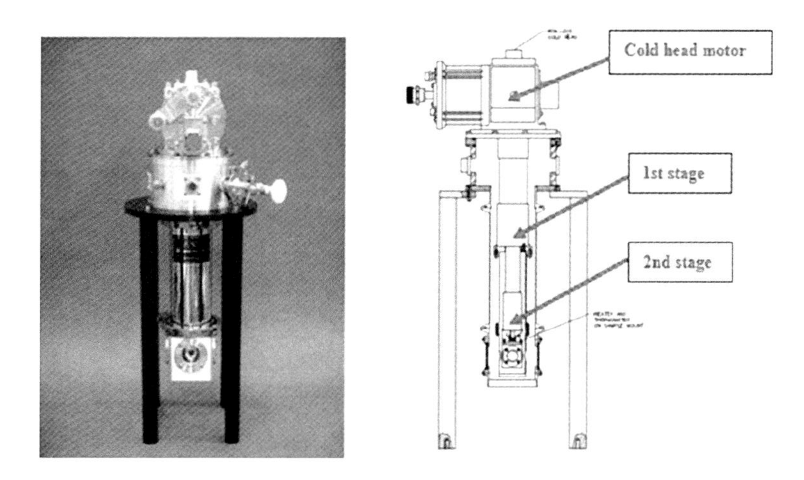

FIGURE 8.2 Basic 4 K cryogen-free system. (Courtesy of Janis Research Company.)

in Example 8.4, while the structures and operations of most 4 K systems are quite straightforward.

Example 8.1: Low Vibration System with Sample in Vacuum

Figure 8.3 illustrates a low vibration 4 K system with the sample cooled in vacuum.

The experimental cryostat is usually seated on an optical table with vibration isolation legs to minimize vibrations from the floor of the laboratory to the experimental surface on the optical table. The cold head of the cryocooler is only connected to the cryostat by a flexible rubber bellows. The cold head is connected to a stand that is mounted on the floor of the laboratory. The cold-head support stand does not make contact with the experimental surface of the optical table or the cryostat. The cold head does not make physical contact with the rest of the cryostat. The soft rubber bellows greatly reduce the vibrations transmitted from the cold head to the cryostat. Actual vibrations have been demonstrated to be less than 15 nm at Janis with proper installation.

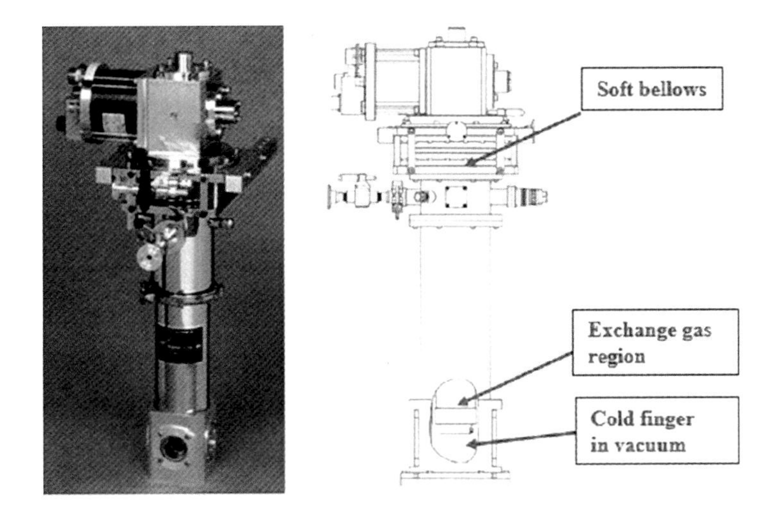

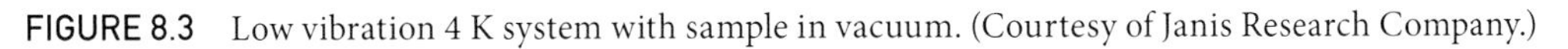
FIGURE 8.3 Low vibration 4 K system with sample in vacuum. (Courtesy of Janis Research Company.)

The cold stage of the cryocooler does not make direct contact with the experimental or any other part of the cryostat that makes rigid physical contact with the sample. The space between the cold head and the cryostat (with the "cold finger") is filled with approximately 1 atm of helium exchange gas. The helium exchange gas is cooled by the cold head and cools the cryostat only through convection since no physical contact between the cold head and the cryostat exists. The sample is installed on the "cold finger" in the vacuum side and cooled purely by thermal conduction with the cold finger.

Example 8.2: Optical System with Large Sample Space

Figure 8.4 illustrates a custom designed optical system with large sample space.

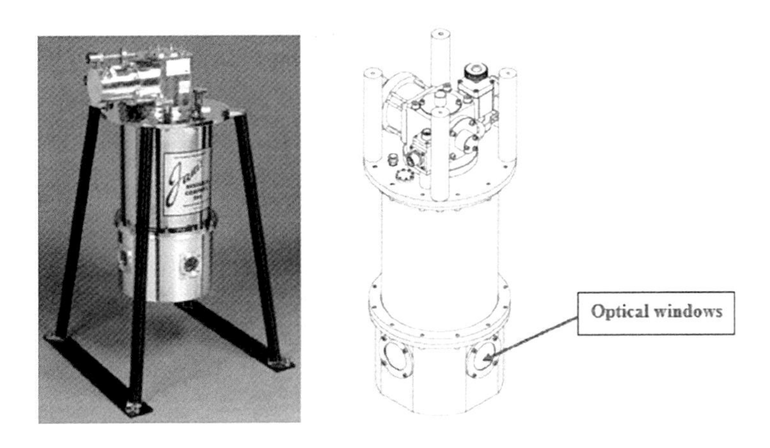

FIGURE 8.4 4 K cryogen-free optical system. (Courtesy of Janis Research Company.)

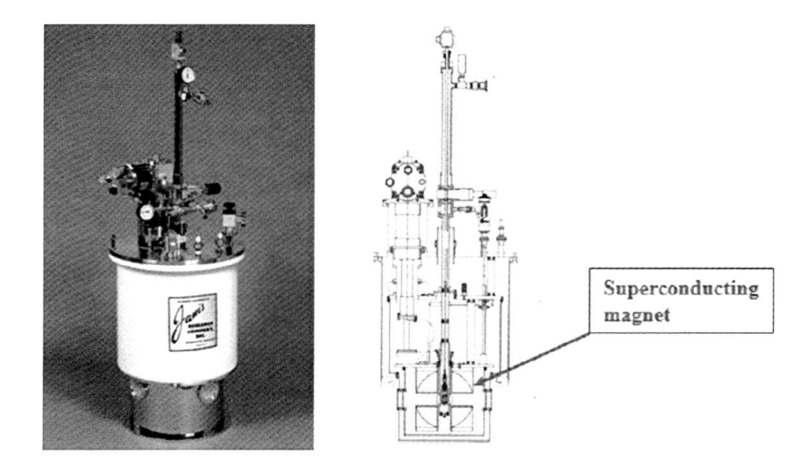

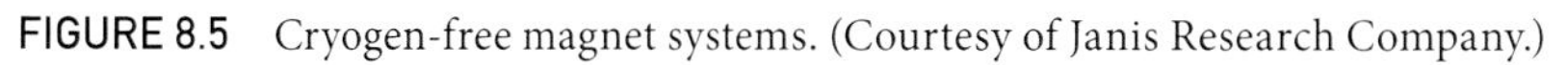

FIGURE 8.5 Cryogen-free magnet systems. (Courtesy of Janis Research Company.)

Example 8.3: Cryogen-Free Magnet Systems with Sample in Exchange Gas

A superconducting magnet is rigidly attached to the second stage of a two-stage cryocooler through highly conductive thermally links.

Non-optical magnet systems with vertical bores are often integrated with a top-loading cryogen-free cryostat as shown in Figure 8.5. Typical systems of this type can be operated between 1.5 K and 300 K or higher.

Example 8.4: Cryogen-Free Mossbauer Cryostat with Sample in (Helium) Exchange Gas

Mossbauer spectroscopy is a type of resonance spectroscopy using gamma rays to probe the electronic and magnetic structure of the sample in transmission geometry. As with many other measurements, Mossbauer spectroscopy is usually performed at low temperatures to improve the signal-to-noise ratio. Features that cannot be observed at room temperature are easily recorded at cryogenic temperatures. At non-resonant energies, the gamma rays pass through the sample mostly undisturbed. When the sample is exposed to gamma radiation at the resonant energy, the gamma radiation is strongly absorbed. Samples examined using Mossbauer spectroscopy are often in the form of a powder or solid sample of a shape that is difficult to thermally anchor to a cold finger in a vacuum. Samples of this type can more effectively be cooled using thermal convection through an exchange gas or dynamic flow of cold vapor. Therefore, cryostats that cool the sample using thermal convection in an environment of static exchange gas or a dynamic flow of vapor are generally better suited for Mossbauer spectroscopy. A cryogen-free cryostat that cools the sample in static exchange gas is simpler to manufacture and therefore lower in price than a cryogen-free cryostat that would cool the sample in a flowing vapor. [NOTE: Helium cryostats are also very commonly used for Mossbauer spectroscopy. Sample environments in flowing vapor and static exchange gas are available]. The following discussion will focus on cryogen-free systems cooling the sample in static exchange gas.

All types of spectroscopy require a source of radiation, a sample under investigation, and a detector to measure the radiation after the radiation interacts with the sample. As Mossbauer spectroscopy utilizes gamma rays to interact with the sample, the source of the gamma radiation is a radioactive isotope. In order to find the resonant frequency where the gamma rays are absorbed by the sample, the frequency of the incident radiation must be scanned over the range of interest. Radioactive isotopes emit radiation at a single, fixed frequency (or energy), not over a broad range that would be useful for scanning a sample to find the resonant energy. Since the energy of the source itself cannot be varied, the frequency is scanned by oscillating the position of the source while keeping the sample position fixed, resulting in a Doppler shift of the gamma ray frequency. Precisely knowing the velocity of the source relative to the sample is of supremely important to collecting a useful Mossbauer spectrum and a very important detail to keep in mind when designing a cryostat for use in Mossbauer spectroscopy.

The velocity of the source relative to the sample is typically measured in units of millimeters/second. A typical commercially available Mossbauer spectrometer has a resolution of 2 mm/sec. The actual velocity of the source is usually precisely known through the mechanics and electronics that drive the motion of the source. Uncertainty in the velocity measurement is mostly due to vibration of the sample, which is generally not a quantity that can be measured. The precision of a Mossbauer spectrum is measured by the line width of the absorption. A broad line width is less precise than a sharp, well-defined line width. Vibrations of the sample result in broader line widths and less precise measurements. Therefore, minimizing sample vibrations is crucial to the design of a cryogen-free Mossbauer cryostat.

Cryocoolers are mechanical refrigerators with moving parts and expanding helium gas that result in vibrations of the cold head. Further information regarding the vibrations from the various cold heads available is discussed in other sections of this text. A cryogen-free Mossbauer cryostat must isolate vibrations of the cold head from the sample as much as possible.

A typical cryogen-free Mossbauer system as illustrated in Figure 8.6 includes the following key components:

1. Two-stage GM cold head operating in an inverted orientation (motor on the bottom, second stage on the top). Note that a PT cold head must be operated in the opposite orientation (motor on top and second stage below) and therefore are not suitable for use in this particular cryostat design.
2. Exchange-gas sample tube mounted on the second stage of the GM cold head with heater and temperature sensor.
3. Sample probe with sample holder, temperature sensor, and heater. The sample probe is suspended in the sample tube without making physical contact with the sample tube.
4. Metal or rubber vibration isolation bellows installed between the sample probe and the rest of the cryostat.
5. Vibration isolation stand supporting the sample probe independently from the rest of the cryostat and from vibrations through the floor of the laboratory in which the system is installed. The vibration isolation stand is intended

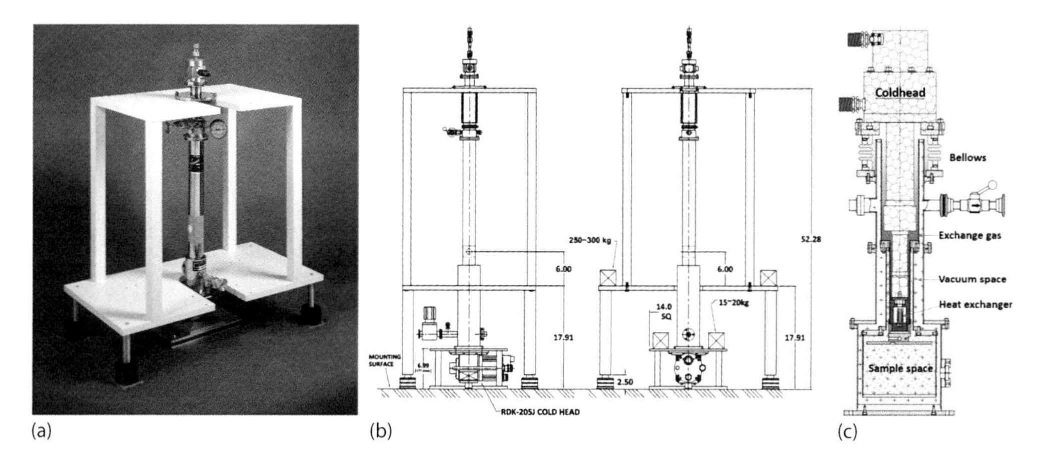

FIGURE 8.6 (a) Image of cryogen-free Mossbauer cryostat with sample in (helium) exchange gas, (b and c) sketch of cryogen-free Mossbauer cryostat with sample in (helium) exchange gas.

to isolate vibrations that would otherwise be communicated from the floor of the laboratory to the sample probe. Common sources of the vibrations are the compressor for the cryocooler, vacuum pumps, and vibrations that are simply present in the building. Reducing these vibrations as much as possible is vital to building a cryostat that will be suitable for Mossbauer spectroscopy. The vibration isolation stand includes pneumatic vibration isolation feet on all four legs along with ~300 kg of ballast in the form of lead bricks that can also serve as part of the radiation shielding. The spectrometer is mounted directly onto the vibration isolation stand. The vibration isolation stand does not make physical contact with the rest of the cryostat.

6. Mylar windows to transmit gamma rays on the outer vacuum jacket, aluminized Mylar windows on the radiation shield, and cold Mylar windows on the sample tube.

The only physical connection between the sample probe and the rest of the cryostat that is directly in contact with the cold head (and therefore feeling all of the vibrations from the cold head) is the metal or rubber bellows between the sample probe and the rest of the cryostat. Rubber bellows offer better vibration isolation than metal bellows. However, proper installation of a system with the metal bellows can still result in vibrations low enough to be less than the 2 mm/s resolution of the Mossbauer spectrometer. Although metal bellows do not isolate vibrations as well as rubber bellows, metal bellows do have two distinct advantages over rubber bellows. Metal bellows never require replacement, while rubber bellows will eventually degrade over time and need to be replaced. Also, some applications, including operation above 150 K, require the sample tube to be evacuated. Mylar is permeable to helium gas at temperatures above ~150 K. Therefore, the sample tube needs to be under vacuum at temperatures above ~150 K or the vacuum between the sample tube and outer vacuum jacket will be lost. Rubber bellows will collapse under vacuum and lose most of their vibration isolation qualities, while metal bellows perform essentially the same whether the sample tube is under vacuum or with exchange gas at 1 atm of pressure.

Example 8.5: Cryogen-Free Micromanipulation Probe Stations

Cryogen-free micro micromanipulated probe stations as shown in Figure 8.7 provide the researcher with vacuum and cryogenic probing capabilities for many applications, such as Microelectromechanical systems (MEMS), nanoscale electronics, superconductivity, ferroelectrics, material sciences, and optics.

Similar to the cryogen-free Mossbauer cryostat, the cryocooler is in the upside-down position; a GM cryocooler instead of a PT refrigerator needs to be employed.

Example 8.6: Special Cryogen-Free Low-Vibration CRYOSTATION

Figure 8.8 illustrates a cryogen-free low-vibration CRYOSTATION crossover premium (xp) platform operated from 1.7 to 350 K and its performance.

Key performances data

Temperature Range	~1.7–350 K	
Temperature Stability	<2 mK	peak to peak at base T (1.7K)
	<200 mK	1.7–15 K
	<50 mK	15–350 K
Vibration Stability	<20 nm	peak to peak (measured on platform *x*-axis)
Cooling Power at Base Temperature	>20 mW	

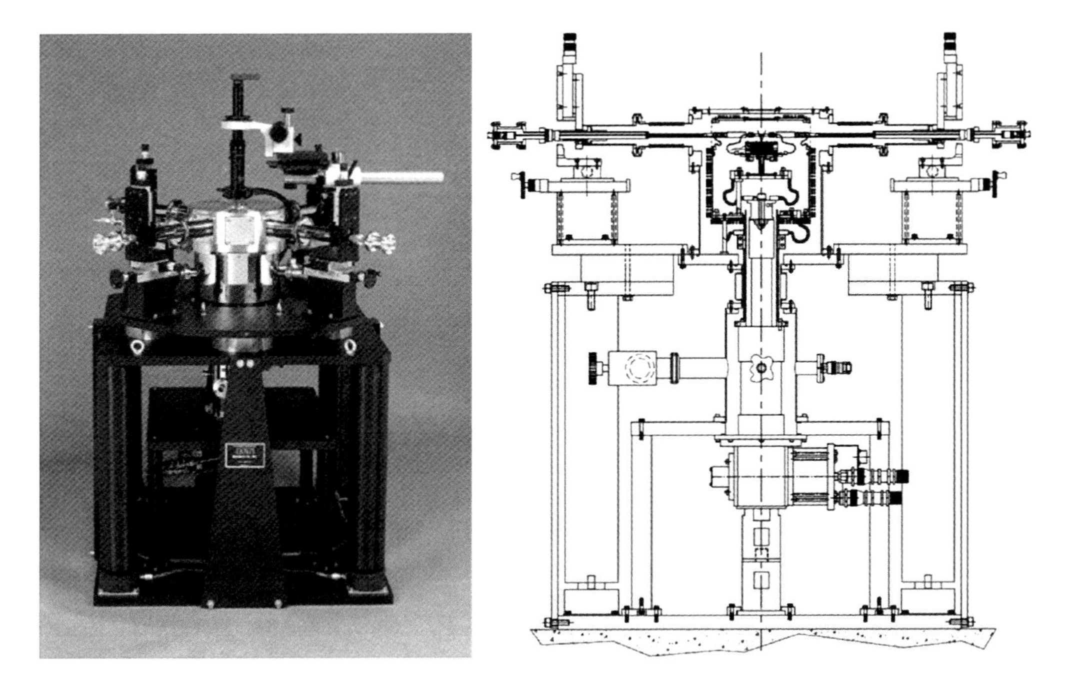

FIGURE 8.7 Cryogen-free micromanipulation probe stations. (Courtesy of Janis Research Company.)

FIGURE 8.8 Cryogen-free low vibration CRYOSTATION crossover premium (xp) platform operated from 1.7 to 350 K. (Courtesy of Montana Instruments; https://www.dropbox.com/sh/2lohhxm3yk3ufci/AADhHqEhcr9_U4qwSmh1cLn6a?dl=0.)

8.1.1 Notes Regarding Cryogen-Free Systems

1. The PT cryocooler in general generates less vibration than the GM cryocooler of similar cooling power and size. GM cryocoolers typically have ~20 μm (peak-to-peak) vibration along the axis of the cold head and ~5 μm perpendicular to the axis of the cold head. PT cryocoolers have 5–15 μm of vibration along the axis of the cold head and <5 μm perpendicular to the axis of the cold head.

   Vibrations in both GM and PT cold heads are primarily due to the valve motors and expansion of the helium gas inside the cold heads. Most PT coolers are available with a "remote valve motor" option that separates the valve motor from the cold head with a flexline instead of having the valve motor directly mounted onto the cold head. In a PT cooler, all of the moving parts are contained in the valve motor. Separating the valve motor from the cold head gives the user the ability to isolate some of the vibrations from the valve motor to the cold head. PT cryocooler cold heads do not have any internal mechanically moving parts to add to vibration levels of the cold head. However, expansion of the compressed helium gas inside the cold head does cause vibration at the second stage, and it happens to both the PT and GM cold heads. GM cold heads also have "displacers" that resemble pistons in each stage of the cold head. The displacers oscillate with the expansion of helium gas from high pressure to low pressure. Motion from the displacers also contributes to the vibrations of GM cold heads. PT cold heads do not have displacers. The cooling process inside of a PT cold head is purely due to the expansion of helium gas. The mechanics of GM and PT cold head has been discussed in great details in Chapters 3 and 4.
2. In order to further reduce the vibration induced from the rotary valve motor on the PT cold heads, a "remote valve option" is available for all Cryomech PT systems, that is, the rotary valve is mounted externally from the cold head for further vibration reduction (Figure 8.9).

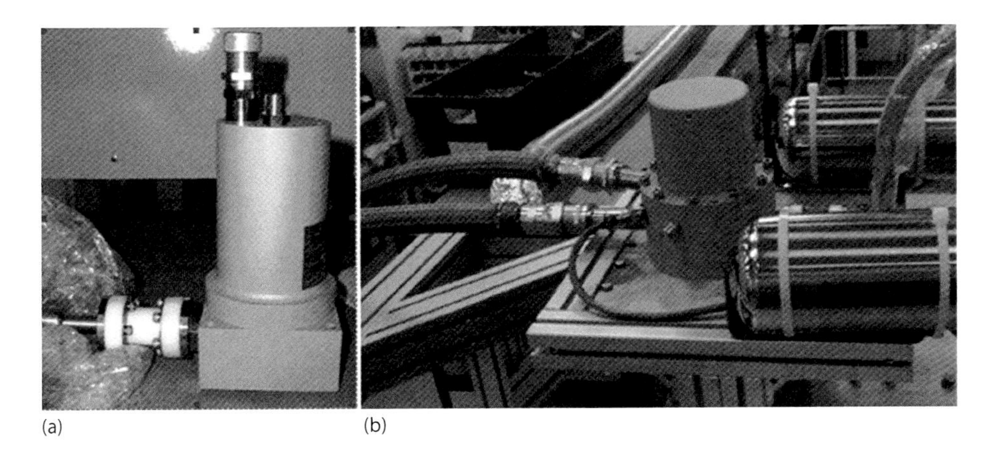

FIGURE 8.9 (a and b) Remote motor and its mounting. (Courtesy of Janis Research Company.)

3. The PT cold head should be oriented with the second-stage pointing down. Cooling power and base temperature of a PT cold head will be reduced if the cold head tilts more than 35° from the vertical position.

   The performance of a GM cold head is maximized when the cold head is operated vertically. The performance is the same with the cold head oriented such that the second stage is pointing upward or downward. The cooling power is likely to be reduced when tilted away from the vertical axis to a maximum of ~15% when horizontal in some (but not all) GM cold heads (Figure 8.10).

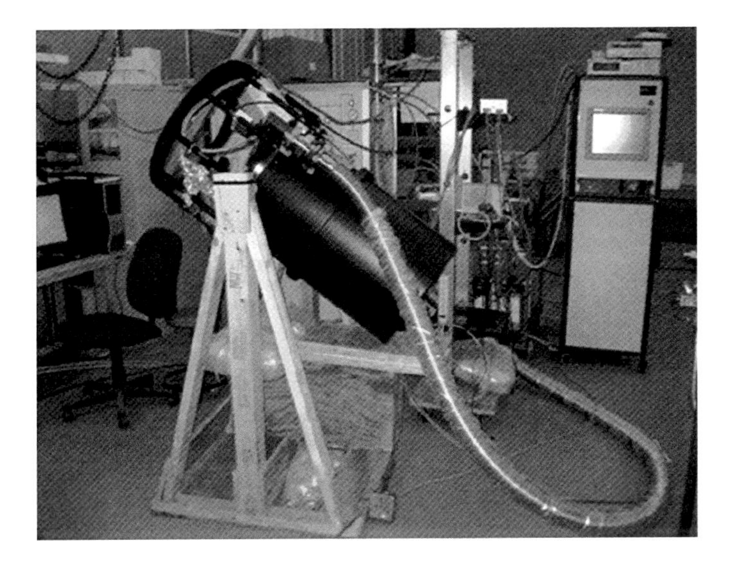

FIGURE 8.10 Pulse-tube cryocooler performance test at tilted position. (Courtesy of Janis Research Company.)

8.2 1.5 K Cryogen-Free Systems

The development of commercial cryocoolers reached a breakthrough point when the cryocoolers reached the base temperature below 4.2 K, the temperature that was previously accessible only by using liquid helium.

The development of 4 K cryocoolers is significant for two reasons:

- 4 K cryocoolers provide temperatures low enough for helium condensation.
- 4 K cryocoolers provide a platform from which much lower temperatures can be realized with additional steps.

The following sections discuss cryogen-free systems with a base temperature of approximately 1.5 K.

Systems to reach below 1 K will be discussed in separate sections.

8.2.1 Helium-4 Phase Diagram

Refer to **Section 2.2** and we start from Helium-4 phase diagram (Figure 8.11) in zero magnetic field (Figure 8.12).

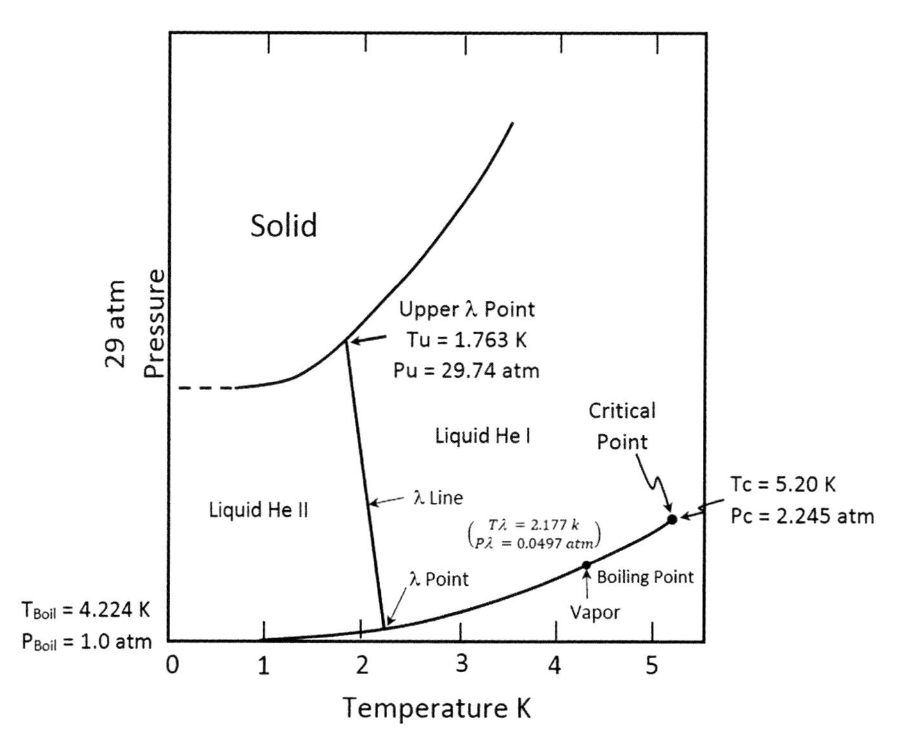

FIGURE 8.11 Phase diagram of Helium-4.

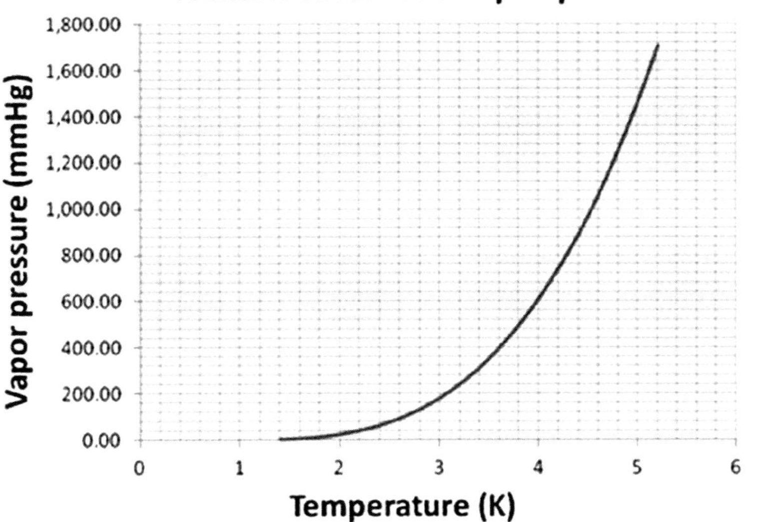

FIGURE 8.12 Saturated vapor pressures of liquid helium as a function of temperature. (From Donnelly, R.J., and Barenghi, C.F., *J. Phys. Chem. Ref. Data*, 27, 1998, 1217–1274.)

8.2.2 Single-Shot-Type 1.5 K Cryogen-Free System

Gaseous helium can be liquefied directly by the second stage of a cryocooler (either GM cryocooler or PT cryocooler) under proper head pressure. The liquid helium can be further cooled down to 1.5 K or below by connecting a vacuum pump to the sample tube and pumping on the condensed LHe to reduce its saturated vapor pressure. This type of system is usually called a "1.5 K cryostat."

Figure 8.13 illustrates a typical single-shot-type 1.5 K cryogen-free cryostat with a dedicated sample tube. The system includes the following main components:

1. Cryocooler
   The 1.5 K single-shot cryostat may include either a GM cryocooler or a PT cryocooler. The temperatures of the first and second stage are typically 40 K and 2.8 K, respectively.
2. Cryostat
3. Sample tube and sample probe

The sample tube is composed of four sections made of different materials.

Section #1 is made of stainless steel. The stainless-steel section starts at the top of the cryostat (room temperature) and terminates at the second stage of the cryocooler. The first stage of the cryocooler is thermally anchored to the middle section of the sample tube to intercept the conductive heat load from the room temperature top section of the sample tube.

Section #2 is made of oxygen-free high-conductivity (OFHC) copper. The top of this section is thermally connected to the second stage of the cryocooler. For systems built on a GM cooler, the thermal link is usually an OFHC copper plate bent into

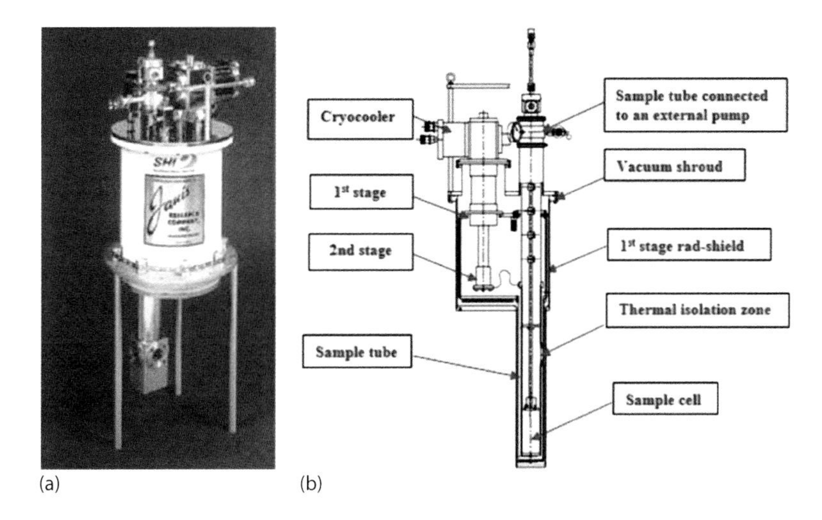

FIGURE 8.13 (a) Single-shot-type optical 1.5 K cryogen-free system. (b) Single-shot-type non-optical 1.5 K cryogen-free system. (Courtesy of Janis Research Company).

the shape of a "U" to allow for thermal expansion and contraction with temperature change. For systems built using a PT cooler, the thermal link is usually made with OFHC copper braids. Pressurizing the sample tube to ~3 psig with helium exchange gas when the sample tube is at base temperature will allow the helium exchange gas to condense into liquid. Once a volume of liquid helium has been condensed in the sample tube, a vacuum pump (typically a large-capacity single-stage rotary vane pump or scroll pump) can be used to pump on the condensed liquid helium to reduce the vapor pressure above the condensed helium and therefore reduce the temperature to 1.5 K. The hold time at 1.5 K will depend on the dimensions of the condensing zone, cooling power of the cryocooler, capacity of the vacuum pump, and experimental heat load. A typical system can hold a temperature of 1.5 K for about 8 hours after condensing for about 8 hours.

Section #3 Consists of a short section of stainless-steel tube, which serves as a thermal isolation zone between the second stage and the section #4, also known as the "sample area" where the condensed liquid helium is collected.

Section #4 is made of OFHC copper and located at the bottom of the sample tube.

Depending on the temperature and pressure inside the sample tube, the sample tube can provide different environments to the sample, such as vacuum, helium vapor, or immersed in liquid helium.

A sample probe is used to load the sample into the sample tube of the cryostat.

The sample tube will remain the base temperature until the liquid helium runs out. Liquid helium needs to be recondensed inside the sample tube before the system can reach the base temperature again.

Cooling power at certain temperatures, defined as the maximum heat load applied by the user while the system can hold a fixed temperature, is another important performance specification of great interest to the users. Cooling power is a function of temperature.

Since the *base temperature* is defined as the lowest temperature a system can reach *without* any applied power from the user, the cooling power at the base temperature is zero.

Since the cooling mechanism of the single-shot-type 1.5 K cryogen-free system comes from the heat of evaporation of liquid helium when it is being pumped, the cooling power depends on how fast liquid is being removed by the vacuum pump. Cooling power is calculated by multiplying the molar latent heat of liquid helium at a given temperature with the throughput (i.e., moles of liquid helium pumped out per unit time).

The user may take the following approach and estimate the cooling power of single shot type 1.5 K cryogen-free system:

Step 1: Get the properties of liquid helium at its saturated vapor pressure.

1. Refer to Table 8.1 and Figure 8.13 for the liquid Helium-4 saturated vapor pressures as a function of temperature

Table 8.1 Saturate Vapor Pressure of Liquid Helium

	Temperature [K]	Pressure [atm]	Density [g/cm³]	Entropy [J/g K]	C_{sat}[J/g K]
	1.4	0.002835	0.1450	0.1.32	0.780
	1.6	0.007487	0.1451	0.284	1.572
	1.8	0.01641	0.1453	0.535	2.810
	2.0	0.03128	0.1456	0.940	5.18
$T\lambda$	2.172	0.04969	0.1462	—	—
	2.2	0.05256	0.1461	1.671	3.16
	2.4	0.08228	0.1453	1.898	2.25
	2.6	0.1219	0.1442	2.068	2.10
	2.S	0.1730	0.1428	2.225	2.22
	3.0	0.2371	0.1411	2.382	2.42
	3.2	0.3156	0.1393	2.542	2.67
	3.4	0.4100	0.1372	2.706	2.95
	3.6	0.5220	0.1348	2.875	3.28
	3.8	0.6528	0.1321	3.050	3.68
	4.0	0.8040	0.1290	3.234	4.19
	4.2	0.9772	0.1254	3.429	4.88
T_{Nbp}	4.224	1.000	0.1250	3.454	4.98
	4.4	1.174	0.1213	3.640	5.86
	4.6	1.397	0.1163	3.873	7.44
	4.S	I.64S	0.1101	4.144	10.5
	5.0	1.929	0.1011	4.495	19.1
T_c	5.201	2.245	0.0696	5.589	—

Source: Ventura, G. and Risegari, L., *The Art of Cryogenics Low-Temperature Experimental Techniques,* Elsevier, Linacre House, Jordan Hill, Oxford, UK, First edition 2008 Copyright © 2008 Elsevier Ltd. All rights reserved.

Note: 1 atm = 760 torr = 1.013 bar = 1.013 × 103 Pa.

Table 8.2 The Recommended Values for the Latent Heat (L) of Liquid Helium as a Function of Temperatures at the Saturated Vapor Pressure

Temperature (K)	L (Joule/mole)	Temperature (K)	L (Joule/mole)	Temperature (K)	L (Joule/mole)
0.00	58.83	1.95	93.16	3.35	93.63
0.05	60.87	2.00	93.07	3.40	93.41
0.10	61.91	2.05	92.80	3.45	93.14
0.15	62.95	2.10	92.27	3.50	92.84
0.20	64.00	2.18	90.75	3.55	92.48
0.25	65.04	2.18	90.75	3.60	92.08
0.30	66.08	2.18	90.75	3.65	91.63
0.35	67.13	2.18	90.75	3.70	91.13
0.40	68.17	2.18	90.74	3.75	90.58
0.45	69.21	2.18	90.74	3.80	89.97
0.50	70.24	2.18	90.74	3.85	89.31
0.55	71.28	2.18	90.74	3.90	88.59
0.60	72.31	2.18	90.73	3.95	87.82
0.65	73.33	2.18	90.73	4.00	87.00
0.70	74.35	2.18	90.73	4.05	86.13
0.75	75.37	2.18	90.73	4.10	85.21
0.80	76.38	2.20	90.87	4.15	84.22
0.85	77.38	2.25	91.15	4.20	83.19
0.90	78.37	2.30	91.43	4.25	82.11
0.95	79.36	2.35	91.71	4.30	80.98
1.00	80.33	2.40	91.98	4.35	79.80
1.05	81.30	2.45	92.24	4.40	78.57
1.10	82.26	2.50	92.50	4.45	77.27
1.15	83.21	2.55	92.74	4.50	75.86
1.20	84.14	2.60	92.97	4.55	74.32
1.25	85.06	2.65	93.18	4.60	72.59
1.30	85.97	2.70	93.38	4.65	70/64
1.35	86.87	2.75	93.56	4.70	68.44
1.40	87.73	2.80	93.71	4.75	65.94
1.45	88.56	2.85	93.85	4.80	63.11
1.50	89.35	2.90	93.96	4.85	59.90
1.55	90.09	2.95	94.05	4.90	56.28
1.60	90.77	3.00	94.11	4.95	52.22
1.65	91.38	3.05	94.14	5.00	47.67
1.70	91.91	3.10	94.14	5.05	42.59
1.75	92.36	3.15	94.11	5.10	36.95

(Continued)

Table 8.2 (*Continued*) The Recommended Values for the Latent Heat (L) of Liquid Helium as a Function of Temperatures at the Saturated Vapor Pressure

Temperature (K)	L (Joule/mole)	Temperature (K)	L (Joule/mole)	Temperature (K)	L (Joule/mole)
1.80	92.72	3.20	94.05	5.15	29.34
1.85	92.98	3.25	93.94	—	—
1.90	93.13	3.30	93.80	—	—

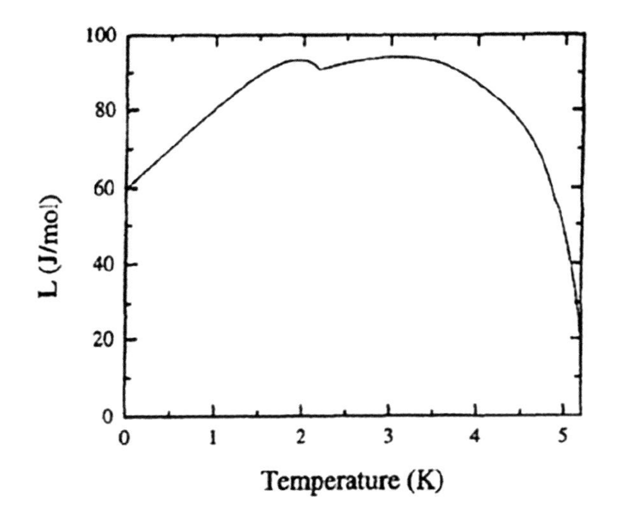

FIGURE 8.14 The recommended values for the latent heat (L) of liquid helium as a function of temperatures at the saturated vapor pressure. (From Donnelly, R.J., and Barenghi, C.F., *J. Phys. Chem. Ref. Data*, 27, 1998, 1217–1274.)

2. Refer to Table 8.2 and Figure 8.14 for the latent heat of helium as a function of temperature under its saturated vapor pressure

Step 2: Decide the temperature you need to reach and maintain.

Step 3: Determine the pump to be used.

Get the performance specifications of that pump, including the ultimate vacuum, the pumping speed as a function of inlet pressure, etc. Make sure that the ultimate pressure (vacuum) of the selected pump is lower than the saturated vapor pressure of the temperature you need to reach.

As an example, the performance curve of Edwards model XDS-35i scroll pump is illustrated in Figure 8.15, and it is a proper and cost-effective candidate for the 1.5 K cryogen-free system of discussion.

Step 4: Calculate the throughput.

Step 5: Estimate the intrinsic heat load to the sample.

Step 6: Calculate the net cooling power.

Multiply the latent heat by the throughput, and minus the intrinsic heat load, you get the net cooling power of the system.

The previously mentioned approach to estimate the net cooling power is based on the assumption that the inlet pressure of the pump is equal to the saturated pressure of

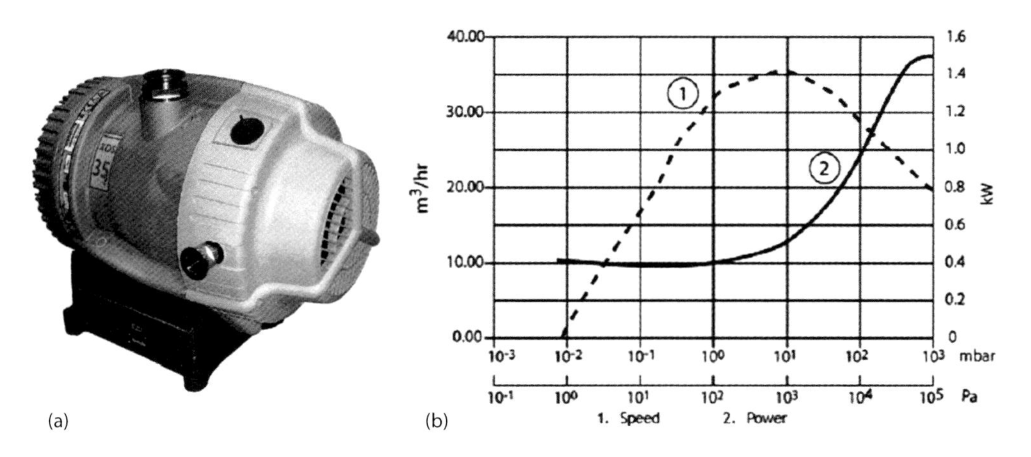

FIGURE 8.15 (a) Model XDS-35i scroll pump. (b) Model XDS-35i scroll pump performance. (Courtesy of Edwards Vacuum.)

the liquid helium. This approach needs some correction when a major pressure gradient exists inside a long pumping line with smaller conductance. Correction can be made by installing a vacuum gauge at the inlet port of the pump and calculating the throughput based on the **true** inlet pressure of the pump. This approach also gives the pressure gradient along the pumping line.

8.2.3 Continuous Flow-Type 1.5 K Cryogen-Free System [5]

Longer (or unlimited) holding time is highly desirable for many users and can be realized by a so-called continuous flow (or recirculating) system as illustrated in Figure 8.16.

Similar to the single-shot-type system, helium gas is introduced into the system from a high-pressure gas bottle as shown on Figure 8.17. The gas will be liquefied by the cryocooler. The condensed liquid helium will be further cooled down by reduction of its vapor pressure (i.e., pumping on the condensed LHe).

An alternative approach is to install a fixed amount such as a few Standard Temperature and Pressure (STP) liters of helium gas inside the system. The helium gas is first liquefied by the cryocooler and then circulated with a hermetic pump. Instead being vented to the air, the helium vapor is sent back to the system for recondensation and recirculating.

The continually operating system includes the following main components:

1. Vacuum jacket and first-stage radiation shield.
   The second stage (below 3 K) radiation shield can be installed as an option.
2. Two-stage PT or GM cryocooler with cold head, compressor, and high-pressure gas lines.
3. Heat exchangers.
   a. First-stage heat exchanger, which is a cylindrical bobbin made of OFHC copper (Figure 8.18)
      A 0.125" diameter copper capillary is wrapped around the bobbin and then attached to it with solder.

FIGURE 8.16 (a) General continuous-flow-type cryogen-free system. (b) Continuous-flow-type 1.5 K cryogen-free system. (Courtesy of Janis Research Company.)

Room temperature helium gas enters the cryostat from the top and is cooled down to the first-stage temperature inside the copper capillary.

Multiple bobbins can be installed since the capillary should be long enough (typically a few dozens of inches) for a full thermal exchange between the incoming helium gas and the first stage of the cryocooler.

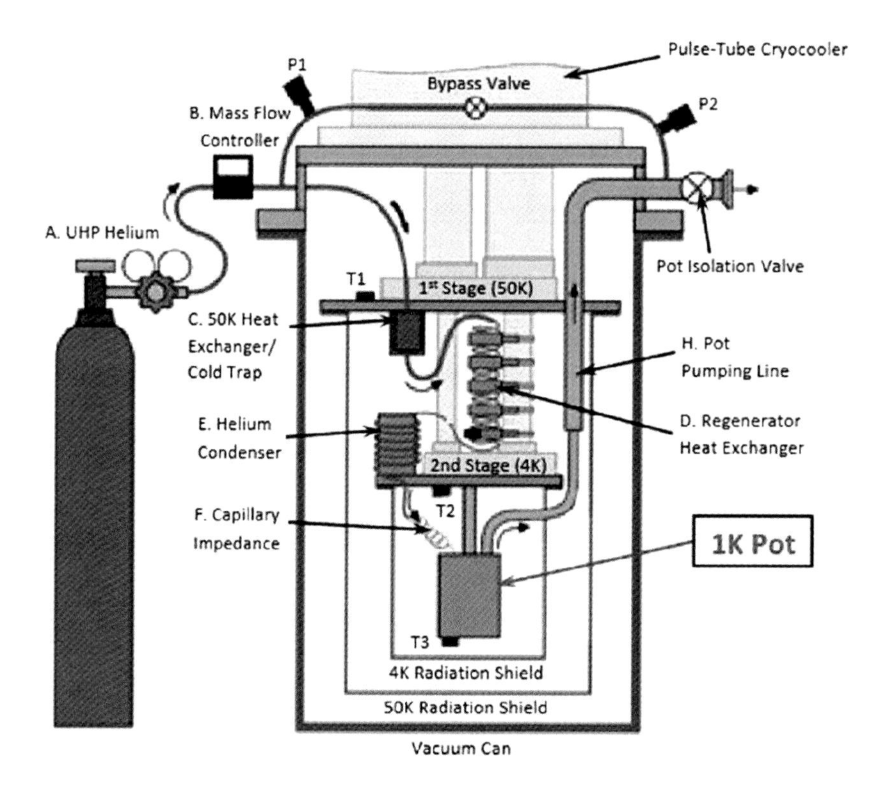

FIGURE 8.17 Continuous-flow-type 1.5 K cryogen-free system with helium gas filled from a high-pressure gas bottle. (From DeMann, A. et al., *Cryogenics*, 73, 60–67, 2016.)

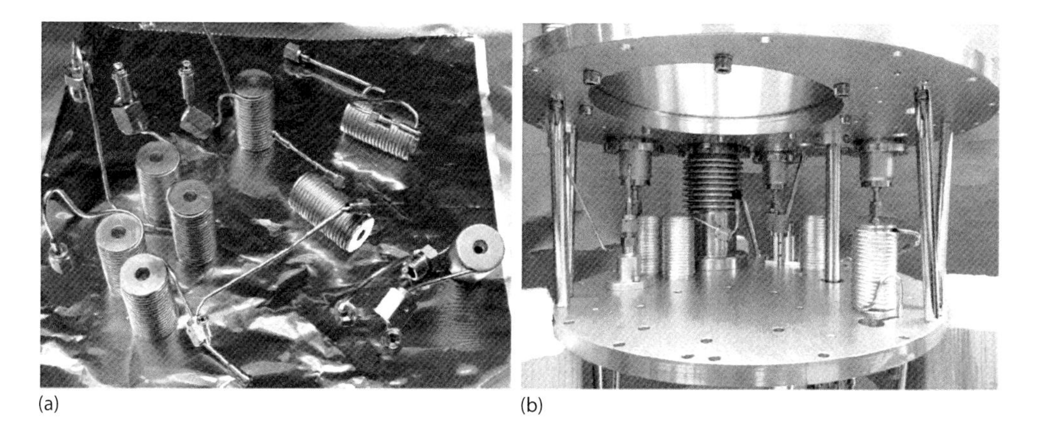

FIGURE 8.18 (a) Demountable heat exchangers. (b) Heat exchangers installed on the second-stage cold plate. (Courtesy of Janis Research Company.)

b. Second-stage regenerator heat exchanger on the PT cooler or GM cooler

 The PT cryocooler has an accessible second-stage regenerator. A stainless-steel precooling line (with approximate diameter of 0.93″) can be attached/soldered onto the second-stage regenerator as a heat exchanger. Excessive cooling power from the second-stage regenerator dramatically improves the helium condensation efficiency. Refer to **Section 6.2.6** for more details (Figure 8.19).

FIGURE 8.19 Distributed heat exchanger installed on a Cryomech pulse-tube second stage refrigerator. (Courtesy of Janis Research Company.)

This heat exchanger is extremely important for an optimized operation of the recirculating 1.5 K system. The heat exchanger utilizes the excessive cooling power of the second-stage regenerator and cools the incoming helium gas from the first-stage temperature (around 40 K) to 6 K or lower [7]. Assuming that the pressure of the returning helium gas is 1.0 atm, approximately 80% of the enthalpy of helium is removed by the heat exchanger when the helium gas is cooled from 40 K to 6 K. Therefore, the retuning helium gas adds very little heat load to the second stage, which allows smooth condensation of the helium. In this case, the helium gas can be condensed below 1.0 atm so that no compressor is necessary. This approach reduces the noise, vibration, and cost.

c. Second-stage heat exchanger

Heat exchangers (including copper bobbin and copper capillaries) similar to the heat exchangers mounted on the first-stage cold plate are installed on the second-stage cold plate.

The precooled Helium gas condenses into liquid in the second stage heat exchanger.

4. Condensing impedance and JT condensing unit

A cryogenic flow impedance is necessary to restrict the flow and generate a proper head pressure for efficient liquid condensation in the second-stage heat exchanger. The impedance can be a fixed impedance or a variable impedance with a needle valve as described **Section 6.2.5**.

The impedance controls the helium flow rate into the 1 K pot and achieves a balance between the amount of the incoming helium to the 1 K pot and that being pumped out from the 1 K pot.

Joule–Thomson effect takes place after the impedance and makes the condensed liquid helium colder. The cold liquid helium enters the 1 K pot and is depressurized there by an external pump for the base temperatures.

5. 1 K pot

   The 1 K pot is a canister usually made of OFHC copper. Liquid helium is collected inside this canister. The typical 1 K pot has capacity around 100 cm^3. A 1 K pot with much larger capacity is not uncommon when long "single-shot" holding time is needed.

   A pumping tube is installed between the 1 K pot at top of the cryostat. Superfluid film flow is usually not a major concern in this type of cryostat, and no film burner is needed.

   However, it is not uncommon to install an orifice at the joint between the bottom of the pumping line and the top of the 1 K pot for superfluid restriction.

6. Mini-Gas Handling System (GHS) with hermetic pump, helium storage, cold trap, and accessories.

   The helium gas needs to be sent back to the cryostat (called "circulation") for recondensation.

   A GHS is needed to complete the circulation loop.

Notes:

- The hermetic pump should have the leak rate less than 1.0×10^{-7} cc.atm/sec. Any air leak will cause blockage inside the cryogenic impedance.
- Oil-free pump, such as the Scroll pump, is highly desirable.
- In order to speed up the helium condensation at the second stage, a compressor (pressure booster) can be, but not necessarily, installed to enhance the head pressure.
- Liquid nitrogen temperature charcoal trap is often included even when an oil-free pump is used because we always have air diffusions through joints. This cold trap is kept at 78 K by being submerged in liquid nitrogen during the operation. Helium gas will pass the cold trap and all air (if any) will be trapped by the charcoal.

  As an example, the Janis Research Company model PTSHI-950-LT cryostat employs the Cryomech model PT-415 PT cryocooler. The system can be operated over a temperature range from ~1.5 K to 300 K.
- The heat load from the returning helium needs to be considered for the cooling power estimate of the continuous flow type 1.5 K system.

  Assuming that the cryocooler absorbs all the heat of condensation when the gaseous helium condenses into the liquid phase at 4.2 K, that is, what enters the 1 K pot after the condensing impedance or JT condensing unit, is all *4 K helium liquid*. Part of the 1 K pot cooling capacity is needed to cool the 4 K liquid helium down to its own temperatures of the 1 K pot (e.g., 1.5 K).

  Listed in Table 8.3 and illustrated in Figure 8.20 is the liquid helium heat capacity from 1.0 K to 5.0 K. With a known quantity of the throughput from the pump, the user can estimate the power needed to cool the incoming 4 K liquid helium to the 1 K pot temperature. Deduct this amount from the total cooling power for the net cooling power estimate.
- Typical cooling power of a Janis 1.5 K continuous flow cryostat is illustrated in Figures 8.20 and 8.21.

Table 8.3 Heat Capacity of Liquid Helium at Saturated Pressure

Temperature (K)	Heat Capacity (Joule/mole. K)	Temperature (K)	Heat Capacity (Joule/mole. K)	Temperature (K)	Heat Capacity (Joule/mole.K)
1.0	0.4154	2.10	28.64	2.8	9.374
1.1	0.7658	2.15	36.89	3.0	9.944
1.2	1.297	2.1760	55.69	3.2	10.75
1.3	2.057	2.1763	57.84	3.4	11.78
1.4	3.092	2.1765	60.68	3.6	13.00
1.5	4.468	2.1767	66.40	3.8	14.37
1.6	6.285	2.1769	45.69	4.0	15.96
1.7	8.678	2.1780	32.64	4.2	17.88
1.8	11.81	2.2	16.73	4.4	20.3
1.9	15.90	2.3	10.78	4.6	23.6
2.0	21.28	2.4	9.515	4.8	29.8
2.05	24.59	2.6	9.066	5.0	44.7

Source: Donnelly, R.J. and Barenghi, C.F., *J. Phys. Chem. Ref. Data*, 27, 1998.

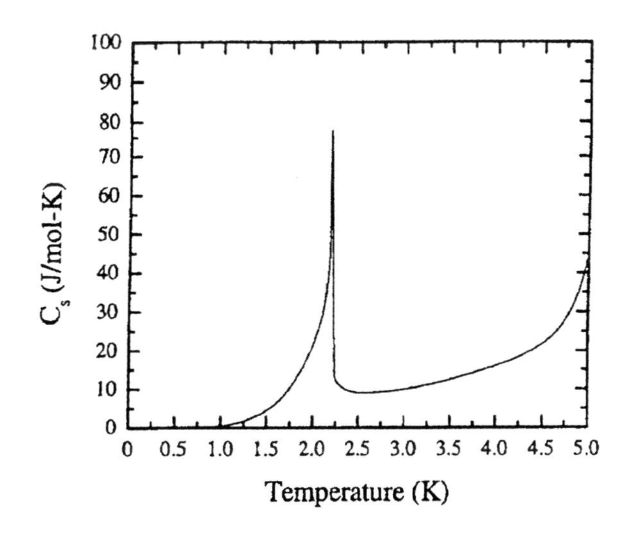

FIGURE 8.20 Specific heat of liquid helium. (From Donnelly, R.J., and Barenghi, C.F., *J. Phys. Chem. Ref. Data*, 27, 1998, 1217–1274.)

8.2.4 Alternative Continuous-Flow-Type Cryogen-Free System

The advanced cryocooler has provided a cooling platform for helium liquefication. Sub-4 K continuous-flow-type cryogen-free systems with various designs were developed upon this platform.

Figure 8.22a depicts a Janis Research Company open-loop model ST-500 cryostat. The conventional operation is to transfer liquid helium into the ST-500 cryostat from

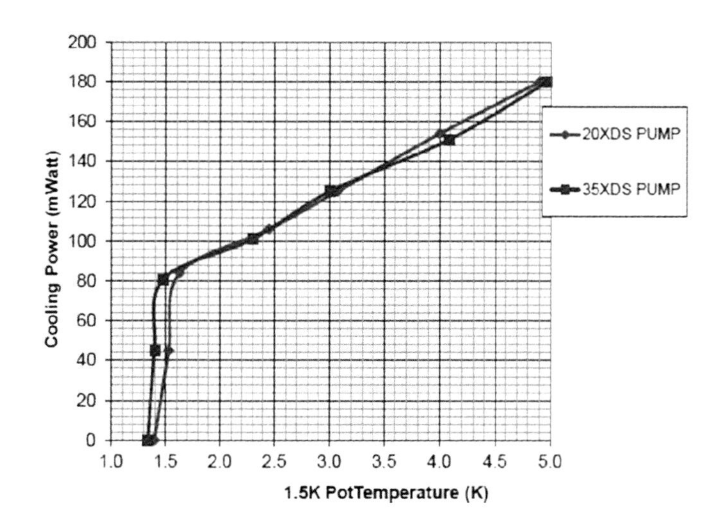

FIGURE 8.21 Typical cooling power. (Courtesy of Janis Research Company.)

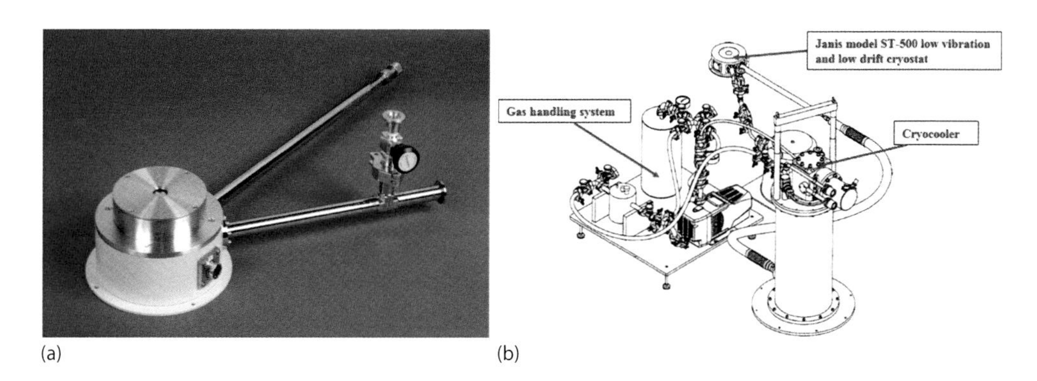

FIGURE 8.22 (a) Model ST-500 open-loop cryostat. (Courtesy of Janis Research Company.) (b) Model ST-500 is turned into closed-loop cryostat. (Courtesy of Janis Research Company.)

a helium storage dewar for cooling. The helium is then vented to the atmosphere. This cryostat can be turned into a closed-loop system as shown in Figure 8.22b.

A fixed amount of helium is confined in the system. Helium gas is first liquefied by a two-stage cryocooler, then recirculated with a hermetic pump (e.g., Edwards model XDS scroll pump) and cools the cryostat below 2 K.

This approach not only eliminates the need of liquid helium but also physically decouples the cryocooler from the cryostat. Vibration measurements were performed on the sample holder with a laser vibrometer. The vibration contributed from the cryocooler was not detected. This continuous flow sub-4 K "cryogen-free" experimental platform of low vibration can be applied in the fields of microphotoluminescence, micro-Raman, STM, and other vibration-sensitive techniques.

Acknowledgments

The contributor would like to send a special acknowledgment to his Janis colleague, Dan Logan, for his tremendous help. Dan is an expert of 4 K systems, either conventional "wet" system or cryogen-free one. It is fair to say that Dan is truly the co-contributor of this chapter.

Thanks to Xi Lin for his important statement regarding the importance of the cryogen-free 1.5 K system described in this chapter.

Sincere acknowledgment goes to the contributor's Janis colleagues:

Thanks to Vladimir Luppov for his review of this chapter and comments for improvement.

Thanks to Lewis Bobb for providing all the cryocooler system drawings.

Last but not the least, special thanks goes to Ann Carroll for her constant help with software difficulties, reference searches, request of copyrights, etc.

References

1. Ventura, G. and Risegari, L.: *The Art of Cryogenics Low-Temperature Experimental Techniques*, Elsevier, Linacre House, Jordan Hill, Oxford, UK, 1st edition 2008 Copyright © 2008 Elsevier Ltd. All rights reserved.
2. Donnelly, R.J. and Barenghi, C.F., The observed properties of liquid helium at the saturated pressure, *Journal of Physical and Chemical Reference Data*, 27(8), 1998.
3. http://www.cryomech.com.
4. http://www.shicryogenics.com/products/4k-cryocoolers.
5. Paine, C.T., Naylor, B.J., Prouve. T., *Cryogenics*, NASA Tech Briefs NPO-48355, 2013.
6. DeMann, A., Mueller, S., and Field, S., 1 K cryostat with sub-millikelvin stability based on a pulse-tube cryocooler, *Cryogenics*, 73, 60–67, 2016.
7. Wang, C., Extracting cooling from pulse tube and regenerator in a 4 K pulse tube cryocooler, *Cryocoolers 15*, ICC Press, Boulder, CO, 2009, pp. 177–184.

9. Cryogen-Free 3He Systems

Zuyu Zhao

Helium-3 (3He) is practically the best cryogen for reaching temperatures below 1 K.

As an isotope of Helium-4 (4He), 3He has two protons and one neutron in its atomic nucleus.

Natural abundance of 3He on the earth is approximately 0.00013% [1], and most of 3He on earth comes from the beta decay of tritium (also called Hydrogen-3).

Tritium has a half-life of approximately 12 years, and it decays into 3He by beta decay per Equation 9.1:

$$(3/1)\mathrm{T}^{®} \rightarrow (3/2)\mathrm{He}(1+) + \mathrm{e}^{-} + \nu_{-e}\,(\text{antineutrino}) \tag{9.1}$$

Tritium (and Helium-3) is produced in a nuclear reactor by the neutron activation of Lithium-6.

$$(6/3)\mathrm{Li} + \mathrm{n} \rightarrow (4/2)\mathrm{He} + (3/1)\mathrm{T} \tag{9.2}$$

9.1 3He Phase Diagram

Figure 9.1 is the 3He phase diagram in zero magnetic fields. It shows the location of three different phases, that is, solid, liquid, and gas, as a function of pressure (P) and temperature (T).

The following unique properties of 3He are illustrated:

- 3He does not solidify under normal (1.0 atm) pressure even at zero kelvin.
  The minimum pressure for 3He to solidify is 29.3 bar (2.93 MPa) near 0.32 K.
- ***Melting curve***: Also called *solid–liquid coexistence curve*. This curve illustrates the P–T relationship when the solid and liquid phases coexist. It has a pressure minimum of 29.3 atm at 0.32 K.
- ***Curve of evaporation***: Also called *liquid–vapor coexistence curve*. This curve illustrates the saturated vapor pressure as a function of temperature of saturated liquid 3He.
- Gaseous phase to liquid phase transition takes place at 3.15 K under normal (i.e., 1.0 atm) pressure.
  Lower temperatures can be realized along the *curve of evaporation* by reducing the saturated vapor pressures of liquid helium 3He.
- 3He has no triple point.
- 3He has a critical temperature of 3.315 K, that is, 3He does not have a liquid phase above the critical temperature, even under high pressures.
  A cryocooler that can reach below 3 K is needed for direct condensation of 3He.
- 3He becomes *superfluid* below 2.49 mK along the melting curve.

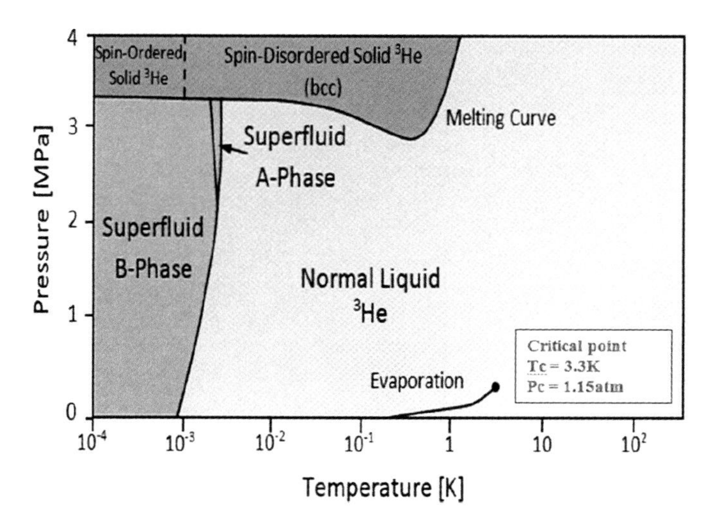

FIGURE 9.1 Phase diagram of 3He.

9.2 Introduction to Selected Properties of Liquid 3He

This section presents a brief summary on some commonly used thermodynamic properties of liquid 3He for pragmatic engineering applications, such as saturated vapor pressure (SVP), density, specific heat, latent heat, and thermal conductivity.

Most of our presentation will focus near the temperatures range under the SVP of liquid 3He.

9.2.1 Saturated Vapor Pressure

The SVP of liquid 3He as a function of temperature is shown in Table 9.1 [2].

Table 9.1 Saturated Vapor Pressure of Liquid Helium-3 as Function of Temperature

Temperature (K)	Pressure (mmHg)	Temperature (K)	Pressure (mmHg)
3.200	762.5225	0.750	1.9939
3.150	723.2165	0.700	1.3490
3.100	685.3463	0.650	0.8701
3.000	613.7502	0.600	0.5292
2.900	547.4773	0.550	0.2990
2.800	486.2437	0.500	0.15379
2.700	429.7685	0.470	0.09749
2.600	377.8150	0.450	0.06992
2.500	330.1802	0.420	0.04040
2.400	286.6534	0.400	0.02690
2.300	247.0504	0.370	0.01364
2.200	211.1909	0.350	0.00819
2.100	178.8990	0.340	0.00623
2.000	150.0011	0.330	0.00466
1.900	124.3243	0.320	0.00344
1.800	101.6952	0.310	0.00249
1.700	81.9384	0.300	0.00177
1.600	64.8763	0.290	0.00123
1.500	50.3283	0.280	0.00084
1.400	38.1103	0.270	0.00055
1.300	28.0343	0.260	0.00036
1.200	19.9074	0.250	0.00022
1.100	13.5318	0.240	0.00013
1.000	8.7030	0.230	0.00008
0.950	6.8029	0.220	0.00004
0.900	5.2097	0.210	0.00002
0.850	3.8959	0.200	0.00001
0.800	2.8334		

Source: International Temperature Scale of 1990 [ITS-90].

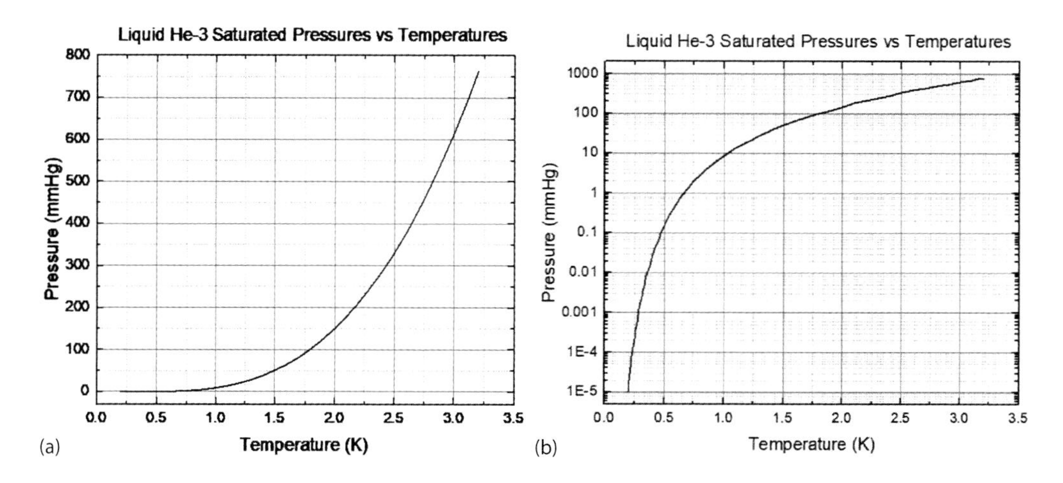

FIGURE 9.2 (a) Plot of saturated vapor pressure of liquid 3He as function of temperature in linear scale. (b) Saturated vapor pressure of liquid 3He as function of temperature plotted in semi-logarithm scale.

Temperature and pressure have a one-to-one correspondence on the curve of evaporation. Lower temperatures can be achieved by reducing the SVP of liquid 3He (Figure 9.2).

The SVP of liquid 3He is very low when the temperature reaches the millikelvin range. Due to the lack of an easy and cost-effective approach to reach such low pressures, the lowest temperature for most 3He cryostats is practically limited between 250 and 300 mK.

9.2.2 Density

The density of liquid 3He is a function of both temperature and pressure. Table 9.2 [3–5] has listed the volume density of liquid 3He as a function of temperature under its SVP. 3He is usually condensed between 1.2 K (wet system) and 3 K (dry system) in most of 3He cryostats. Since the volume density of liquid 3He at its condensing temperature is larger than that at the base temperature (say, 300 mK) this information can be used to determine the volume of the 3He pot needed to contain the volume of liquid 3He when it is first condensed (Figure 9.3).

Extensive research has been done to study the density, or the expansion coefficients, of liquid 3He as a function of temperatures under fixed pressures, as well as a function of pressures under certain temperatures.

Table 9.3 lists the thermal expansivity of liquid 3He from 10 to 600 mK over the pressure range from zero to 3.0 MPa.

Liquid 3He reaches a density maximum approximately near 0.5–0.6 K under its SVP as depicted in Figure 9.4 where the density of liquid 3He is normalized to its value at 1.2 K under different pressures. The density maximum is suppressed as the pressure increases.

This feature has a direct impact on 3He system design due to the so-called hydrostatic effect on the 3He performance [7].

Consider a cylinder containing saturated liquid 3He. The top layer of liquid 3He cools down first when the vapor pressure of the liquid 3He is reduced. The cold layer sinks since it usually has higher density. At the same time, the warmer layer moves to the top

Table 9.2 Volume Density of 3He as a Function of Temperature under Its Saturated Vapor Pressure (SVP)

Temperature (K)	Density (gram/cm^3)	Mole Volume (cm^3)
0.0	0.08235	36.43
1.0	0.08185	36.65
1.2	0.08147	36.82
1.4	0.08093	37.07
1.6	0.08029	37.36
1.8	0.07924	37.86
2.0	0.07801	38.46
2.2	0.07645	39.24
2.4	0.07448	40.28
2.6	0.07200	41.67
2.8	0.06882	43.59
3.0	0.06462	46.43
3.1	0.06193	48.44
3.2	0.05861	51.59

Source: Rives, J.E. and Meyer, H., *Phys. Rev. Lett.*, 7, 217, 1961; Grilly, E.R., *J. Low Temp. Phys.*, 4, 615–635, 1971; Gibbons, R.M. and Nathan, D.I., *Thermodynamic Data Of Helium-3*, Air Products, Technical Report AFML-TR-67-175, October 1967.

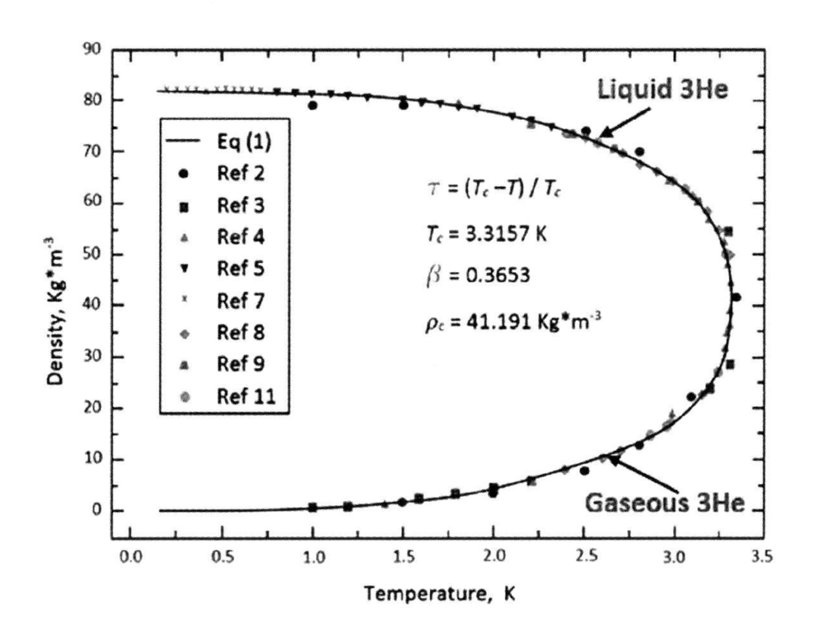

FIGURE 9.3 Plot of volume density of liquid 3He as a function of temperature under its saturated vapor pressure. (Huang, Y.H. et al., June 5, 2018, private communications.)

Table 9.3 Thermal Expansivity (α) of Liquid 3He, Listed as $-10^3\alpha$ (K^{-1})

	Pressure (MPa)										
T (mK)	0	0.3	0.6	0.9	1.2	1.5	1.8	2.1	2.4	2.7	3.0
10	1.62	1.53	1.47	1.42	1.39	1.36	1.35	1.34	1.33	1.33	1.33
20	3.25	3.06	2.91	2.81	2.73	2.68	2.62	2.60	2.58	2.61	2.68
30	4.88	4.55	4.26	4.04	3.89	3.82	3.80	3.79	3.78	3.81	3.90
40	6.45	5.91	5.46	5.13	4.90	4.81	4.79	4.81	4.82	4.85	4.92
50	7.82	7.08	6.50	6.05	5.76	5.65	5.63	5.65	5.69	5.72	5.75
60	8.98	8.09	7.39	6.86	6.52	6.37	6.32	6.34	6.36	6.39	6.43
80	10.73	9.68	8.78	8.17	7.73	7.45	7.32	7.27	7.27	7.31	7.37
100	11.92	10.77	9.76	9.02	8.53	8.17	7.96	7.84	7.82	7.86	7.96
120	12.62	11.42	10.34	9.55	9.00	8.60	8.35	8.17	8.12	8.18	8.30
140	13.02	11.74	10.64	9.84	9.24	8.82	8.53	8.35	8.30	8.35	8.46
160	13.16	11.84	10.77	9.94	9.32	8.88	8.59	8.41	8.35	8.38	8.50
180	13.12	11.78	10.73	9.89	9.27	8.83	8.53	8.35	8.28	8.31	8.44
200	12.90	11.58	10.55	9.71	9.09	8.66	8.38	8.21	8.14	8.16	8.28
250	11.60	10.48	9.56	8.84	8.33	7.97	7.76	7.66	7.61	7.63	7.70
300	9.68	8.88	8.24	7.73	7.38	7.15	7.03	6.97	6.98	7.04	7.13
350	7.50	7.10	6.79	6.62	6.47	6.38	6.35	6.38	6.44	6.52	6.64
400	5.08	5.34	5.53	5.66	5.73	5.78	5.87	5.96	6.05	6.15	6.27
450	2.78	3.82	4.50	4.93	5.20	5.37	5.58	5.72	5.88	6.00	6.14
500	0.70	2.40	3.64	4.32	4.77	5.09	5.34	5.55	5.76	5.93	6.10
600	N/A	0.00	2.09	3.13	3.86	4.33	4.70	5.04	5.34	5.63	5.89

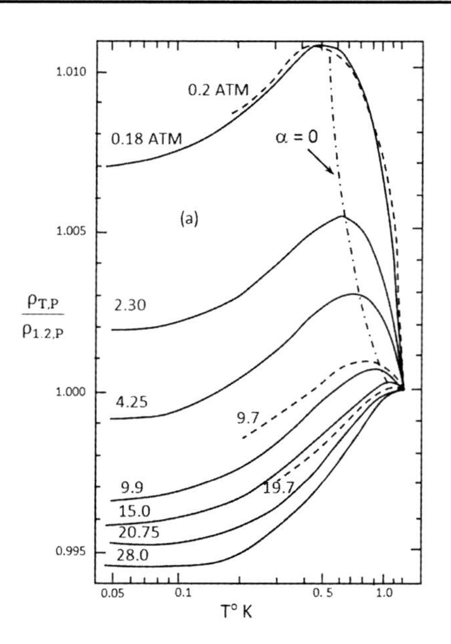

FIGURE 9.4 Liquid 3He density maximum near 0.5 K under "lower" pressures. (From Boghosian, C. et al., *Phys. Rev.*, 146, 1966.)

and gets cooled when the vapor pressure is further reduced. This convection cooling process will slow down and even stop when the liquid 3He reaches 0.5 K because the liquid 3He that is cooled below 0.5 K at a later stage will remain above the liquid that was cooled to 0.5 K due to their density difference.

Conduction cooling will become the only cooling mechanism in this case.

High-field magnets usually have small room temperature bores. All 3He cryostats that fit into such magnets have long cylindrical 3He pots (often made of **stainless steel** to minimize the eddy current) with small diameters (typically 17 mm). Conduction cooling becomes highly inefficient in that case since stainless steel is a very poor thermal conductor at that temperature range. Convection remains the main approach of heat exchange.

Hydrostatic pressure will make the convection cooling very inefficient as reported by Takacs et al. [7].

The SVP of liquid 3He at 0.5 K is approximately 0.2 mmHg, which is equivalent to 3.2 cm hydrostatic pressure of liquid 3He. Attention must be paid to the design if the liquid 3He column is much taller than 3.2 cm because the hydrostatic pressure may generate a 0.5 K layer of liquid 3He. This layer of liquid 3He may stay at the bottom of liquid 3He column for long time before it is cooled down.

Copper wires may be installed inside the 3He pot to improve the conductive heat exchange.

An alternative way to improve the heat exchange is to thin down the wall of the stainless steel 3He pot and then copper plate the outer surface of the thinned-down section. A narrow gap should be left to reduce eddy currents when changing the magnetic field (refer to Figure 9.5).

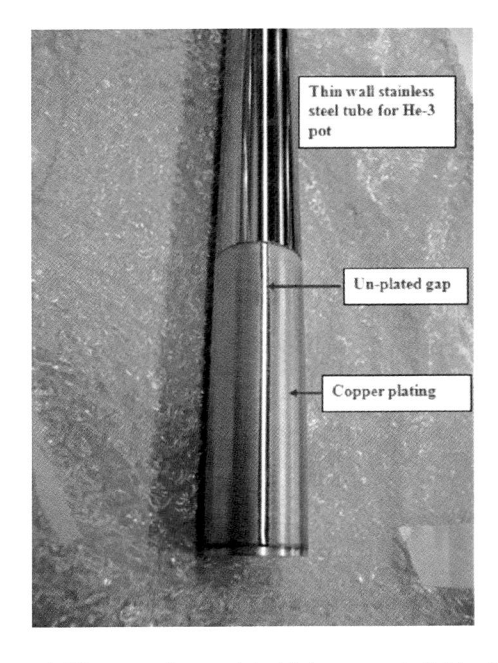

FIGURE 9.5 Design of special 3He pot to be used in high magnetic fields. (Courtesy of Janis Research Company.)

9.2.3 Latent Heat

3He cryostat operation includes two basic steps:

- 3He condensation, that is, change the 3He from the gaseous phase to the liquid phase; and
- Reduce the vapor pressure of the liquid 3He to reach lower temperatures.

Both steps involve a first-order phase transition that obeys the **Clausius–Clapeyron** relation:

$$\mathbf{dP/dT} = (\mathbf{S}_L - \mathbf{S}_V)/(\mathbf{V}_L - \mathbf{V}_V) = \mathbf{L/T}(\mathbf{V}_L - \mathbf{V}_V) \tag{9.3}$$

where:

dP/dT is the slope of the tangent on the liquid–vapor coexistence curve
$\mathbf{S}_L$ is the specific entropy (commonly in units of molar entropy) of 3He liquid at temperature *T*,
$\mathbf{S}_V$ is the specific entropy (commonly in units of molar entropy) of 3He saturated vapor at temperature *T*,
L is the specific latent heat (commonly in units of molar latent heat) of the saturated 3He at temperature *T*,
T is the 3He temperature,
$\mathbf{V}_L$ is the specific volume (commonly in units of molar volume) of 3He liquid at temperature *T*, and
$\mathbf{V}_V$ is the specific volume (commonly in units of molar volume) of 3He saturated vapor at temperature *T*.

Latent heat, also called heat of evaporation, is the amount of energy (or enthalpy) removed from the liquid during the process of turning the liquid phase to the gaseous phase.

Heat of condensation is the amount of energy (or enthalpy) transferred from the gas to the liquid during the process of turning the gaseous phase to the liquid phase.

Specific units in moles are often used.

The latent heat and heat of condensation are equal to each other in value at any specific temperature.

Since the 3He cryostat operation involves the liquid–vapor transition only, we are mostly interested in the latent heat under the SVP of the liquid, that is, the latent heat along the liquid–vapor coexistence curve as shown in Figure 9.1.

Latent heat of liquid 3He under its SVP is shown in Table 9.4. Recall that the latent heat of helium at its critical temperature is zero (Figure 9.6).

The latent heat is often used to estimate the performance of 3He systems. Thus, for a known amount of liquid and a known heat load into the 3He liquid, the user can estimate the holding time of the 3He cryostat at certain fixed temperatures.

Table 9.4 Latent Heat of Liquid 3He under SVP

Temperature (K)	Latent Heat of Saturated Liquid 3He (Joule/mole)	Temperature (K)	Latent Heat of Saturated Liquid 3He (Joule/mole)
0.1	22.12	1.4	43.2
0.2	24.0	2.2	46.3
0.3	26.06	2.6	44.8
0.4	28.13	3.0	38.0
0.54	30.8	3.2	24.7
0.8	35.2	3.3	14.1
1.0	38.1	3.315	0

Source: Greywall, D.S., *Phys. Rev. B*, 27, 1983.

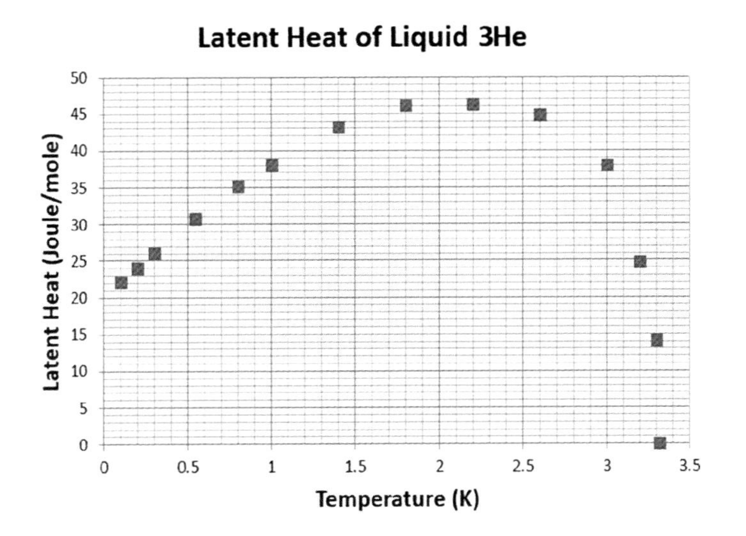

FIGURE 9.6 Plot of latent heat of liquid 3He under SVP. (From Huang, Y.H., *Phys. Procedia*, 67, 582–590, 2015.)

9.2.4 Specific Heat

The specific heat of liquid 3He is a function of temperature and pressure, and we are most interested in the specific heat of liquid 3He under the SVP pressure as listed in Table 9.5 and Figure 9.7.

9.2.5 Thermal Conductivities

Based on the measured data from the saturated liquid 3He, Huang et al. developed the following empirical formula in the temperature range of 3 mK to 3.3175 K (3He critical temperature) [14] (Figure 9.8).

Table 9.5 Specific Heat of Liquid 3He as a Function of Temperature under Its Saturated Vapor Pressure

Temperature (K)	Specific Heat (C_v) of Saturated Liquid 3He (Joule/mole K)	Temperature (K)	Specific Heat (C_v) of Saturated Liquid 3He (Joule/mole K)
0.005	0.1139	0.45	3.2433
0.01	0.2278	0.50	3.2973
0.02	0.4514	0.60	3.4170
0.04	0.8780	0.70	3.5717
0.06	1.2562	0.80	3.7621
0.08	1.5788	0.90	3.9799
0.10	1.8582	1.00	4.2152
0.15	2.4085	1.20	4.7057
0.20	2.7203	1.60	5.6909
0.25	2.9090	1.80	6.1664
0.30	3.0329	2.00	6.2220
0.35	3.1211	2.50	7.6413
0.40	3.1876	N/A	N/A

Source: Greywall, D.S., *Phys. Rev. B*, 27, 1983; *Phys. Rev, B*, 31, 2675, 1985; *Phys. Rev. B*, 33, 7520, 1986; Halperin, W.P., and Pitaevskii, L. P. (Eds.), *Helium Three*, Elsevier, Amsterdam, the Netherlands, 1990, http://ltl.tkk.fi/research/theory/helium.html; Dobbs, E.R., *Helium Three*, Oxford University Press, 2000.

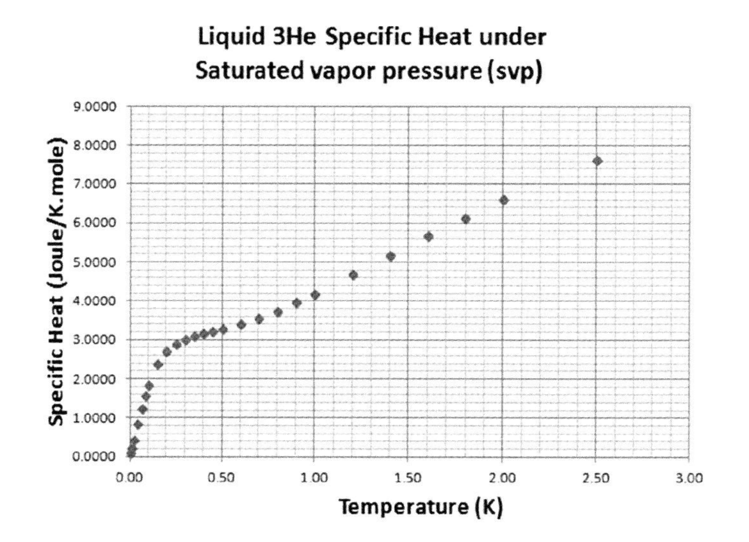

FIGURE 9.7 Plot of specific heat of liquid 3He as a function of temperature under its saturated vapor pressure. (From Greywall, D.S., *Phys. Rev. B*, 27, 1983; *Phys. Rev, B*, 31, 2675, 1985; *Phys. Rev. B*, 33, 7520, 1986; Halperin, W.P., and Pitaevskii, L. P. (Eds.), *Helium Three*, Elsevier, Amsterdam, the Netherlands, 1990, http://ltl.tkk.fi/research/theory/helium.html; Dobbs, E.R., *Helium Three*, Oxford University Press, 2000.)

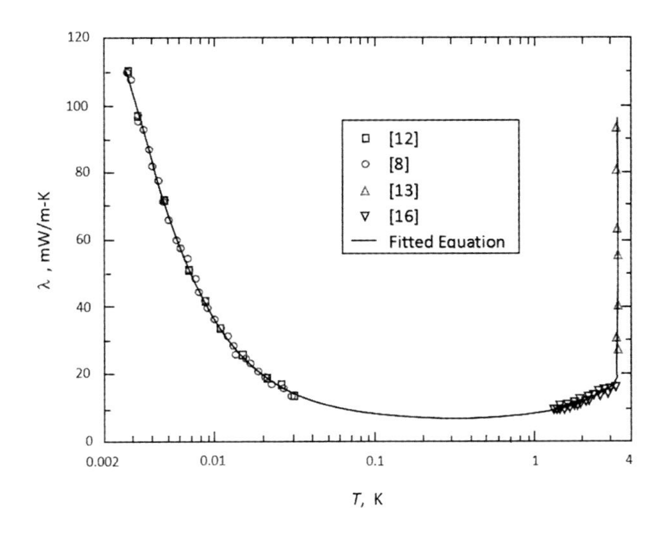

FIGURE 9.8 Plot of thermal conductivity of saturated liquid 3He. (From Huang, Y.H. et al., *AIP Conf. Proc.*, 1434, 1849, 2012.)

$$\lambda_{\text{sat}} = C_1 + C_2\tau + C_3/\tau + C_4\tau^2 + C_5/\tau^2 + C_6\tau^3 + C_7/\tau^3 \tag{9.4}$$

where:

λ_{sat} is the thermal conductivity of saturated liquid 3He, that is, under the SVP; and

τ is the non-dimensional temperature, $\tau = \ln(1 - T/T_c)$, where T_c is the 3He critical temperature 3.3157 K.

Least-squares fit to Equation 9.4 upon the reference data collected by the authors give the coefficients:

$C_1 = 1.00054 \times 10^1$

$C_2 = -8.81 \times 10^{-3}$

$C_3 = 7.83247 \times 10^{-5}$

$C_4 = -2.16839 \times 10^{-3}$

$C_5 = 5.0133 \times 10^{-8}$

$C_6 = -1.97967536 \times 10^{-4}$

$C_7 = 3.61 \times 10^{-11}$

9.2.6 Viscosity

Multiple measurements on the viscosity of saturated liquid 3He, η, were performed [15,16]. It is approximately proportional to T^{-1} above 1 K, then becomes a function of T^{-2} below 1 K. It increases rapidly when the temperature decreases in the mK range as illustrated in Figure 9.9.

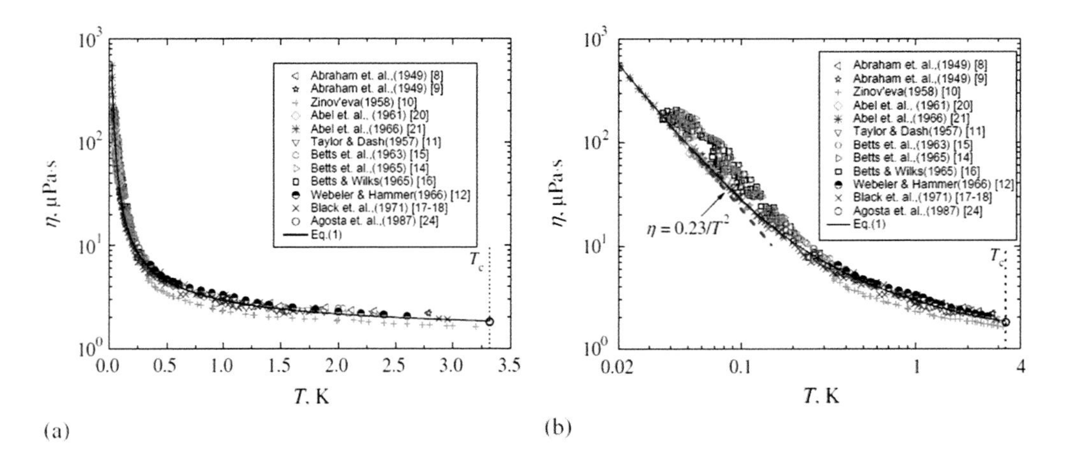

FIGURE 9.9 Viscosity of saturated liquid 3He: (a) Semi-Logarithmic scale. (b) Logarithmic scale. (From Huang, Y.H. et al., *AIP Conf. Proc.*, 1434, 1849, 2012.)

Viscous liquid 3He will have difficulties flowing in the traditional long and narrow counterflow heat exchangers at very low temperatures. Solutions will be discussed in Chapter 11: Cryogen-free Dilution Refrigerator.

Huang et al. did extensive surveys on the measured data [14] and fit the following formula of the viscosity of saturated liquid 3He below its critical temperature.

$$\eta_{sat} = C_1 / T^2 + C_2 / T^{1.5} + C_3 / T + C_4 \tag{9.5}$$

where η_{sat} is the viscosity of saturated liquid 3He (in Pa s),

T is the temperature (in K),

$$C_1 = 2.897 \times 10^{-7} \text{ Pa s K2},$$

$$C_2 = -7.02 \times 10^{-7} \text{ Pa s K1.5},$$

$$C_3 = 2.012 \times 10^{-6} \text{ Pa s K},$$

$$C_4 = 1.323 \times 10^{-6} \text{ Pa s}.$$

Huang et al. also did extensive surveys on other thermodynamic property data of liquid 3He during the past a few decades [8,14,17]. They have developed empirical equations and programs for data simulation that cover a wide range of pressures and temperatures. Interested readers can refer to *Yonghua Huang, Guobang Chen, Vincent Arp. Thermal Property Package of Helium-3: He3Pak. version 2.0.1, 2016. Shanghai Jiao Tong University and Zhejiang University* as well as the ULR link: http://www.htess.com/he3pak.htm regarding the program.

Readers may also refer to the book "***Helium Three***" by E. R. Dobbs [13] for an excellent summary on the properties of 3He.

9.3 Cryogen-Free 3He Refrigerators

Due to its simplicity, a 3He refrigerator using 3He as coolant is one of the most popular methods to reach temperatures below 1.0 K. Similar to the 1.5 K system (pumped 4He), the basic working principle of the 3He refrigerator is to reduce the vapor pressure of liquid 3He to reach lower temperatures. The main difference is that the coolant is 3He this time.

Liquid 3He is not commercially available. The refrigerator needs to first generate liquid 3He by itself before the evaporative cooling. In other words, all 3He refrigerators include two-step operations: Step 1 is to condense (i.e., liquefy) 3He, and Step 2 is to reduce the vapor pressure of liquid 3He and to cool it down to lower temperatures.

Two basic types of 3He refrigerators exist, as illustrated in Figure 9.10, known as the *Single-Shot* system, and the other is the Continuously Circulating system.

The single-shot system often uses charcoal as the (cryo) pump. The refrigerator cools down when the charcoal adsorbs 3He vapor and reduces the SVP of liquid 3He. The cooling process will stop when all the 3He is adsorbed. Then 3He needs to be recondensed.

An external hermetically sealed pump is employed in the continuously circulating 3He refrigerator to reduce the SVP of liquid 3He. The 3He gas is then sent back to the refrigerator for recondensation. It is a recirculating system, and the base temperature can be kept for an unlimited period of time.

9.3.1 Single-Shot 3He Refrigerator

9.3.1.1 Activated Charcoal and Its Basic Properties as an Adsorbent

Activated charcoal is light weight black carbon and ash residue hydrocarbon with water being removed at very high temperatures, as shown in Figure 9.11a.

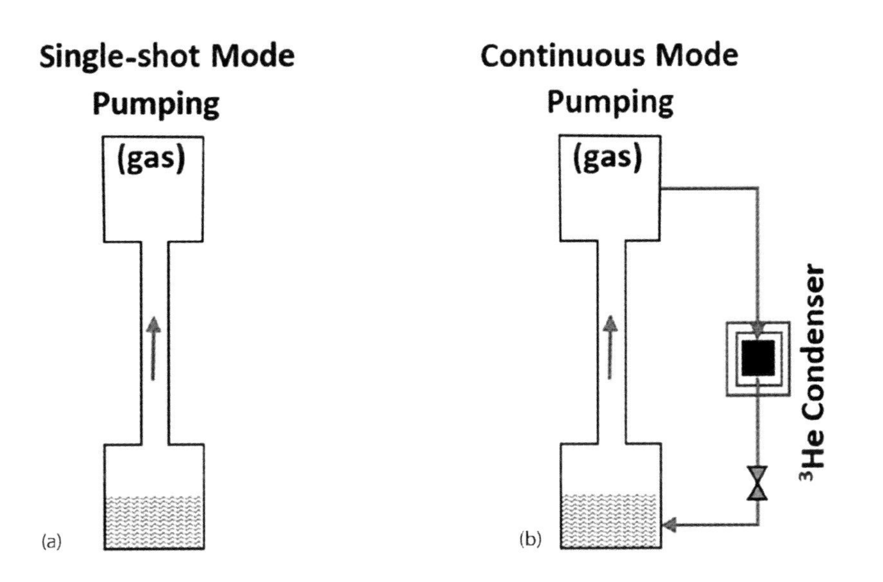

FIGURE 9.10 Two basic steps of 3He refrigerator operation: (a) Single-shot mode. (b) Continuous mode.

FIGURE 9.11 (a) Regular charcoal. (b) Synthetic charcoal.

A similar material called bituminous charcoal is also a good choice at helium temperature.

Synthetic charcoal, as shown in Figure 9.11b, is also commercially available. It is a highly uniform spherical activated carbon with less dust. Due to its small size, synthetic charcoal has a higher filling factor.

All three types of charcoal have enormous surface areas. For example, synthetic charcoal has surface areas of order of a few hundred to thousands of square meters per gram.

Our discussion will be focused on the activated charcoal since it is the most popular adsorbent for helium gas. It is often used as a cryopump at low temperatures.

9.3.1.1.1 Adsorption of 4He Vapor Interaction between the gas molecules and the charcoal comes from the Van der Waals interaction between the oscillating electrical dipoles. It can either be attraction or repulsion, as illustrated by Lennard–Jones potential in Figure 9.12.

The Lennard–Jones model consists of two "parts": a steep repulsive term and smoother attractive term. The Lennard–Jones potential is given by Equation 9.6:

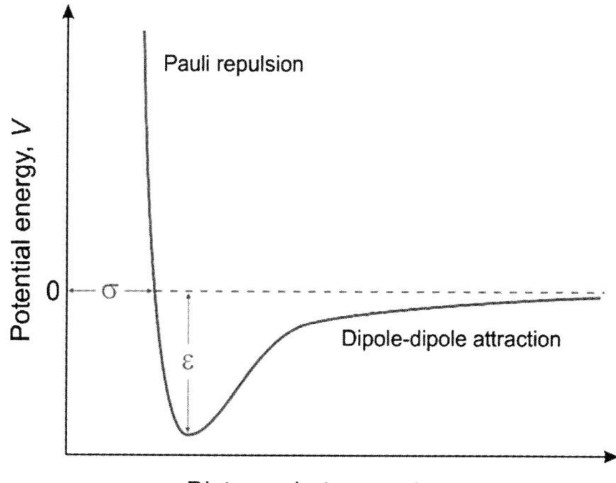

FIGURE 9.12 Lennard–Jones potential.

$$V_{LJ} = 4\varepsilon[(\sigma/\mathbf{r})^{12} - (\sigma/\mathbf{r})^{6}] \tag{9.6}$$

where σ is the distance at which the intermolecular potential between the two particles is zero.

It is also referred to as the ***Van der Waals radius***.

When the distance between the charcoal atom and the helium atom is smaller than σ, the interaction is repulsive. The interaction will become attractive if the distance between the charcoal atom and the helium atom is greater than σ, where a "***potential well***" exists.

ε represents the "depth" of the potential well with the unit of energy. It is a function of temperature and pressure. The helium atom falls into the ***potential well*** when the helium is adsorbed by the charcoal. Helium atoms will be desorbed when the charcoal is heated.

Another important parameter, ξ, is called the "adsorption" of charcoal. It is defined as the amount of helium adsorbed by unit weight of charcoal.

The adsorption increases tremendously when the charcoal temperature decreases and approaches the critical temperature of the gas.

Extensive research to study the adsorption, ξ, of some commercial activated charcoal has been performed on **4He** at different temperature and pressures, which is a good starting point for our later discussion on the activated charcoal adsorption on 3He.

The example illustrated in Figure 9.13 [18] indicates that ξ is a function of temperature and pressure.

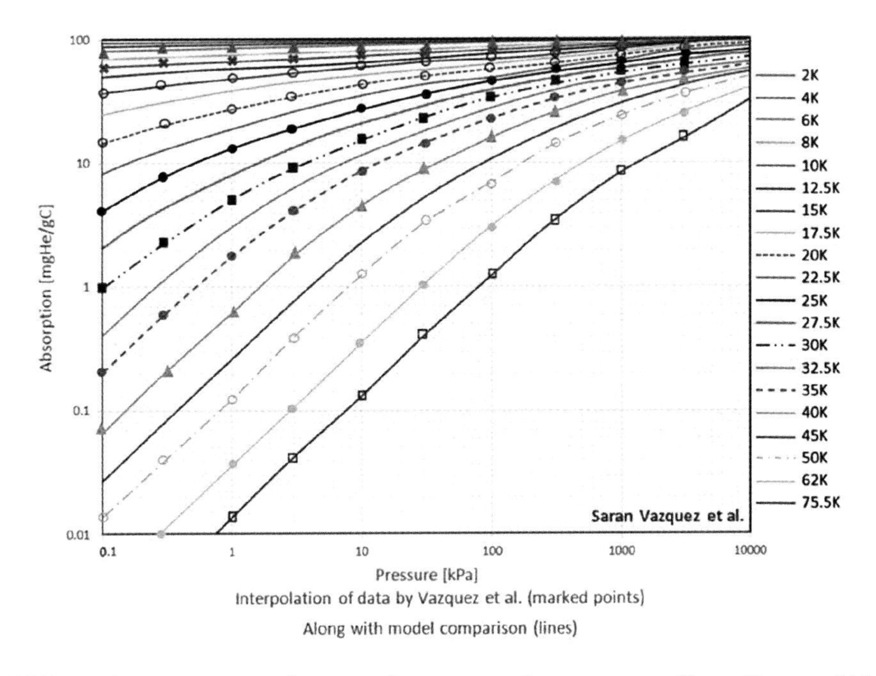

FIGURE 9.13 Adsorption, K, as a function of pressures and temperatures. (From Vazquez, I.M. et al., *Advanced Cryogenic Engineering*, 33, 1013, 1988.)

Adsorption ξ increases when the gas pressure increases. It also increases when the charcoal temperature decreases. ξ almost remains unchanged when the charcoal temperature reaches near 4.2 K. Adsorption drops to a very low level at higher charcoal temperatures. A good compromise appears to be approximately 40–50 K to desorb helium, while this temperature does not place an excessive heat load on the rest of the cryogenic system.

9.3.1.1.2 Typical Examples of 3He Adsorption by the Activated Charcoal

The preceding results have given a useful guidance to estimate the charcoal's adsorption of 3He gas. A good temperature appears to be approximately 40–50 K to desorb 3He gas during the 3He condensation in the single-shot 3He cryostat.

Listed in Table 9.6 and Figure 9.14 for the reader's reference are data of the quantities of 3He gas and the charcoal installed in a several functioning 3He cryostats. All data were taken when the charcoal temperature was kept below 5 K and under the 3He SVP near 300 mK.

The 3He gas quantity is in Normal Temperature and Pressure (NTP) liters for system #3. It is in Standard Temperature and Pressure (STP) liters for all other systems.

Discussion:

1. These examples have shown that approximately 6.1 g of activated charcoal at a temperature below 5 K are needed to adsorb each STP liter of 3He gas. This number can be used as a safe and starting point for the single-shot 3He refrigerator design.

Table 9.6 Quantities of Charcoal and 3He Gas Installed in Some Real Systems

System #	Amount of Activated Charcoal Installed (grams)	Amount of 3He Gas Installed (liters)
#1	8.6	1.51
#2	43	6.0
#3	100	12.4
#4	150	20.0
#5	155	20.0
#6	161	20.0
#7	169	20.0
#8	190	30
#9	213	30
#10	299	45
#11	346	52

Note: #1 system data from [19].
#3 system data is from [20].
#11 system data is from [21].
All other data are from the courtesy of Janis Research Company.
Capacity for all charcoal sorbs were designed with a safety factor of 2.
The 3He gas quantity is in NTP liters for system #3. STP liters are used for all other systems.

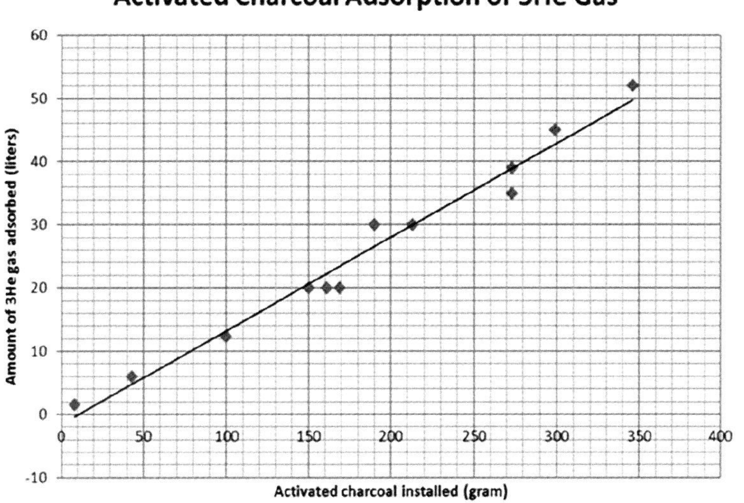

FIGURE 9.14 Quantities of charcoal and 3He gas installed in a few functioning 3He cryostats. All data were taken when the charcoal temperature was kept below 5 K and under the 3He SVP near 300 mK.

This number actually has a safety factor near 2. Some 3He refrigerator designs have pushed to the limit of approximately 3.3 g activated charcoal for each STP liter of 3He.

2. Although the amount of 3He gas ranges from 6 to 52 L only in our examples, there is no reason why it cannot be extended to larger but reasonable quantities of 3He gas.
3. Commercial activated charcoal is available from many vendors. The charcoal used by Janis was originally purchased from Barnebey Sutcliffe Corporation in Columbus, Ohio, with the following empirical parameters:
   - Charcoal granular dimensions: approximately 2.0–3.0 mm;
   - Density of the **packed** charcoal (also referred to as ***packing density***) is approximately 0.35–0.51 g/cm^3 depending how tightly it is packed;
   - Packing factor (the total real volume all the charcoal divided by the volume of the charcoal sorb chamber) is approximately 0.33–0.48; and
   - Recall the density of the ***real*** charcoal is around 2.25 g/cm^3; the above parameters imply that this activated charcoal has approximately 50% porosity.
4. Moisture always gets adsorbed in the charcoal when it is purchased. Moisture should be removed (or "dehydrated") before the charcoal is used as a cryopump (the charcoal may have been located in the vendor's stockroom for years).

   The commercially activated charcoal also has lot of "dust" that should be removed.

   The readers can start from the following steps to remove dust as well the adsorbed moisture:

   - Put the charcoal inside a stainless steel mesh (hole size should be approximately 2.0 mm × 2.0 mm) to remove the smaller particles since we do not want the small charcoal particles to fall into the "charcoal adsorption chamber" (see below).

- Gently blow away the remaining dust with a hair dryer.
- Put the charcoal inside a sealed "charcoal adsorption chamber."
- Heat the charcoal up with a hair dryer or a heat gun.
  At the same time, keep pumping on the chamber until you see a burst of pressure when the adsorbed water evaporates from the charcoal.
- Repeat this process a few times until no burst of pressure is observed.

5. Figure 9.15 shows the adsorption isotherms of 3He on activated charcoal as a function of the 3He vapor pressure at different temperatures. The activated charcoal should be kept near 4.2 K during the operation of the 3He refrigerator.

9.3.1.2 Zeolite and Its Basic Properties as an Adsorbent

Zeolite is a microporous crystalline material with typical pore size in the nanometer range (Figure 9.16).

Zeolite is a very stable solid with a melting point over 1000°C, and it does not burn.

Zeolite can withhold high pressures. It does not dissolve in water or other inorganic solvents and does not oxidize in the air [23].

It is an excellent adsorbent for water moisture and can also be used as an alternative 3He gas adsorbent. For example, Cheng et al. at NASA/Goddard Flight Center installed 2.1-kg Zeolite 5A pellets in a 3He refrigerator for balloon-borne payloads with 2.6 moles of 3He gas in 1996 [24].

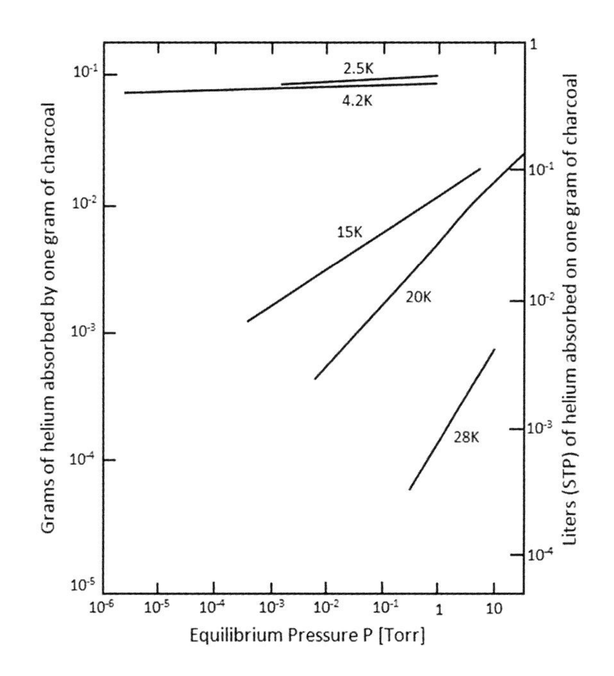

FIGURE 9.15 Shows the adsorption isotherms of 3He on activated charcoal as a function of the 3He vapor pressure. (From *Material and Methods at Low Temperatures*, 3rd edition, Springer-Verlag, Berlin, Germany, 2006.)

FIGURE 9.16 Regular Zeolite.

9.3.1.3 Structures of Typical Cryogen-Free Single-Shot-Type 3He Refrigerator

A typical cryogen-free single-shot-type 3He refrigerator is illustrated in Figure 9.17.

The refrigerator includes the following main items:

1. **Cryocooler**

   A two-stage pulse-tube refrigerator (PTR) is needed for the cryogen-free 3He refrigerator.

   The excessive cooling power from its second-stage regenerator plays an important role for the system performance and will be discussed later. Lower vibration is another reason why PTR is highly desirable.

   The typical temperatures of the first and second stages of PTR are 40 K and 2.8 K, respectively.

   In order to further reduce the vibration, flexible copper braids are commonly employed as thermal links between the cryocooler and the rest of the cryostat as illustrated in Figure 9.17.

   A PTR with a remote rotary valve/motor assembly does generate lower vibration than a PTR with a direct-mount motor. The remote-valve motor has little to no effect on the base temperature (300 mK) of 3He refrigerators. The situation is different for cryogen-free dilution refrigerators operated near 10 mK as discussed in **Chapter 11: Cryogen-Free Dilution Refrigerator**.

   Cryomech PT-400 series PTR adapts the U-type design. Excessive cooling power can be extracted from the distributed heat exchanger installed on the second stage regenerator as described in **Chapter 3: 4 K Regenerative Cryocoolers, Chapter 4: Features and Characteristics of 4 K Cryocoolers**, and **Section 8.2** (Figure 9.18).

2. **Dewar with vacuum shroud and radiation shields**

   A vacuum shroud is needed to provide the high-vacuum space necessary for all cryogenic systems. Aluminum or copper radiation shields are installed on the first

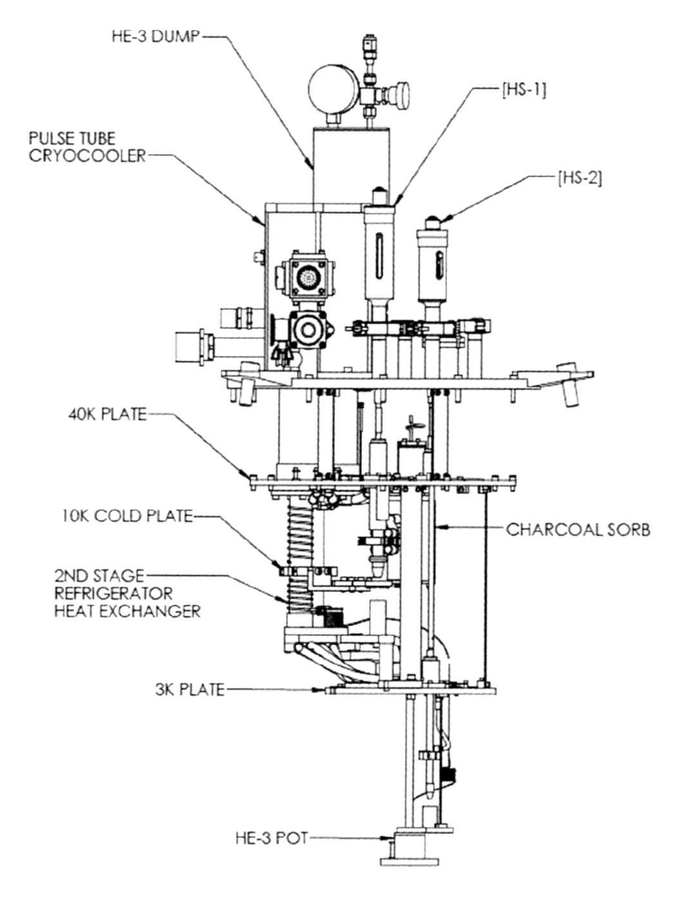

FIGURE 9.17 Typical cryogen-free single-shot-type 3He refrigerator. (Courtesy of Janis Research Company.)

stage and the second stage, respectively. In order to minimize the radiation heat load, multilayer superinsulation needs to be installed on both radiation shields (Figure 9.19).

3. **3He "insert"**

The 3He insert includes the *3He gas, the gas storage bottle, the charcoal sorption pump, the 3He condenser,* the *3He pot,* and their connecting tubes.

a. 3He gas and gas storage bottle

Pure 3He gas is undoubtedly the key item in all 3He refrigerators. Listed in Table 9.7 are the typical specs of commercial pure 3He gas that can be used for the 3He cryostat.

At room temperature, the 3He gas fills the whole space of the 3He insert. The total volume is dominated by the gas storage bottle. A smaller bottle with higher room-temperature pressure helps to improve the 3He gas condensing efficiency, but it has a lower safety factor under high pressures. An optimized compromise could be to keep the 3He space at a pressure around 10 bars at room temperature.

b. Charcoal sorption pump

FIGURE 9.18 (a) Cryomech model PT-415 PT cryocooler. (b) The distributed heat exchanger installed on the second-stage regenerator cryocooler. (Courtesy of Janis Research Company.) (c) Flexible thermal link between the second stage of PTR and the 3He cryostat 3 K cold plate. (Courtesy of Janis Research Company.)

The charcoal sorption pump, also called charcoal sorb, usually is a cylindrical copper chamber filled with activated charcoal. Its functions include releasing the 3He gas for condensation and adsorbing the 3He gas for cooling. A heater and thermometer are installed on the charcoal sorb. The charcoal sorb needs to be controlled at different temperatures to perform different functions.

c. 3He condenser

   The 3He condenser is located below the charcoal sorb. It is thermally connected to the second stage of the PTR (the coldest position of the cryocooler) at a temperature below 3 K. 3He gas will be condensed at the inner surface of the condenser.

FIGURE 9.19 (a) 3He cryostat vacuum shroud. (Courtesy of Janis Research Company.) (b) Multilayer superinsulation installed on the 40 K radiation shield. (Courtesy of Janis Research Company.)

Table 9.7 Typical Specs of Commercial "Pure" 3He Gas

Item	Percent
3He, atomic fraction	99.96
4He, atomic fraction	0.04
Total helium volume fraction	99.999
Nitrogen, volume fraction	2.6×10^{-4}
Hydrocarbons, volume fraction	$<1.0 \times 10^{-4}$
Hydrogen, volume fraction	$<1.0 \times 10^{-4}$
Oxygen, volume fraction	$<1.0 \times 10^{-4}$
Carbon oxide, volume fraction	$<1.0 \times 10^{-4}$
Tritium, atomic fraction	$<4.2 \times 10^{-12}$
Tritium, activity	$<5.4 \times 10^{-11}$ Ci/lt

d. 3He pot

3He pot is usually a cylindrical chamber made of oxygen-free high thermal conductivity copper (OFHC). It is located at the bottom of the 3He insert. The liquid 3He gets collected inside the 3He pot during the condensation.

The bottom of the 3He pot is often used as the sample mount for the user.

4. **Main heat switch and auxiliary heat switch**

The charcoal sorb needs to be operated at different temperatures during different operation stages of the single-shot-type 3He refrigerator. The main heat switch (referred to as ***HS #1***) is the thermal bridge between the charcoal sorb and the 3He cryostat. Its operation is as illustrated in Figures 9.20 and 9.21.

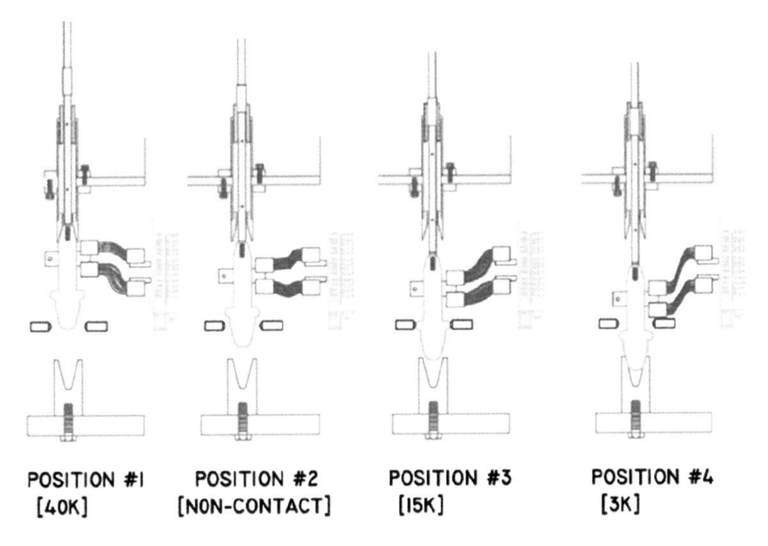

FIGURE 9.20 Main heat switch and its operation. (Courtesy of Janis Research Company.)

FIGURE 9.21 Main heat switch engaged at different operations. (Courtesy of Janis Research Company.)

The heat switch body is made of gold-plated oxygen-free high-conductivity (OFHC) copper, and it is thermally connected to the charcoal sorb with flexible copper braids. The heat switch can be located at different positions with a linear motion manipulator for different functions.

Position #1: 40 K position

The heat switch makes contact to the first-stage cold plate (approximately 40 K). The charcoal temperature will be stabilized at approximately 40 K with **HS #1** located at this position. The heat switch is located at this position either during the 3He condensation or to cool the charcoal sorb down from higher temperatures.

Position #2: Floating (non-contact) position

The heat switch makes no contact with any other item of the cryostat.

The charcoal sorb can be controlled at higher temperature, for example, 70 K, for more effective 3He condensation.

Position #3: 10 K position

This position is located near the middle of the second-stage regenerator as shown in Figures 9.18b and 9.20.

This is an intermediate stage with temperature of approximately 10–12 K. A gold-plated OFHC copper adapting plate is attached to that position, called a *10 K cold plate*. The charcoal sorb will be cooled down to this intermediate temperature at the end of the 3He condensation before the **HS #1** moves to Position #4. This approach will minimize the heat load to the PTR second stage during the cooling process of the charcoal sorb and makes the operation smoother.

The 10 K cold plate also provides an additional thermal anchor stage for wires, which is crucial when the system has large number of wires and coaxial cables.

Position #4: 3 K position

Charcoal sorb is cooled down to below 3 K with the **HS #1** located at this position.

Because most of the cryogen-free 3He cryostats do not have the "Inner Vacuum Can" (IVC), that is, the 3He pot (and the samples) shares the same vacuum with the dewar vacuum shroud, installation of an auxiliary single stage mechanical heat switch, referred to as **HS #2**, (as discussed in **6.3.2 Mechanical Heat Switch**) between the 3He-pot and the 3 K stage of the cryocooler is highly desirable. The 3He pot (and the user's samples) can be precooled much faster when **HS #2** is engaged. It also keeps the 3He pot as cold as possible during the condensation and makes the 3He condensing process more efficient.

9.3.1.4 Operation and Performance of the Cryogen-Free Single-Shot-Type 3He Refrigerator

1. Operation guidance

Operation of the single-shot-type 3He refrigerator consists of two steps: (1) condensation of gaseous phase 3He, and (2) vapor pressure reduction of liquid phase 3He (Figure 9.21).

Following the preceding discussion, the charcoal sorb is heated up to 40 K (**HS #1** at Position #1) or higher temperatures (**HS #1** at Position #2) with the 3He pot kept below 3 K or lower by the cryocooler (**HS #2** engaged). 3He gas is released from the charcoal sorb, condenses inside the condenser, and then falls into the 3He pot. The liquid 3He that drops into the 3He pot initially evaporates when the 3He pot temperature is above the 3He liquefaction temperature. "Convection cooling" or "evaporative cooling" occurs inside the 3He pot.

After the 3He pot reaches the 3He liquefaction temperature under the gas pressure inside the cryostat, the 3He pot will start collecting liquid. The vapor pressure will continue dropping as the 3He pot gets colder and colder (refer to Table 9.1).

The condensing efficiency, defined as the amount of 3He gas condensed into the liquid phase divided by the total amount of 3He gas, is an important performance parameter of the 3He cryostat. All designers must pay close attention to this parameter because it determines the total amount of 3He gas needed for a 3He cryostat when certain holding time is required.

For example, the second stage of PTR usually reaches 2.7 K or lower. The SVP of liquid 3He at 2.7 K is 428.77 mmHg per Table 9.1. 3He will stop condensing when its vapor pressure inside the cryostat reaches that pressure while the cryocooler stays at its base temperature. The designer can estimate the uncondensed amount of 3He, which remains in the "gaseous" phase inside the cryostat. Wet 3He systems use a 1 K pot for 3He condensation that can reach 1.5 K or lower. The 3He SVP at 1.5 K is 50.33 mmHg. **The condensing efficiency of the cryogen-free system is quite lower than that of the wet system.**

Disengage **HS #2** after 3He condensation is completed. Move **HS #1** to Position #1 (40 K) (if its starting position is at Position #2), then to Position #3 (10 K), finally to Position #4 (3 K). The charcoal sorb will reach below 4 K, and the 3He pot will reach its base temperature.

The base temperature of the 3He cryostat depends on the parasitic heat load of the system.

The holding time depends on the amount of 3He gas installed as well as the total heat load to the 3He pot.

The typical temperature stability at the base temperature can be maintained within +/− 1.0 mK even without PID control as shown in Figure 9.22. The 3He pot temperature usually drops a few millikelvins just before it becomes dry.

2. **Temperature control**

There are two basic approaches to control the 3He pot temperatures above the base temperature. An automated temperature controller is necessary for each operation.

The first approach is to set the desired 3He pot temperature at the temperature controller, and then select the PID setting. The temperature controller will turn on the heater and stabilize the 3He pot temperature at the set point.

The advantage of this approach is that the 3He pot can be stabilized precisely at any desired temperature. The disadvantage is that the applied heating will make the holding time shorter.

The second approach is to control the 3He vapor pressures by setting the charcoal sorb temperature between 5 K and 40 K. The advantage is that this approach does not apply additional heating to the 3He pot, and the holding time will be longer (Note: it is equivalent to 3He condensation if the user sets the charcoal sorb above 40 K, which offers unlimited "holding time.") The disadvantage is that it is difficult to stabilize the 3He pot at certain desired temperatures accurately.

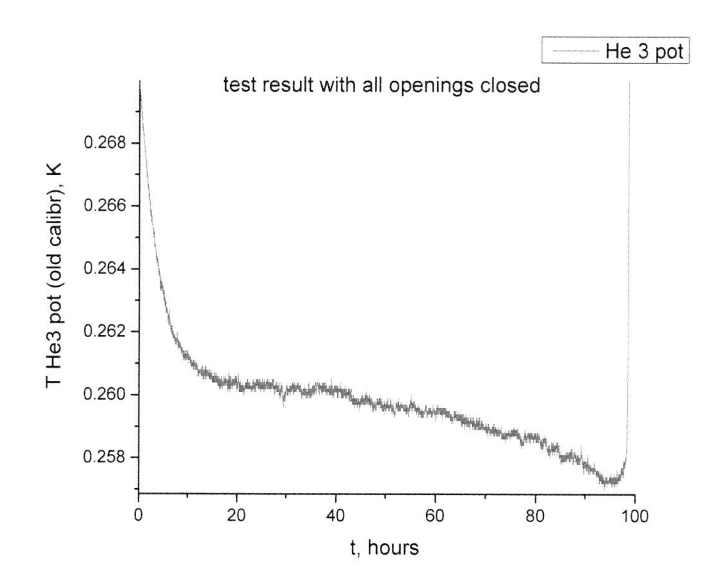

FIGURE 9.22 Typical temperature stability at base temperature without PID control. (Courtesy of Janis Research Company.)

3. **Cooling power**

   If the 3He pot holds a certain stable temperature when a fixed amount of power is applied to it, this applied power is defined as the **cooling power** of this 3He refrigerator at the **specific temperature**. Recall that when the 3He pot temperature is stabilized, the 3He system is always being operated under the liquid 3He SVP (Figure 9.23).

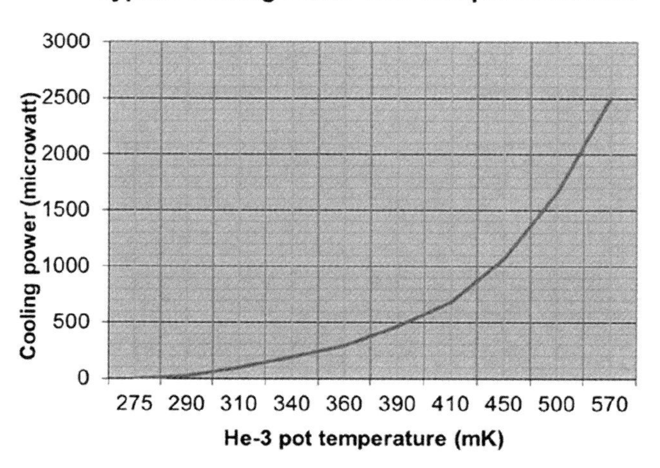

FIGURE 9.23 Typical cooling power of single-shot 3He refrigerator with sample in a vacuum. (Courtesy of Janis Research Company.)

Based on our previous discussion, the total cooling power of 3He cryostat is expressed by Equation 9.7:

$$Q = n_3 L_3 \tag{9.7}$$

where:

$n_{3\,(\text{mole/second})}$ is the throughput of 3He vapor adsorbed by the charcoal sorb, and
$L_{3(\text{Joule/mole})}$ is the mole latent heat of 3He liquid under its SVP.

Higher heat load on the 3He pot will result in higher throughput, which in turn will generate more heat in the charcoal sorb during the adsorption process. When the heat load reaches a few milliwatts, a regular single-shot-type 3He cryostat containing a limited amount of charcoal is not able to maintain a stable charcoal sorb temperature below 5 K. The adsorption of the 3He system will decrease per Figure 9.15, and the base temperature of the 3He pot will increase.

A recirculating (also called continuous flow) 3He refrigerator should be considered when high cooling powers are required.

9.3.2 Recirculating 3He Refrigerator

The single-shot 3He cryostat has two intrinsic (although not fatal) disadvantages: (1) limited holding time, and (2) limited cooling power. For example, a single-shot-type 3He refrigerator cannot practically keep the sample temperature in the millikelvin range for experiments with larger heat loads. A typical example is for the angle-resolved photoemission spectroscopy (ARPES) application [25] where the 3He pot (and the sample) are exposed to room-temperature radiation.

A recirculating 3He refrigerator is the solution where the charcoal sorb is replaced by hermetic pump(s) located outside the 3He refrigerator, as depicted in Figure 9.24.

1. **Hermetic pump**
   Pumps employed in the recirculation 3He refrigerator need to have the following features:
   a. Being hermetic
      The system, including the pump, must be "***truly***" vacuum tight for two reasons:
      - No 3He gas should be lost during the recirculation due to the high cost of 3He gas.
      - No air or other impurities should get into the recirculation loop. Air and impurities will freeze and block the condensing impedance and stop the 3He circulation.

   Special attention should be paid to pumps with a shaft seal. A careful leak check must be performed when the pump is running. The total leak rate of the whole system should be near 1.0E-8 liter.torr/sec. or lower.
   b. With proper performance features
      The following features should be considered when pumps are selected for the recirculation systems:
      - Be able to reach pressures lower than the SVP of liquid 3He at the desired base temperature for the experiment;

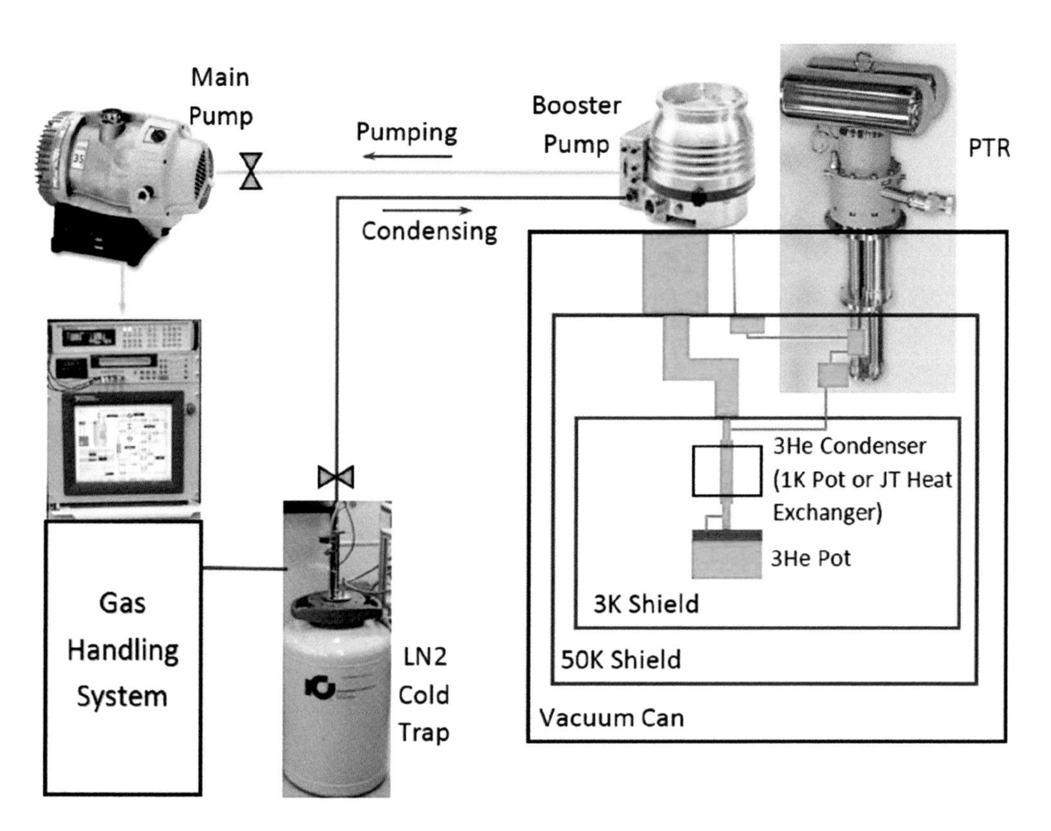

FIGURE 9.24 Working principle of the continuously circulating 3He refrigerator.

- Having sufficient throughput, n_3, to provide the required cooling power; and
- An "oil-free" pump is highly desirable. If not, the lubricant oil must have a very low vapor pressure and degas rate. Synthetic oil is a good choice.

Only one pump is needed for "smaller" systems with higher base temperatures and lower cooling powers. Both the mechanical ("Roots") pump, such as the Pfeiffer model A100L pump, and the Scroll pump, such as the Edward model XDS35i pump as illustrated in Figure 9.25, are good candidates.

A booster pump, such as a Roots blower or a high-speed turbo pump is needed when higher throughputs, that is, larger cooling powers, are needed. These pumps must be backed by a "backing" pump (often referred to as the "main pump" in the recirculating systems).

Good candidates are the Pfeiffer model OKTA-1000/2000 Roots blower or Pfeiffer model HiPace series turbo pumps as illustrated in Figure 9.26.

2. **3He Condensing Unit**
   a. Integrated 1 K system

      Effective condensation of warm gaseous 3He at room temperature into liquid 3He within the circulating loop is challenging for the recirculating cryogen-free 3He refrigerator. The ***3He Condensing Unit*** is responsible for this function.

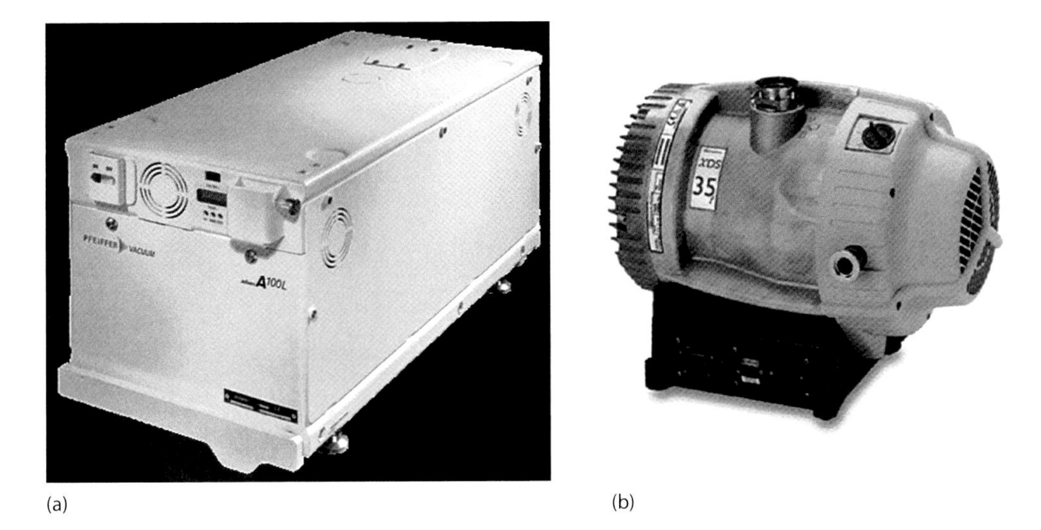

FIGURE 9.25 (a) Pfeiffer model A100L pump. (Courtesy of Pfeiffer Vacuum.) (b) Edward model XDS35i Scroll pump. (Courtesy of Edwards Vacuum Engineering Company.)

FIGURE 9.26 (a) Pfeiffer model OKTA-1000/2000 Roots blower. (Courtesy of Pfeiffer Vacuum.) (b) Pfeiffer model HiPace series turbo pump. (Courtesy of Pfeiffer Vacuum.)

A separate "1 K pot" can be installed on the second stage of the cryocooler. 4He gas from a high-pressure gas bottle is injected into the system and gets liquefied inside the 1 K pot by a cryocooler. An external pump is used to reduce the 4He vapor pressures (this is why the 1 K pot is sometimes referred to as 4He evaporator) inside the 1 K pot and cool down the 1 K pot below 1.5 K.

The 3He gas enters the system through a 3He return line, which can be made of stainless steel capillary with diameter of approximately 0.038–0.062 in. The capillary is coiled in approximately 3–4 turns inside the 1 K pot. The 3He gas inside the returning line passes the 1 K pot and condenses into liquid.

As an alternative, a dedicated hermetic pump for the 1 K circuit can be implemented and turns the setup into a recirculation 1 K system as described in Section 8.2.

After the 4He gas is liquefied, a hermetic pump will be used to pump on the 1 K pot and circulate the 4He. The 3He gas will condense when it passes the 1 K pot.

The recuperative heat exchanger (or also called JT condenser) is an alternative approach for 3He condensation [26]. As a matter of fact use of JT condenser has become a standard approach in the cryogen-free dilution refrigerator system, which will be discussed in **Chapter 11: Cryogen-Free Dilution Refrigerator**.

b. Condensing impedance

Condensing impedance discussed in Section 6.2.4.2 is an inevitable part of the **3He Condensing Unit**. The impedance is located above the 3He pot and keeps the proper head pressures for the 3He condensation.

The impedance is also a cryogenic Joule–Thomson (JT) valve, and 3He gets liquefied after being further cooled by the JT expansion. The proper impedance value optimizes the 3He circulation rate and stabilizes the liquid 3He level inside the 3He pot.

3. Examples

a. Use of 4He evaporator for 3He condensation

A cryogen-free recirculating 3He refrigerator using a separate 4He stream evaporator to condense 3He has been reported by Burton et al. [27]. Summarized below are the main features of that system for the reader's reference:

i. Main items

Cryocooler: Cryomech model PT410 with base temperature of 2.7 K
3He refrigerator hermetic pump: Varian SH-100 Scroll pump
Impedance for 3He refrigerator: Range of 10^{12} cm^{-3}
Gas-handling systems for 3He operation with LN cold trap
1 K pot impedance: Range of approximately 10^{12} cm^{-3} to 10^{11} cm^{-3}
4He gas source: High-pressure pure 4He gas bottle

ii. Performance of 3He refrigerator

Typical circulation rate: 1.5 mmole/s when the pressure drop across the impedance is ~ 101 kP

Base temperature: 0.4 K
Cooling power: 1.5 mWatt at 0.5 K

iii. Performance of the 1 K pot
Typical flow rate: 0.35 mmole/s when the head pressure is 225 kPa
Base temperature: 1.5 K
Cooling power: 20 mWatt at 1.5 K

b. Use of JT condenser for 3He condensation
Successful and optimal operation of using JT condenser for 3He condensation requires a good understanding on the 3He JT inversion curve [28,29].

A prototype of highly custom designed "hybrid" recirculating 3He refrigerator as illustrated in Figure 9.27 has been developed at Janis Research Company. This cryostat is designed to be UHV compatible for ARPES applications. That system contains both a closed-loop recirculating 1 K pot system (also called 4He evaporator) as well as a JT condenser. The system has two separate 3He condensing paths. One goes to the recirculating 1 K pot (referred to as 1K operation mode), and the other one is connected to the JT condenser (referred to as JT operation mode). Both the 1 K pot and the JT condenser are precooled below 3 K by the second stage of a Cryomech model PT415 pulse tube refrigerator. The user is able to change the condensing modes during the operation.

It is a common understanding that the JT operation mode usually generates less noise than that of the 1 K operation mode. This feature is highly desirable for the vibration-sensitive experiments. With a 3.6 × 19.5 mm slot facing room temperature this system has reached base temperature near 546 mK with JT mode operation and 506 mK with 1 K mode operation, respectively.

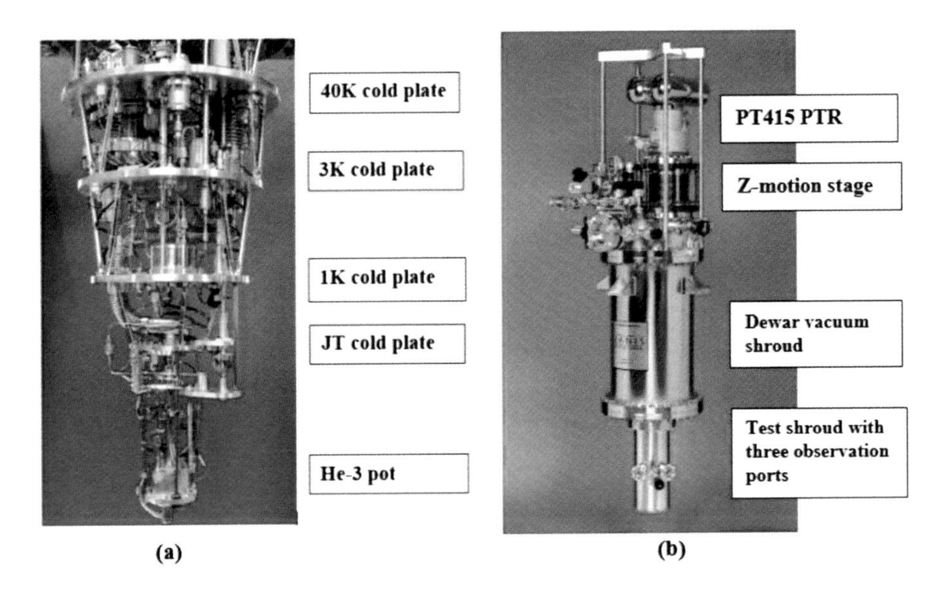

FIGURE 9.27 (a) Cryogen-free "hybrid" re-circulating 3He insert with integrated cryogen-free recirculating 1K pot, (b) cryogen-free "hybrid" re-circulating 3He cryostat. (Courtesy of Janis Research Company.)

Acknowledgments

The contributor acknowledges Munir Jirmanus for reviewing this chapter for its scientific integrity. Sincere acknowledgment goes to Yong-hua Huang for his helpful discussion on 3He properties.

The contributor acknowledges his colleagues at Janis Research Company for their help:

Thanks to Ashley Huff for her helpful comments.

Thanks to Jeonghoon Ha and Sam Gonzales for their final proofreading. Last but not the least, thanks to Ann Carroll for her constant help with software difficulties and reference searches.

References

1. Rosman, K.J.R., and Taylor, P.D.P. The 1997 report of the IUPAC Subcommittee for Isotopic Abundance Measurements, *Pure Appl. Chem.*, 71, 1593–1607 (1999).
2. International Temperature Scale of 1990 [ITS-90].
3. Rives, J.E., and Meyer, H. Density of Liquid Helium-3 Between 0.045 and 1.3°K, *Phys. Rev. Lett.*, 7, 217 - Published 15 September (1961).
4. Grilly, E.R. Pressure-volume-temperature relations in liquid and solid ^{3}He, *J. Low Temp. Phys.*, 4(6), 615–635 (1971).
5. Gibbons, R.M., and Nathan, D.I. *Thermodynamic Data of Helium-3*, Air Products, Technical Report AFML-TR-67-175, October (1967).
6. Boghosian, C., Horst M., and Rives, J.E. Density coefficient of thermal expansion, and entropy of compression of liquid Helium-3 under pressure below 1.2°K, *Phys. Rev.*, 146(I) (1966).
7. Takacs, I., Takacs, J., Reiff, W.M., and Ramsden, J.D. Hydrostatic pressure effect on the minimum temperature of a helium-3 cryostat, *Rev. Sci. Instrum.*, 57(10) (1986).
8. Huang, Y.H. A deeper look into the thermodynamic perfection of the Debye equation of state of helium-3, 25th International Cryogenic Engineering Conference and International Cryogenic Materials Conference in (2014), ICEC 25 - ICMC 2014, *Phys. Procedia*, 67, 582–590 (2015).
9. Greywall, D.S. Specific heat of normal He-3, *Phys. Rev. B*, 27(5) (1983).
10. Greywall, D.S. He 3 melting-curve thermometry at millikelvin temperatures, *Phys. Rev, B*, 31, 2675 (1985).
11. Greywall, D.S. 3He specific heat and thermometry at millikelvin temperatures, *Phys. Rev. B*, 33, 7520 (1986).
12. Halperin, W.P., and Pitaevskii, L. P. (Eds.) *Helium Three*, Elsevier, Amsterdam, the Netherlands (1990). http://ltl.tkk.fi/research/theory/helium.html
13. Dobbs, E.R. *Helium Three*, Oxford University Press (2000).
14. Huang, Y.H., Fang, L., Wang, X.J., Wang, R.Z., and Xu, L. Thermal conductivity of helium-3 between 3 mK and 300 K, *Advances in Cryogenic Engineering, AIP Conference Proceedings*, 1434, 1849 (2012).
15. Betts, D.S., Osborne, D.W., Welber, B., and Wilks J. The viscosity of liquid helium 3, *The Philosophical Magazine: A Journal of Theoretical Experimental and Applied Physics*, Series 8, 8(90) (1963).
16. Black, M.A., Hall, H.E., and Thompson, K. The viscosity of liquid helium 3, *J. Phys. C*, 4(2) (1971).
17. Huang, Y.H., Chen, G.B., Li, X.Y., and Arp, V. Density equation for the saturated 3He, *Int. J. Thermophys.*, 26(3), 729–741 (2005).
18. Vazquez, I.M., Russell, M.P., Smith, D.R., and Radebaugh R. Helium adsorption on active carbon at temperature 4 and 76 K, *Adv. Cryog. Eng.*, 33, 1013 (1988).
19. Ventura, G., and Risegari, L. *The Art of Low Temperature Experimental Technologies*, Elsevier, Oxford, UK (2008).

20. Palumbo, P., Aquilin, E., Cardoni, P., P. de Bernardis, A. De Ninno, Martinis, L., Masi, S., and Scaramuzzi, F. Balloon-borne 3He cryostat for millimeter bolometric photometry, *Cryogenics*, 34(12) (1994).
21. Masi, S., Aquilini, E., Cardoni, P., de Bernardis, P., Martinis, L., Scaramuzzi, F., and Sforna, D. A self-contained 3He refrigerator suitable for long duration balloon experiments, *Cryogenics*, 38(3) (1998).
22. Pobell, F. *Material and Methods at Low Temperatures*, 3rd edition, Springer-Verlag, Berlin, Germany, 2006.
23. Tominaga, H. Zeolite no Kagaku to Ouyou, Kodansha Scientific (1987).
24. Cheng, E.S., and Meyer, S.S. A high capacity 0.23K He-3 refrigerator for balloon-borne payloads, *Rev. Sci. Instrum.*, 67(11) November (1996).
25. Xingjiang Zhou, X.J. *Lecture: Introduction to ARPES* (2013).
26. Daunt, J.D., and Lerner, E. A closed-cycle Joule-Thomson liquefier and cryostat for 3He, *Cryogenics*, December (1970).
27. Burton, J.C., Cleve, E. Van, Taborek, P. A continuous 3He cryostat with pulse-tube pre-cooling and optical access, *Cryogenics*, 51 (2011).
28. Maytal, B.-Z., and Shavit, A. On the integral Joule-Thomson effect, *Cryogenics*, 34 (1994).
29. Maytal, B.-Z. 3He Joule-Thomson inversion curve, *Cryogenics*, 36(4) (1996).

10. Cryogen-Free Adiabatic Demagnetization Refrigerator (ADR) System

Charlie Danaher

10.1 Cooling Approaches below 100 mK

There are several methods of achieving temperatures below 100 mK. In choosing the best method, several factors should be considered. To name a few: (a) desired base temperature; (b) whether the cooling must be continuous or can be "one-shot"; (c) desired cooling power; (d) desired cooling energy (which is the product of cooling power and time); (e) system cooling density, or amount of cooling that can be achieved per volume or mass; (f) sensitivity to gravity; and (g) sensitivity to magnetic fields.

One of the most common methods of achieving temperatures below 100 mK is to utilize magnetic cooling—employing the magnetocaloric effect (MCE)—with an adiabatic demagnetization refrigerator (ADR). Use of the ADR as a refrigerator was first described by Peter Debye in 1926 [1] and William Francis Giauque in 1927 [2].

ADRs are attractive due to their high thermodynamic efficiency, the fact that their temperature can be precisely controlled without dissipating any heat, and their independence of gravity. The fact that the MCE is reversible leads to the high thermodynamic efficiency of the ADR [3].

ADRs are typically used in the temperature range from a few Kelvin down to a few millikelvin, and, depending upon heat load, common ADRs can offer cooling durations anywhere from a few minutes to on the order of a week [4]. From a system architectural standpoint, it's important to note that ADRs can be used in stages. In this section, we will discuss two-stage ADRs, as well as Helium-3 backed (3He), single-stage ADRs. We'll also address continuous-ADRs (CADRs), which offer, as the name suggests, continuous cooling. Typical ADRs are one-shot, meaning that they need to be regenerated after their cooling energy capacity is exhausted.

Usually the cooling source for an ADR is either a liquid helium bath (at ~4.2 K) or a cryocooler (at ~3 K). The cooling source can also be a pumped He-4 bath (at ~1 K) or a pumped 3He bath (~0.3 K). However, more stages can be employed, thereby raising the required temperature of the cooling source.

Much of the physics that governs an ADR is independent of the number of stages. Therefore, we will review this information and then describe the different types of cryostats that employ these designs.

10.2 Review of Entropy and the Principle of Adiabatic Demagnetization

10.2.1 Refrigeration Process of Adiabatic Demagnetization

Entropy is often explained as the measure of disorder in a thermodynamic system. ADRs employ materials that have two relevant sources of entropy: lattice (or thermal) and magnetic. In a simple sense, an ADR works by converting a change in magnetic state into a change in temperature, and then using that change in temperature to heat or cool something [5].

From a thermodynamic standpoint, any system that can be manipulated, so as to bring about a change in its entropy state, can be used for refrigeration. Accordingly, whenever such a system moves from a higher entropy state to a lower one, it will be releasing heat—to a heat sink, for instance. And, vice versa, when it moves from a lower state to a higher one, it absorbs heat—from the object being cooled.

In a vapor-based refrigeration cycle, the gas changes entropy states as it flows around the system circuit, giving off heat at one point and absorbing heat at another location. In an ADR system, the refrigerant material changes entropy states without moving and alternates between being cooled and being heated.

ADR cryostats produce this cooling by cycling the magnetic field applied to a paramagnetic refrigerant material. This material is commonly referred to as a "salt pill" since many of the materials used are paramagnetic salts. When the salt pill is subjected to a magnetic field, its magnetic moments (also referred to as dipoles, or spins) will tend to align, becoming more orderly, with a corresponding decrease in magnetic entropy. During this process, assuming that the system is insulated, an increase in the magnetic state results in an increase in temperature. If this heat is then removed while the magnetic field is applied, a lower entropy state is generated for the overall system. This reduced entropy state corresponds to an increase in cooling capacity, which can be subsequently spent intercepting heat flowing toward an apparatus held at low temperature. More simply known as "cooling" it. Thus, a simplified way to understand an ADR is to make an equation between entropy and cooling capacity, where a decrease in entropy corresponds to an increase of cooling capacity, and vice versa.

To assist in this discussion about entropy, we will be referring to a series of graphs that show the relationship between magnetic field strength, normalized entropy, and temperature. For example, let's suppose we have an ADR cryostat and the paramagnetic salt is at 3 K, with no magnetic field present. We can identify this as being, thermodynamically, at point A in Figure 10.1. The first step in the ADR process then is to generate a magnetic field. In this example, we use a 4T field. If this magnetization process happens with the salt pill thermally connected to a cooling source, instead of resulting in an expected temperature rise as the field is increasing, the generated heat is conducted away, and the temperature would, ideally, be held constant during the magnetization process. This step is called isothermal magnetization, and, again, ideally, would be a direct path between points A and B in Figure 10.1.

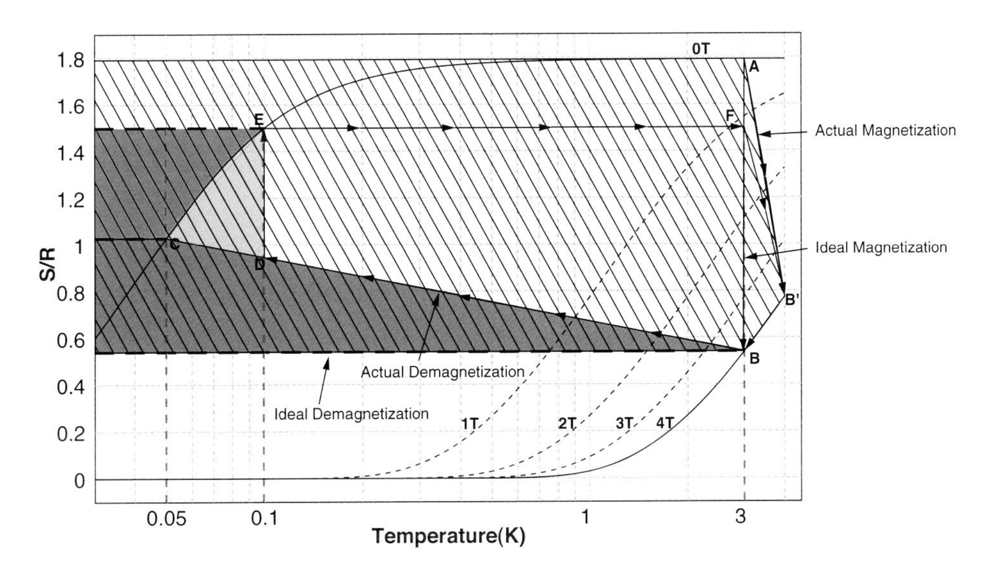

FIGURE 10.1 Entropy as a function of temperature and magnetic field.

However, due to the thermal resistances in the ADR system, the magnetization process is not actually isothermal, and as a result, there will always be a temperature rise during the "isothermal" magnetization. And instead of following a straight vertical line from point A down to B, upon reaching 4T, the system will end up at a temperature higher than where it started. We will call this point B′. In practice, when the system reaches B′, it is given time to equilibrate, so that the remaining heat can be conducted out of the ADR system and into the heat sink. Therefore, while equilibrating, the system state will move down the 4T curve, from the higher temperature (B′), and approach point B. The temperature of point B is called the "launch temperature." It's important to note that the amount of cooling energy available at low temperature is very much dependent upon the launch temperature.

The adiabatic cooling process occurs when, after thermal equilibration at full field is sufficiently complete, the salt pill is disconnected from the cooling source (by disengaging the heat switch, refer to Figure 10.2) and the magnetic field is reduced. This adiabatic cooling process is shown in Figure 10.1 as the line going from point B to C.

When isolated from the cooling source, reducing the magnetic field causes the temperature of the paramagnetic material to fall toward its "ordering" temperature. In assessing this cooling process, one will naturally be curious about what base temperature an ADR cryostat can achieve following demagnetization. One common ADR salt material, ferric ammonium alum (FAA), can achieve a base temperature near 22 mK [6]. See Figure 10.3 for an example of an FAA salt pill, as seen after crystal growth, and before the pill housing is sealed, so as to prevent dehydration.

Returning to Figure 10.1, after having reached point C, as the system rests at the base temperature and heat begins to flow into the cold salt pill, the system will naturally follow the zero field path from point C to point E. Note that the amount of cooling energy that is available to keep the device cool, while the system warms and follows the zero field line, from point C to E, is represented by the area shaded red. Specifically, in the situation shown, the red area represents the amount of cooling energy available as the system warms from 50 to 100 mK.

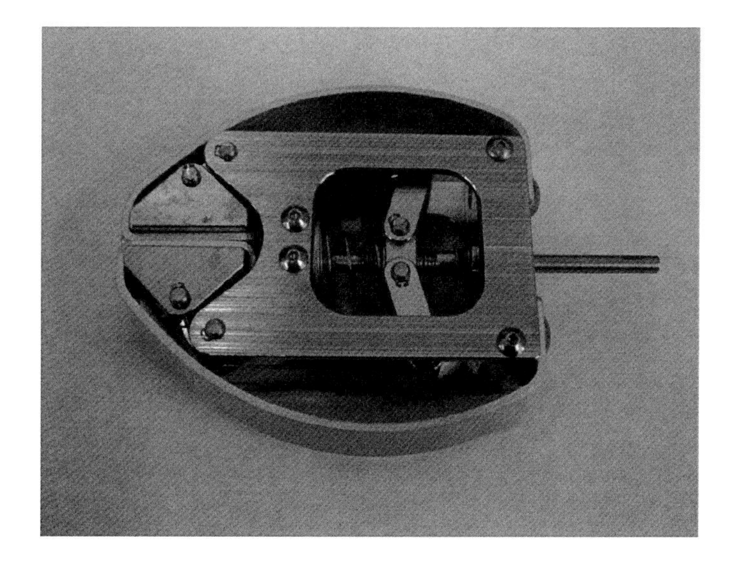

FIGURE 10.2 HPD's mechanical heat-switch vise. (Courtesy of HPD.)

FIGURE 10.3 FAA salt pill during construction. (Courtesy of HPD.)

However, recall that, whenever possible to do so, it is always more efficient to employ cooling at a higher temperature, and is therefore preferable. Following this reasoning, and by studying Figure 10.1 more thoroughly, we can see that additional cooling energy is available if, instead of reducing the magnet current to zero, we only reduce it until reaching the desired temperature, and then regulate the magnet's current to maintain said target temperature. The additional cooling energy made available by following the process just described is shown in the area shaded green. Again, the example shown here is for a regulation at 100 mK. Stepping back and reviewing the graph, now it becomes intuitive that by raising that regulation temperature we can access even more cooling energy.

The beauty of an ADR is that its temperature can be regulated, in a straightforward manner, anywhere between the system's base temperature and its launch temperature. Further, this precise regulation can be accomplished without employing a heater, thereby realizing greater thermodynamic efficiency than if a heater was used.

So, as described above, to regulate at a desired temperature, following the equilibration at full field (the soak period), the magnet current is reduced until the desired temperature is achieved. In this example, we stop at 0.1 K, shown as point D. Then, to maintain a constant temperature of 0.1 K, a control system is utilized whereby the magnet current is slowly reduced (at a rate matching the heat flowing into the system) [7]. During this isothermal process, the system state goes from point D to E. In order to provide further cooling at 0.1 K, upon reaching point E, the system would need to be regenerated. If a regeneration cycle is not performed, the system would travel along the zero field line, approaching the temperature of the cryocooler (or whatever is generating the launch temperature).

Most ADRs offer a "single-shot" cooling approach. This means that once the pill's entropy has increased (or, said another way, the cooling energy has decreased) to the point where the system can no longer provide cooling to maintain the desired temperature, and the magnet current has decayed to zero (or, in the case, where the temperature was not being controlled, and the temperature of the salt material was naturally warming and has warmed to the maximum desired temperature), the system then needs to be regenerated.

To regenerate the system, the process of reducing the magnetic entropy—and thereby creating a magnetic entropy credit, or a cooling capacity—is simply repeated.

The first step in the process of regenerating the ADR, the ramp up of the magnet current, is shown by the line segments going from points E to F, and from F to B, in Figure 10.1. At point F (the point at which the salt pill reaches the temperature of the cryocooler), the heat switch is closed so as to engage the cryocooler to ensure that the rest of the magnetization is as isothermal as possible.

During regeneration, the superconducting magnet is energized up to its rated current. The speed at which the magnet is energized should be governed by the following: the rate that heat is being released from the paramagnetic salts, so as to not overwhelm the cryocooler or otherwise excessively disturb the temperature of the system, and the back electromotive force (EMF) voltage being generated by the magnet due to its inductance. Usually the limiting factor for the back EMF voltage is that which will not cause the protection diodes to activate. It should be noted that different current-style magnets (such as 10 amp versus 20 amp) that generate the same field can have very different inductances, due to the different number of turns in the magnet coil. For instance, a given ramp rate will generate more back EMF voltage when using a 10-amp magnet than a 20 amp one, since the 10-amp version has more coil turns.

When considering how much heat is being expelled from the salt pills during the magnetization step of the cycle (heat of magnetization), by comparing the entropy values of points A and F, we can see that the amount of heat being released during a regeneration cycle (as shown in Figure 10.4) is substantially less than that released during an

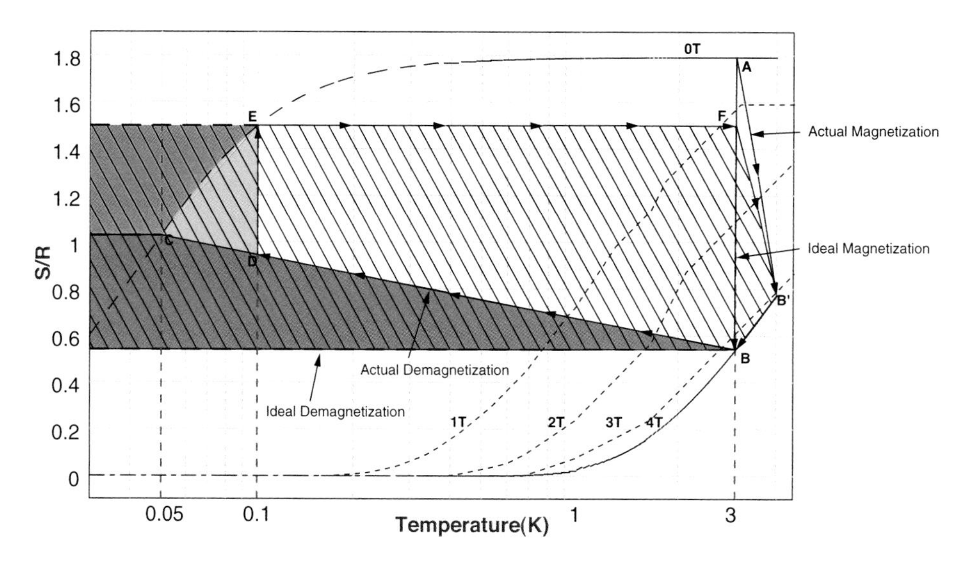

FIGURE 10.4 Entropy graph, showing normalized entropy as a function of temperature and magnetic field.

initial magnet cycle (as shown in Figure 10.1). This comparison can be made easily by comparing the different amounts of heats of magnetization, shown in both graphs as the hatched areas. So as far as how the rate of heat release limits the ramp-up speed, it's clear that the magnet can be ramped up much more quickly on a regeneration cycle than during the initial one. Unlike the rate at which the magnetic field is decreased, the rate which the magnetic field is increased does not have any effect on the final performance of the cooling cycle.

10.2.2 Analogy between Vapor Cycle Refrigeration and a Magnetic Cycle

Since for most people the magnetic refrigeration cycle is a somewhat foreign concept, comparing it to the commonly understood vapor-cycle refrigeration process can be helpful. Figure 10.5 illustrates the steps of the two analogous cycles and their corresponding stages.

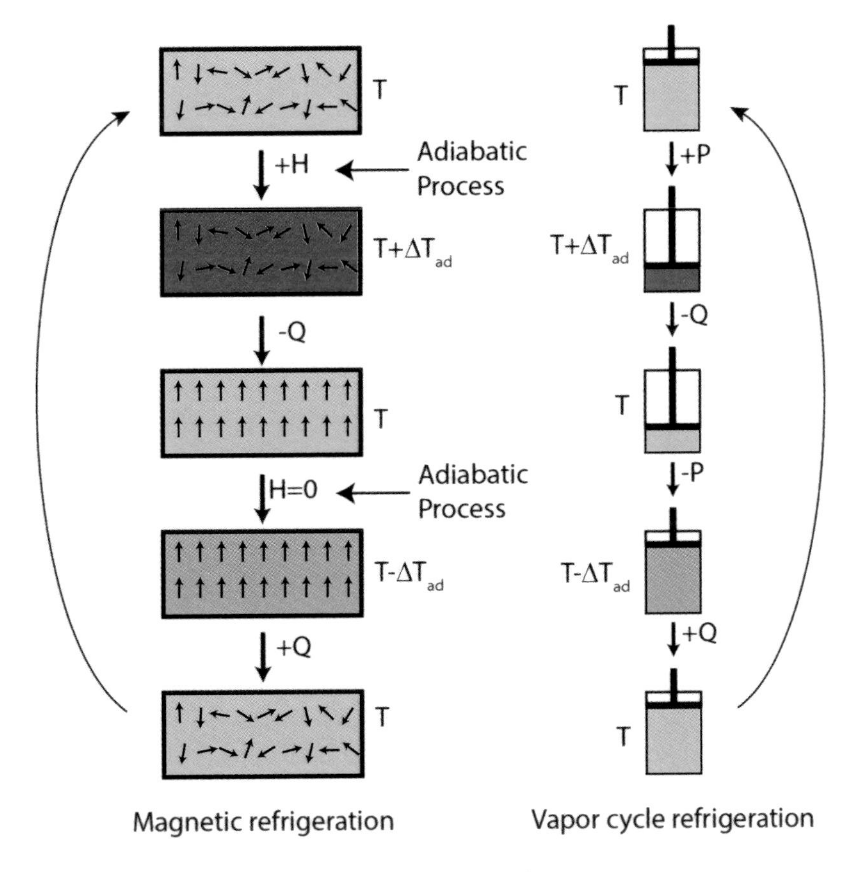

FIGURE 10.5 Analogy between magnetic and vapor-cycle refrigeration.

10.3 The Physics behind the ADR Cycle

Let's step back and look at the physics and some formulae that govern adiabatic demagnetization.

Refrigeration by adiabatic demagnetization derives from the fact that the entropy $S(B, T)$ of a paramagnetic salt depends on the temperature of the salt and the magnitude of an external magnetic field applied to the salt [8].

Expressing it another way, as stated by NASA ADR expert Peter Shirron, "At its most basic level, the magnetocaloric effect (MCE) is a phenomenon in which certain materials warm or cool as they are exposed to increasing or decreasing magnetic fields. In paramagnetic materials, the effect originates in the interaction of an external magnetic field with the electronic magnetic moments of unpaired outer-shell electrons, and results in the entropy having a strong dependence on magnetic field and temperature" [9]. We use the term "electronic" here to distinguish it from "nuclear" magnetic moments. However, because in this forum we will only concern ourselves with the moments of electrons, we will drop the "electronic" qualifier and just refer to them as magnetic moments [10].

The entropy of a magnetocaloric material is the sum of three independent constituents: (1) the magnetic entropy S_m, (2) the lattice entropy S_l, and (3) the entropy of the conduction electrons S_e:

$$S = S_m(B,T) + S_l(T) + S_e(T) \text{[11]} \tag{10.1}$$

However, S_l and S_e can be neglected at low temperatures, leaving only S_m [12].

Magnetic moments impart a total angular momentum, J, on each ion. The magnetic moments have $2J + 1$ possible orientations. In the absence of any applied magnetic field, each of these orientation states has the same energy and will therefore be equally occupied, contributing

$$S_m = R \times \ln(2J+1) \tag{10.2}$$

to the magnetic entropy [13], where R is the ideal gas constant (8.314 J/mol K).

Because in an ADR the temperature of the salt material changes when subjected to a magnetic field, ADRs are conveniently employed to lift heat from an object at a low temperature T_{cold} to a heat sink at T_{hot}, using the Carnot cycle. During the initial magnet cycle (that is, in going from point A to B, in Figure 10.1), the heat absorbed by the heat sink is:

$$Q_{\text{hot}} = T_{\text{hot}} \times (S_A - S_B) = T_{\text{hot}} \times \Delta S_{\text{hot}} \text{ [14]} \tag{10.3}$$

Q_{hot} is the heat of magnetization and, referring to Figure 10.1, is depicted by the hatched area in the rectangle left of the line AB [15]. Notice how much larger the heat of magnetization is than the amount of cooling energy that's available at regulation temperature (the amount of cooling available at regulation being the combined areas shaded red and green).

During the regulated cooling process, the refrigerant absorbs heat

$$Q_{\text{cold}} = T_{\text{cold}} \times (S_E - S_D) = T_{\text{cold}} \times \Delta S_{\text{cold}} [16] \tag{10.4}$$

If T_{cold} is 100 mK, Q_{cold} is represented in Figure 10.4 as the sum of the red and green areas.

As mentioned, before a magnetic field is applied to the salt, the spacing of the magnetic energy states is essentially zero (except near and below the ordering temperature). However, when an external magnetic field is applied, the magnetic moments will tend to align with the field, resulting in increases in the energy differences between the states and reducing the occupation of the higher energy states (those whose spin is counter to the applied field). Assuming that the magnetic moments are completely independent, we find:

$$\frac{S_m}{R} = x\coth(x) - (2J+1)x\coth(x(2J+1)) + \ln\left(\frac{\sinh(x(2J+1))}{\sinh(x)}\right) \quad (10.5)$$

where x equals $U_b gB/2K_b T$ [17], U_b is the Bohr magneton, K_b the Boltzmann constant, and g the Landē factor.

Because magnetic moments interact with each other, it is useful to replace B with an effective field B_{eff}, such that

$$B_{eff} = \left(B^2 + b^2\right)^{0.5} \quad (10.6)$$

where b is the internal or background field resulting from nearby moments in the salt [18].

The Landē factor, g, is defined as:

$$g = 1 + \frac{J(J+1) + S(S+1) - L(L+1)}{2J(J+1)} \quad (10.7)$$

where J is the total angular momentum, L is the orbital angular momentum, and S (not to be confused with entropy) is the spin angular momentum [19].

However, in iron group transition metal ions, where spin–orbit interactions are not the main type of cooling, $g = 2$ [20].

It becomes obvious then that it is useful to choose a salt with a high total angular momentum, J, and Landē factor, g. If we look at Table 10.1, we can see the properties for various salts, including the Landē factor, g [21].

For temperatures below 0.5 K, the only choices for paramagnetic cooling are the various hydrated salts, where the primary trade-off is between spin density (the number of magnetic moments per volume) and ordering temperature. Salts with higher magnetic moment density will have more cooling capacity for a given salt volume, but they will also have higher limits on the lowest temperatures that can be reached [23].

The lowest temperature that can be reached by a particular salt is a function of its zero field entropy.

Here are a few other useful facts about the magnetocaloric effect.

Table 10.1 Total Angular Momentum and Landē Values for Various Paramagnetic Salts

Refrigerant	Chemical Composition	J	g	T_0 (K)	N (cm^{-3})
CMN	$Ce_2Mg_3(NO_3)_{12}24H_2O$	1/2	2	0.0015	1.65×10^{21}
CCA	$CrCs(SO_4)_2 12H_2O$	3/2	2	0.01	2.09×10^{21}
CPA	$CrK(SO_4)_2 12H_2O$	3/2	2	0.009	2.21×10^{21}
FAA	$Fe(SO_4)_2NH_4 12H_2O$	5/2	2	0.026	2.14×10^{21}
MAS	$Mn(SO_4)_2(NH_4)_2 6H_2O$	5/2	2	0.17	2.79×10^{21}
DGG	$Dy_3Ga_5O_{12}$	15/2	8	400	1.28×10^{22}
GGG	$Gd_3Ga_5O_{12}$	7/2	2	0.38	1.26×10^{22}
GLF	$GdLiF_4$	7/2	2	0.38	1.34×10^{22}

Source: Wikus, P. et al., *Magnetocaloric materials and the optimization of cooling power density*, pg 5.

- The magnetization and heat capacity of the system do not change during adiabatic changes of the magnetic field [24].
- To the degree which the demagnetization process is adiabatic (which, in reality, it is not), we find that the magnetization and entropy remain constant during the demagnetization process, or, in other words, that the populations of the various energy levels do not change when the field is being reduced. Rather, the only thing that changes, when the field is changed, is the differences in the energy levels [25].
- As the demagnetized salt absorbs heat, the populations of its energy levels change, which means that the spins of the paramagnetic ions flip. The ions can absorb more heat if the energy change per spin is larger, that is, if their energy separation (which is proportional to the field B_f) is larger [26].

When using an ADR, it's often useful to know what sort of hold time one can expect. We can figure the hold time by employing previously derived terms [27].

$$t_{\text{hold}} = \left(\left(T_{\text{load}}\right)/\left(\dot{Q}_{\text{load}}\right)\right)\times\left(S\left(T_{\text{load}}, 0\right) - S\left(T_{\text{sink}}, B_{\max}\right)\right) \tag{10.8}$$

10.4 Design Aspects of ADR

10.4.1 What Variables Affect Salt Pill Base Temperature and Hold Time?

Given what we've discussed, we can now effectively review base temperature and hold time. The theoretical base temperature that can be achieved for a given refrigerant (a particular salt pill material of a given size) is primarily determined by three things: (1) the "launch" temperature of the system, that is, the temperature of point B on the entropy graphs in Figures 10.1 and 10.4; (2) the magnetic field strength of the regeneration, that is, the magnetic field strength of point B on the entropy graphs in Figures 10.1 and 10.4; and (3) the ballast mass of the experimental setup. Here it's worth repeating that different salt materials have different ordering temperatures.

If we refer to Figure 10.1, we can see how these three factors correspond to those previously identified. The launch temperature is the temperature of points *A* and *B*. Next, one can then see how the magnetic field strength defines point *B*. Finally, the ballast mass of the system is what will determine the size of the blue triangular area, with that triangle defining the difference between the theoretical demagnetization and the actual demagnetization.

The amount of time that a system can maintain a given temperature (i.e., how long it can be regulated, also known as the "hold time"), is a function of the size of the salt pill and the rate of steady-state heat flowing into the paramagnetic material. The contributors to this heat load are typically: conduction through the support system, conduction through the experimental wiring, blackbody radiation from warmer surfaces, and energy dissipated by the experiment.

In order to generate an elegant design solution, one must identify the requirements of the ADR system. Naturally, those requirements are dependent upon the expected functions of the system. And when determining these functions, we inevitably must consider the overall instrument into which the ADR system fits. For instance, sometimes the most important specification of the system is for it to occupy the least amount of volume. Other times, it may be to have as low a mass as possible. Another important specification may be that the operating current should be as low as reasonably possible.

Once these priorities have been determined and a set of requirements identified, a design analysis can be performed to inform the final design. Part of this analysis is weighing competing specification goals and making necessary trade-offs. A thorough treatment of this topic is described in Shirron's 2014 paper, "Applications of the Magnetocaloric Effect in Single-Stage, Multi-Stage and Continuous Adiabatic Demagnetization Refrigerators."

10.4.2 Optimizing the Magnet Cycle

Users can obtain better performance from the system by considering the following effects.

10.4.2.1 Ballast Mass and How It Affects Base Temperature

As described in the adiabatic process, when reducing the magnetic field, the paramagnet cools toward its "ordering temperature." However, in practice, an ADR system will not ever actually reach the salt's ordering temperature because of the ballast mass that is necessarily included in the system. To review, ballast mass is anything that is thermally attached to the paramagnetic material, but that is not the salt itself.

For the sake of this review, let's identify a ***subsystem*** as that which includes all of the objects that are thermally connected to a particular salt pill, and that are, upon opening the heat switch, thermally isolated from other parts of the cryostat. Let's call this subsystem an "ADR stage." Because the salt is the only cooling source in an ADR stage, during the demagnetization phase, some of the salt's cooling energy is "spent" cooling the rest of the stage.

Reviewing Figure 10.1, the blue triangle represents the amount of cooling energy that is spent by the salt in cooling the ballast mass of the ADR stage, as the system cools from point *B*. It should be noted that the shape and size of the blue triangle is not precise and has been simplified for discussion sake.

Since an ADR is often used to cool an instrument—an experimental setup, for instance—we should also consider this instrument as ballast mass. Since some experiments can be fairly substantial in size and weight, they can have appreciable impact on the base temperature that the ADR can achieve.

When analyzing the impact the ballast mass will have on the ADR's performance, there are a couple things to keep in mind. The first is the material makeup of the ballast mass. What's important is the specific heat of the materials included. For instance, one will want to be aware of the amount of heat that must be removed from all ballast mass in order to cool it from the ADR's launch temperature (point B in Figure 10.1, typically around 3 K) down to the anticipated regulation temperature. And, therefore, the user will want to know the specific heat of those materials in that temperature range. The second issue to consider is how well the ballast mass is thermally coupled to the cooling path. (Cooling path means the portions of the ADR stage that are made of high thermal conductivity material, such as copper, through which heat is conducted on its way between the salt pill and the heat sink.)

If any portion of the mass is poorly coupled to the cooling path, it will take a relatively long time to be thoroughly cooled. This delay may mean that more than one ADR cycle is required to achieve an acceptable base temperature within a reasonable amount of time. In addition to possibly requiring additional ADR cycles to get sufficiently cold quickly, material that is poorly coupled to the cooling path will suffer from inefficient cooling and result in unnecessary entropy generation (or diminished cooling capacity). This is because being poorly coupled to the cooling path means that there is a delay in heat transfer. And this delay means that, for a given demagnetization rate, the heat absorbed by the paramagnetic material will occur when the paramagnet is colder than it would have been had the mass been better coupled. This increase in entropy is on the order of the ratio of the temperature of the mass being cooled, to the temperature of the salt material [28].

Poorly coupled masses are those that are either poor thermal conductors themselves (such as non-metallic components), or they have a poor conduction aspect ratio (such as a long, skinny component), or the conduction path connecting them to the cooling source is long and skinny, or, finally, they have a resistive interface between them and the conduction path.

10.4.2.2 Achieve a Good Soak

Since in practice the salt pill does not remain isothermal during the magnetization cycle (rather it warms up), in order to maximize the cooling energy delivered from the regeneration cycle, after the "mag up" step, the salt material must be allowed to fully cool back down to the heat sink temperature. The user can help insure this by allowing adequate equilibration time while the magnet is at full field.

10.4.2.3 Demagnetize the System Slowly

Because the salt pill assembly has finite thermal conductivity, any heat flow will result in a delta T across the conduction path (between the salt and components being cooled). This means that, (as described above in the ballast mass section) during the entire demagnetization process, the salt material absorbing the heat is colder than the part of the system from which the heat is originating. Therefore, slowly demagnetizing the system results in a smaller delta *T*, thereby allowing the heat from the experimental load to flow into the paramagnetic material while that material is at a higher temperature (than it would otherwise have been, if demagnetizing more quickly), where, thermodynamically, it has more cooling capacity. Additionally, decreasing the magnetic field slowly reduces the amount of eddy currents that develop in metallic parts. In sum, slowly demagnetizing the system generates less entropy, which the salt must absorb, than does a faster rate. Recall that using less of the salt's entropy capacity means more cooling capacity is available. However, there is a limit to slowing the demagnetization rate. Continuing to extend the demagnetization time will yield diminishing returns and, eventually, the parasitics of the system will outweigh any additional savings from reducing the delta *T* in the conduction path of the salt pill–ballast mass assembly.

10.4.2.4 Reduce the Launch Temperature

Although usually the launch temperature of an ADR system, based on the system design, is relatively fixed, sometimes it can be improved by paying close attention to the system setup, such as reducing conduction or radiation loading coming from the intermediate cooling stage. Often, the conduction losses can be reduced by improving the effectiveness of the thermal intercepts at higher temperatures, which results in a lower parasitic heat load to the ADR's heat-sink stage, yielding a lower launch temperature. For a given salt pill mass and a given magnetic field, a lower launch temp will yield higher cooling energy.

10.4.2.5 Subsequent Magnet Cycles

In practice, the experimental setup, and other components that contribute to ballast mass, contains materials that are weakly coupled to the salt pill and the heat switch's conduction path. This means that their thermal equilibration may be delayed by several minutes, or even hours. As a result, improvements in performance are often realized between the first and subsequent demagnetization cycles. A governing phenomenon here is that since heat flow is a function of delta *T*, as the temperatures of different components in a mechanical assembly asymptotically approach each other (such as during the initial cooldown of the cryostat, when the ADR and the attached experiment are approaching the cryocooler temperature), the heat flow rate falls accordingly. But when a demagnetization occurs, a relatively large delta *T* is generated, (where the components needing to be cooled are near, say, 3 K, and a new cooling source appears, near 30 mK), thereby providing a strong motivation for the remaining heat to flow. In this manner, the vast majority of any excess heat lingering in the ballast mass is drawn into the salt material, and, once there, can be readily removed via the heat-switch path. It should be noted here that, in most cases, allowing more time for thermal equilibration prior to executing a magnet cycle can have the same result as a subsequent magnet cycle.

Depending upon the design of the cryogenic cooling, various salts can be employed to achieve different temperatures. When selecting a particular paramagnetic salt for an ADR application, it is important to consider the following characteristics of the salt: ordering temperature, heat capacity, cooling density.

10.5 Fundamental Distinctions between Single-Shot ADRs and CADRs

In order to provide continuous cooling, ADR systems can be devised using multiple magnets, salt pills, and heat switches. CADRs are significantly more complex than single-shot ADRs but offer the great advantage of being able to maintain their regulation temperature indefinitely, therefore providing continuous cooling. CADRs are discussed in detail in [29–31].

First let's establish that, from an architectural standpoint, a CADR is always a combination of two or more single-stage ADR units [32]. See Figure 10.6. And for review, in using the term "single-stage ADR" we mean a grouping of a salt pill, a superconducting magnet, and a heat switch. In constructing a CADR, these single-stage ADR building blocks can be configured in either a parallel or series configuration, or both parallel and series, as shown in Figure 10.7. Here are some distinctions between parallel and series configurations.

- In a parallel design, two identical stages are used, whereas in a series approach, the stages can differ in choice of salt material, salt pill size, magnet size, etc.
- In a parallel design, both component ADRs operate over the full temperature range of the ADR. Whereas in a series design, each component ADR operates over a portion of the ADR system's temperature range. This difference will naturally lead to using smaller-field-strength magnets than one would need in a parallel configuration, as well as choosing different salts for the various stages since different salts can be better optimized for the different temperature ranges.
- In a parallel design, the identical stages provide cooling to the cold plate and also reject heat to the heat sink. In a series approach, only the uppermost stage rejects heat to the heat sink.
- As the "continuous" name suggests, both parallel- and series-type ADRs provide continuous cooling. However, they differ in that only parallel-type CADRs continually expel heat to the heat sink. In a series-type CADR, the warmest stage divides its time between cooling the next colder stage and expelling heat to the heat sink.

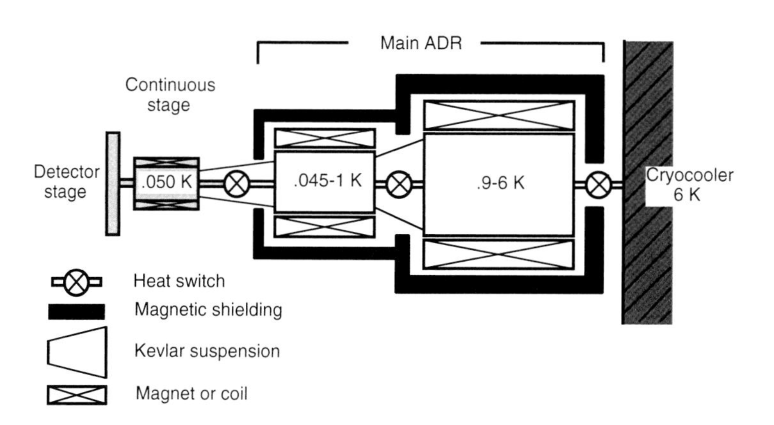

FIGURE 10.6 A three-stage, series-type CADR. (Courtesy of NASA GSFC, https://cryo.gsfc.nasa.gov/ADR/adv_ADR/adv_ADR.html.)

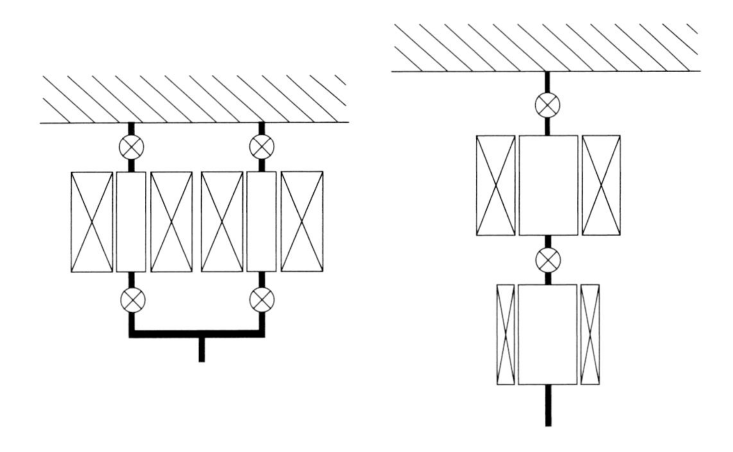

FIGURE 10.7 Comparison of parallel and series configurations of two-stage CADRs. (Courtesy of NASA GSFC; Shirron, P., *Applications of the Magnetocaloric Effect in Single-Stage, Multi-Stage and Continuous Adiabatic Demagnetization Refrigerators*, pg 15.)

- Since a CADR's performance is determined by the heat rate it can reject to the heat sink, a similarly sized parallel system will provide about twice the cooling power since it rejects heat nearly 100% of the time, where in a series configuration that value is about 50%.
- Parallel CADRs require two heat switches per stage, while series types can be implemented with one heat switch per stage. See Figure 10.8.

From a design standpoint, there are several distinctions between single-shot ADRs and CADRs. Here we will identify and discuss them.

Heat switch conductance ratio: Where in a single-shot ADR one of the main goals is to have the hold time be as long as possible, it follows then that the parasitic heat load from the heat switch in its ***off*** (open) state should be as small as possible. For instance, a mechanical heat switch that has zero conductance in its open state is most ideal for

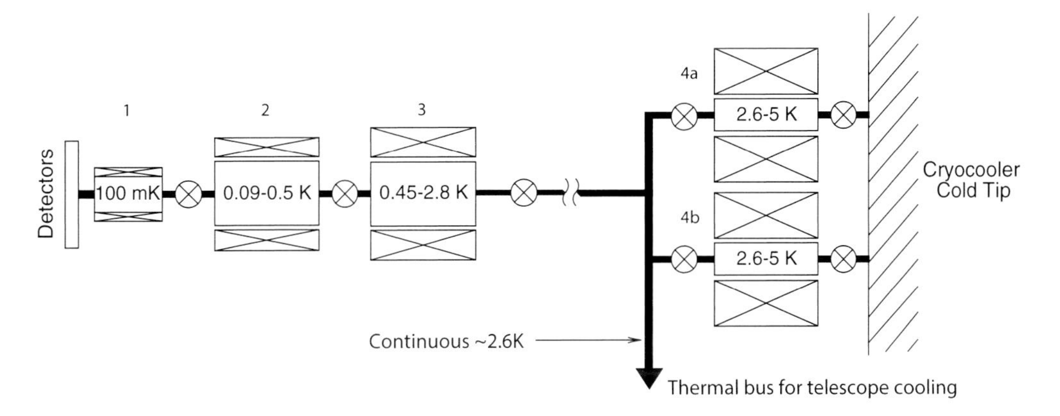

FIGURE 10.8 Schematic of a system containing series and parallel CADR units. (Courtesy of NASA GSFC; Shirron, P., *Applications of the Magnetocaloric Effect in Single-Stage, Multi-Stage and Continuous Adiabatic Demagnetization Refrigerators*, pg 17.)

a single-shot system. But since in a CADR, in order to maximize cooling power, the regeneration time of any particular stage wants to be relatively short (compared to its cooling period), the figure of merit of the heat switch in a CADR is the ratio of its ***on*** (closed) versus ***off*** (open) conductance values. A conductance ratio of about ten thousand is achievable [35]. If a conductance ratio of several thousand can be achieved in a CADR switch, we can reason that the parasitic load from the switch will not limit the CADR's performance, in any real sense.

Thermal conductance of the salt pills: For a salt pill of a given size and given salt material, the conductance of the salt pill in a CADR needs to be much greater than that of a single-shot system. This means that in CADRs, it's necessary to budget a significant percentage of the salt pill's volume and mass to the thermal bus. From a design standpoint, the goal is to minimize the thermal resistance between any part of the salt crystal and the thermal bus termination on the outside of the salt pill. When studying the salt pill matrix, this in turn means that it's desirable to reduce the distance that any part of the salt crystal is from the closest element of the thermal bus. While in a single-shot ADR the percentage of the salt pill volume dedicated to the thermal bus is around 2% or 3%, in a CADR that value approaches 30% [36].

Thermal conductance of the conduction path between the salt and the cooling source: Because the regeneration rate of any CADR stage is not only limited by the rate at which heat can be extracted from the salt material, but it also requires getting that heat conducted all the way to the cooling source, the conduction path between the pill's thermal bus and the cooler is likewise also of utmost importance.

Number of heat switches required: Where a single-shot ADR can be operated with as few as one heat switch, CADRs require at least two switches. The total number of switches required is a function of the system design.

10.6 Description of Common Commercial ADRs

10.6.1 A Common Commercial Two-Stage ADR

The HPD Model 102 Denali was one of the first commercially available dry (liquid cryogen-free) ADRs. Developed in partnership with NIST, the original purpose of the Model 102 was implementation onto a scanning electron microscope (SEM). This SEM requirement motivated the unique rectangular shape (Figure 10.9).

In the Model 102, a superconducting magnet is used to generate magnetic fields of 4 T. In the bore of the magnet is positioned the paramagnetic salt pills, as shown in Figure 10.10. The magnet is cooled by the cryostat's 3 K stage. The salt pills are supported from the 3 K stage using Kevlar thread in a patented suspension design [37].

A critical feature of the Model 102 is the patented suspension design, where the tension in the Kevlar threads is maintained by utilizing metals that have differential coefficients of thermal expansion, as shown in Figures 10.11 and 10.12.

From a thermal standpoint, each of the salt pills represents one of the ADR stages. To reduce the complexity of the system, a single magnet is employed. This means that the two salt pills share the magnet bore.

Earlier we discussed regulating the magnet current in order to maintain a desired temperature. Because, in a two-stage ADR, both salt pills occupy the same magnet bore,

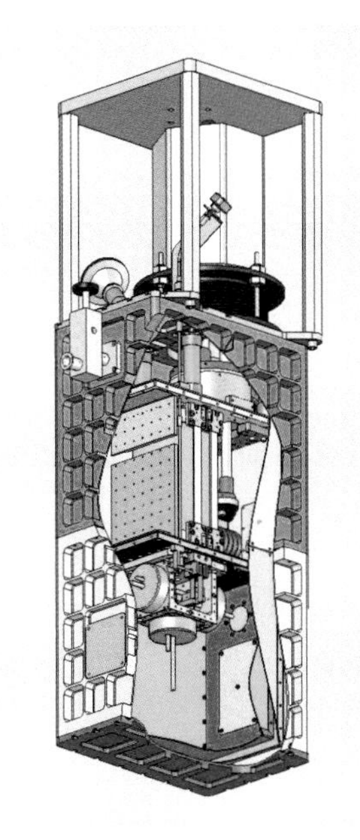

FIGURE 10.9 HPD Model 102 Denali ADR Cryostat.

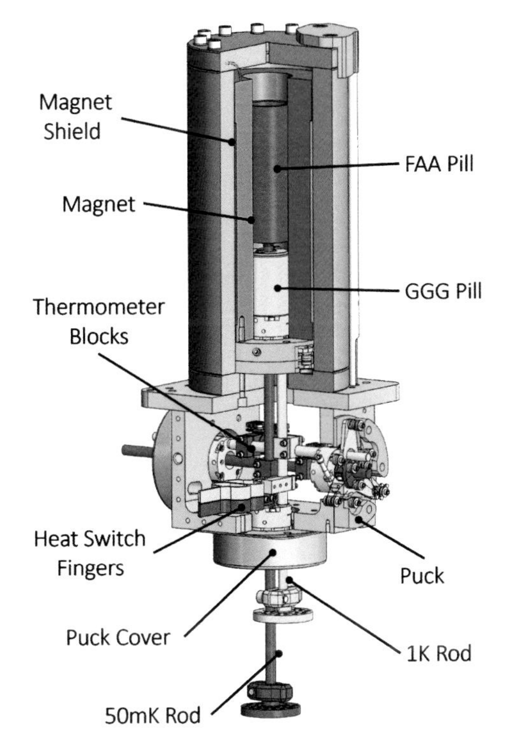

FIGURE 10.10 Cross section of common two-stage ADR subsystem.

FIGURE 10.11 Patented two-stage Kevlar suspension. (Courtesy of HPD; Bartlett, J. et al., *Cryogenics*, 50, 647–652, 2010.)

FIGURE 10.12 Two-Stage ADR in an HPD Model 102 Cryostat, employing the NIST-patented Kevlar suspension. (Courtesy of HPD.)

if we regulate the magnet current to maintain a certain temperature of the FAA stage, the temperature of the other salt pill (in this case, the GGG) will be undefined, and will simply be a function of the magnet current and the pill's entropy state (which is a function of how much heat the GGG salt pill has been absorbing since its last regeneration cycle). As it turns out, this situation is perfectly acceptable, since the GGG stage temperature can vary widely and still serve as an effective thermal intercept. Although having a colder intercept is always preferable, as being colder more effectively protects the next colder stage from parasitic heat flow, which affects the hold time.

10.6.2 A Common Commercial Single-Stage, Helium-3-Backed ADR

The Model 107 K2 cryostat includes a mechanical pulse-tube cryocooler refrigerator with vibration isolation features integrated and generates stage temperatures of 50 and 3 K. Next, a Helium-3 adsorption refrigerator is employed to establish a 300 mK intercept stage, providing 15 J of usable cooling capacity at the stage. The final experimental temperature stage has a user-selectable set point, but it operates nominally at 50 mK and is cooled by the single-stage ADR using an FAA salt pill, which delivers 270 mJ of cooling for the experiment.

The scientific argument for the Model 107 originates with the desire to operate large-scale arrays of superconducting sensors near 100 mK. Applications for such sensors include X-ray materials analysis, nuclear materials accounting, and astrophysics. By employing multiplexing, arrays of thousands of pixels can be used while the number of connections to room temperature is much smaller. But multiplexing requires coax cables that carry GHz frequency signals, and coax cables result in greater thermal load than do lower bandwidth connections. The greater cooling capacity of the Model 107 ADR, due to its ^{3}He stage as well as the larger FAA salt pill, makes it well-suited for such applications.

Images below are of an HPD Model 107 K2 cryostats in service at NIST Boulder (Figures 10.13 and 10.14).

FIGURE 10.13 HPD Model 107 K2 Cryostat, in service at NIST. (Courtesy of Charlie Danaher.)

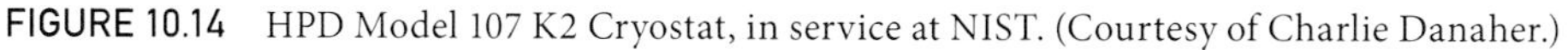

FIGURE 10.14 HPD Model 107 K2 Cryostat, in service at NIST. (Courtesy of Charlie Danaher.)

10.6.3 A Common Commercial Large Volume, Two-Stage ADR

The HPD Model 104 Olympus cryostat has been used extensively at The University of Colorado by the Center for Astrophysics and Space Astronomy (CU-CASA, one of the collaborating institutions working at the South Pole Telescope [SPT]). The most attractive feature of the Olympus cryostat is its large experimental volume. Inside the 3 K radiation shield, scientists can take advantage of a volume measuring 17″ in diameter and 23″ in height. The Model 104 is commonly configured with a two-stage ADR with base temperatures near 30 mK and typical operating temperature around 100 mK (Figures 10.15 and 10.16).

The argument for the Model 104 starts with the science that it supports.

Astronomers and cosmologists are eternally curious about our universe. This curiosity motivates ever more sophisticated observation methods. Land-based telescopes continue to be one of the main instruments employed to study the sky.

The receivers (or cameras)—the part of the instrument where the light is converted to electrical signals—on these telescopes are where many of the advances in hardware are taking place.

Some of the recent work has been concentrated on developing cryogenic polarimeters and multichroic pixels. Scientists have developed large arrays of such detectors, which are extremely sensitive to light coming from the sky.

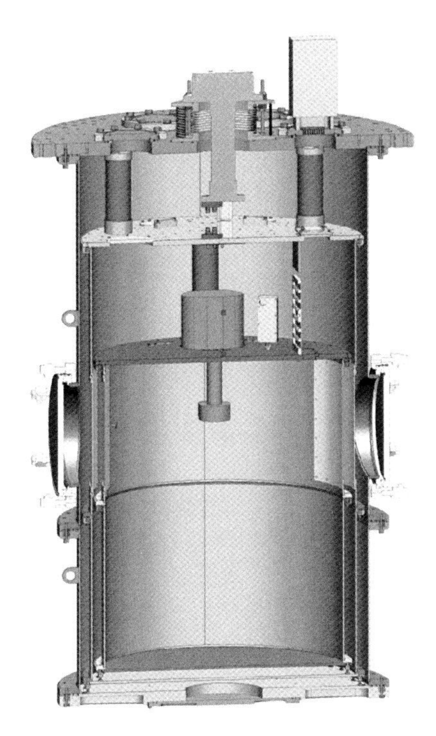

FIGURE 10.15 Section view of HPD Model 104 Olympus.

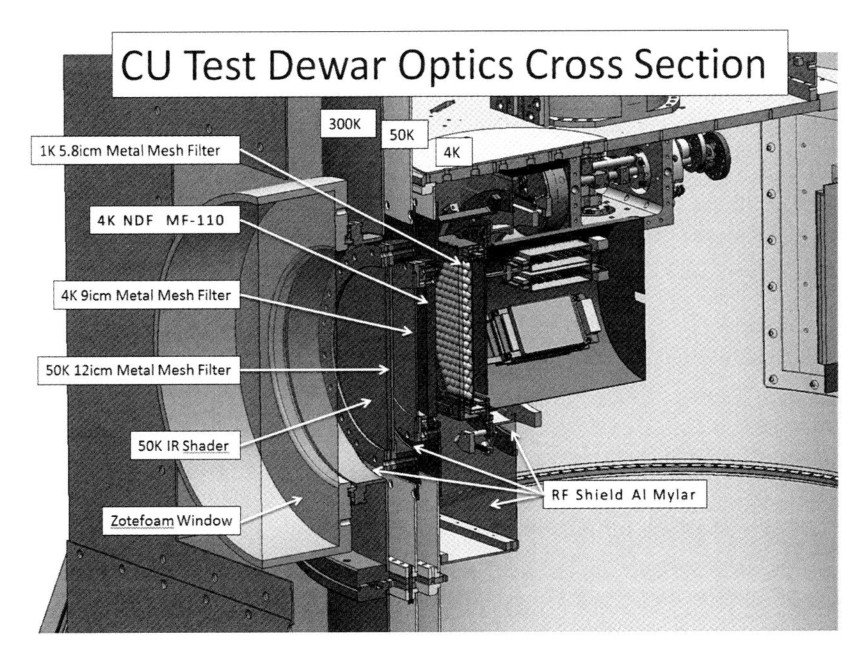

FIGURE 10.16 Section view of Model 104 Olympus cryostat testing 6″ wafers for POLARBEAR-2. (Courtesy of Greg Jaehnig, CU-CASA.)

These advances are pursuing better measurements of the polarization and amplitude of the cosmic microwave background (CMB) radiation. These improved measurements can lead to a more complete understanding of aspects of the universe, such as neutrino physics, inflation (a proposed period of rapid expansion shortly after the big bang), and gravitational waves, to name just a few.

But long before these sensitive detector arrays are deployed in the telescopes around the globe, years of development and testing occur in the laboratory. Prototype components—including cryogenic detectors, feedhorn-waveguide assemblies, and the various electronics—are placed into laboratory cryostats for characterization and testing.

Among the telescope observatories pursuing such research, and that have been either directly or indirectly supported by research at CU-CASA, are:

- *The South Pole Telescope* (*SPT*): A 10-m diameter telescope located at the Amundsen–Scott South Pole Station, Antarctica.
- *The Atacama Cosmology Telescope* (*ACT*): A 6-m telescope on Cerro Toco in the Atacama Desert in the north of Chile.
- *POLARBEAR*: A cosmic microwave background polarization experiment located in the Atacama Desert of northern Chile.

10.6.4 Improved Detector Designs Lead to Compressed Observation Runs

One of the biggest pushes in detector advancement is toward the goal of compressing observation time. If a telescope can collect the same amount of information as previously but in a fraction of the time, or, for a given observation period, if more information can be gathered, such improvement can both reduce exploration costs as well as bring closer a more complete understanding of the universe. Multichroic pixels, having several detectors each, multiply the amount of information gathered, for a given time period, without any cryogenic cost.

The Cosmology Large Angular Scale Surveyor (CLASS) is an array of microwave telescopes currently under construction at Johns Hopkins University that will be deployed to a high-altitude site in the Atacama Desert of Chile as part of the Parque Astronómico de Atacama in 2015. The CLASS experiment aims to test the theory of cosmic inflation and distinguish between inflationary models of the very early universe by making precise measurements of the polarization of the CMB.

CLASS scientists at Johns Hopkins University have employed the Model 104 ADR cryostat to test and guide the development of a new design of detectors that will eventually be employed in the CLASS telescopes (Figure 10.17).

In a just a few years, novel detector arrays, currently being validated in the laboratory, will see their first light at these telescopes. Their observation data will advance our knowledge of the structure and history of the universe. HPD is proud to contribute toward this noble effort by providing physicists with the basic instruments needed to perform such exciting research (Figure 10.18).

FIGURE 10.17 Model 104 hosting optical testing of detectors at Johns Hopkins University. (Courtesy of David Larson, CLASS collaboration.)

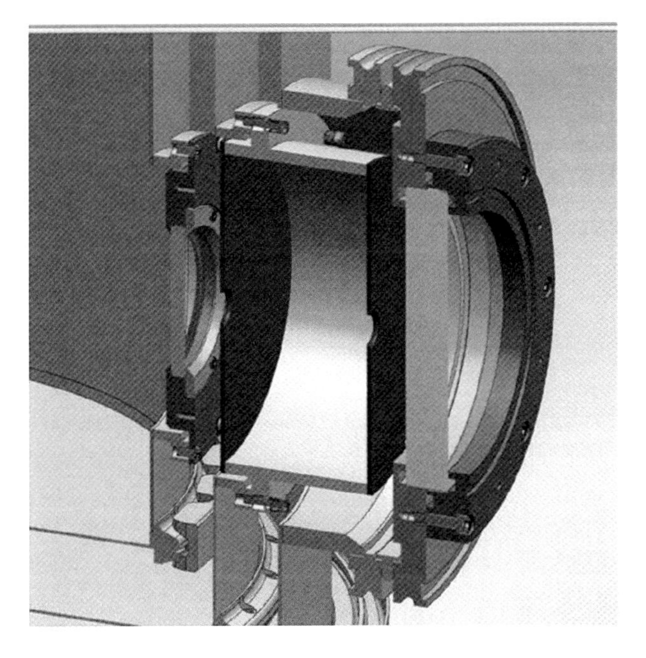

FIGURE 10.18 HPD Model 103 ADR Cryostat with side ports and optics, photo provided by FermiLab, window and filter assembly designed by Greg Derylo.

References

1. Debye, P., Einige Bemerkungen zur Magnetisierung bei tiefer Temperatur, *Ann. Phys.* 81, 386(25): 1154–1160 (1926).
2. Giauque, W. F., A Thermodynamic treatment of certain magnetic effects. A proposed method of producing temperatures considerably below 1° absolute. *J. Am. Chem. Soc.* 19, 1864–1870 (1926).
3. Shirron, P., *Applications of the Magnetocaloric Effect in Single-Stage, Multi-Stage and Continuous Adiabatic Demagnetization Refrigerators,* pg 2.
4. Shirron, P., *Applications of the Magnetocaloric Effect in Single-Stage, Multi-Stage and Continuous Adiabatic Demagnetization Refrigerators,* pg 2.
5. Weisand, J., Cold Facts, 2011.
6. Unpublished results from tests conducted with the HPD 107 K2 Cryostat, when launching from 290 mK and a 4 T field.
7. Wikus, P., et al., *Optimum operating regimes of common paramagnetic refrigerants,* pg 1.
8. Wikus, P., et al., *Optimum operating regimes of common paramagnetic refrigerants,* pg 1.
9. Shirron, P., *Applications of the Magnetocaloric Effect in Single-Stage, Multi-Stage and Continuous Adiabatic Demagnetization Refrigerators,* pg 3.
10. Shirron, P., McCammon, D., *Salt Pill Design and Fabrication for Adiabatic Demagnetization Refrigerators,* pg 1.
11. Wikus, P., et al., *Magnetocaloric materials and the optimization of cooling power density,* pg 1.
12. Wikus, P., et al., *Magnetocaloric materials and the optimization of cooling power density,* pg 1.
13. Shirron, P., *Applications of the Magnetocaloric Effect in Single-Stage, Multi-Stage and Continuous Adiabatic Demagnetization Refrigerators,* pg 3.
14. Wikus, P., et al., *Magnetocaloric materials and the optimization of cooling power density,* pg 3.
15. Pobell, F., *Material and Methods at Low Temperatures,* 3rd edition, Springer-Verlag, Berlin, Germany, 2006, pg 205.
16. Wikus, P., et al., *Magnetocaloric materials and the optimization of cooling power density,* pg 3.
17. Wikus, P., et al., *Magnetocaloric materials and the optimization of cooling power density,* pg 2.
18. Pobell, F., *Material and Methods at Low Temperatures,* 3rd edition, Springer-Verlag, Berlin, Germany, 2006, pg 208.
19. Pobell, F., *Material and Methods at Low Temperatures,* 3rd edition, Springer-Verlag, Berlin, Germany, 2006, pg 207.
20. Wikus, P., et al., *Magnetocaloric materials and the optimization of cooling power density,* pg 5.
21. Wikus, P., et al., *Magnetocaloric materials and the optimization of cooling power density,* pg 4–5.
22. Wikus, P., et al., *Magnetocaloric materials and the optimization of cooling power density,* pg 5.
23. Shirron, P., McCammon, D., *Salt Pill Design and Fabrication for Adiabatic Demagnetization Refrigerators,* pg 3.
24. Pobell, F., *Material and Methods at Low Temperatures,* 3rd edition, Springer-Verlag, Berlin, Germany, 2006, pg 206.
25. Pobell, F., *Material and Methods at Low Temperatures,* 3rd edition, Springer-Verlag, Berlin, Germany, 2006, pg 208.
26. Pobell, F., *Material and Methods at Low Temperatures,* 3rd edition, Springer-Verlag, Berlin, Germany, 2006, pg 208.
27. Shirron, P., *Applications of the Magnetocaloric Effect in Single-Stage, Multi-Stage and Continuous Adiabatic Demagnetization Refrigerators,* pg 7.
28. Shirron, P., *Applications of the Magnetocaloric Effect in Single-Stage, Multi-Stage and Continuous Adiabatic Demagnetization Refrigerators,* pg 10.
29. Tuttle, J., et al., *Development of a space-flight ADR providing continuous cooling at 50 mK with heat rejection at 10 K.*
30. Duval, J. M., et al., *50 mK Continuous Cooling with ADRs Coupled To He-3 Sorption Cooler.*
31. A yet cited CADR source.
32. Shirron, P., *Applications of the Magnetocaloric Effect in Single-Stage, Multi-Stage and Continuous Adiabatic Demagnetization Refrigerators,* pg 14.
33. Shirron, P., *Applications of the Magnetocaloric Effect in Single-Stage, Multi-Stage and Continuous Adiabatic Demagnetization Refrigerators,* pg 15.

34. Shirron, P., *Applications of the Magnetocaloric Effect in Single-Stage, Multi-Stage and Continuous Adiabatic Demagnetization Refrigerators*, pg 17.
35. Bartlett, J., et al., Thermal characterization of a tungsten magnetoresistive heat switch, Cryogenics 50, (2010) 647–652.
36. Shirron, P., McCammon, D., *Salt Pill Design and Fabrication for Adiabatic Demagnetization Refrigerators*, pg 2 & 10.
37. The NIST patent for the two-stage ADR suspension is number 5,934,077.

11. Cryogen-Free Dilution Refrigerator Systems

Zuyu Zhao

11.1 Introduction

A dilution refrigerator (referred to as DR in future discussion) can be operated continuously (i.e., with infinite "holding time") at temperatures of a few to a few dozens of millikelvin. A DR can also provide a few hundred microwatt to a few milliwatts of cooling power at 100 mK.

The evolution of the DR occurred in the following three stages.

11.1.1 1960s–1970s: Realization, Commercialization, and Technology Development [1–4]

The idea of the DR was first proposed by Heinz London in the early 1950s, followed by an improved version in 1962 [2].

Their ideas were quickly realized in the mid-1960s. Examples of systems and pioneers who had developed these systems include, but are not being limited to, the DRs at Leiden University in 1965 [5]; a system developed by Das, de Bruyn, and Taconis in 1965 [6]; the DR at the University of Manchester by Hall, Ford, and Thompson in 1966 [7]; Nagano, B.S., Borisov, N.S., and Limburg, M., Yu in 1966 [8], among others.

John Wheatley was certainly another pioneer of the DR development for his creative contribution in developing copper step heat exchangers, which are the key components for reaching below 40 mK in a DR.

From 1965 to 1967, John Wheatley quickly developed multiple versions of DRs based on his own design and calculations. In his seminar celebrating 50 years of DR's Vilches introduced Wheatley's first few versions of systems as shown in Figure 11.1 [9].

The first version of Wheatley's refrigerator with four small heat exchangers reached 40 mK on the first try. The machine was then completely dismantled. Each component was tested and studied individually for better understanding and improvement. The complete second version started running one month later. The continuous mode reached the base temperature of 20 mK while the single shot mode reached 13.7 mK. Wheatley and Vilches developed the first single-cycle refrigerator as illustrated in Figure 11.1b.

A great deal of effort was focused on the development of heat exchangers during that period. Discrete heat exchangers with large surface areas were a particular topic of focus. More effective sintered silver heat exchangers developed in 1970s replaced copper powder heat exchangers and enabled the systems to reach approximately 2 mK.

Several excellent review articles [10–13] were published in 1970s that reviewed the progress and achievements of DR development during that period. More profound theories on the DR, including quantitative analysis of thermodynamics, were also developed [14].

The first commercial DR was developed in collaboration with Heinz London at Oxford Instruments factory in the UK in 1967 [15]. In the same year, Oxford Instruments won its first Queen's Award.

Rapid progress has been made since then.

11.1.2 1980s–1990s: Mature Technologies Resulted in Rapid Development for Wide-Range Applications

DR technology became mature during these two decades. The DR has become a standard and mature commercially available product.

Highly custom designed DR systems for various types of physical research have been developed. One example is the plastic DRs for very high magnetic fields, along with highly custom designed DRs for Scanning Tunneling Microscope (STM) applications, quantum computing, and many other applications.

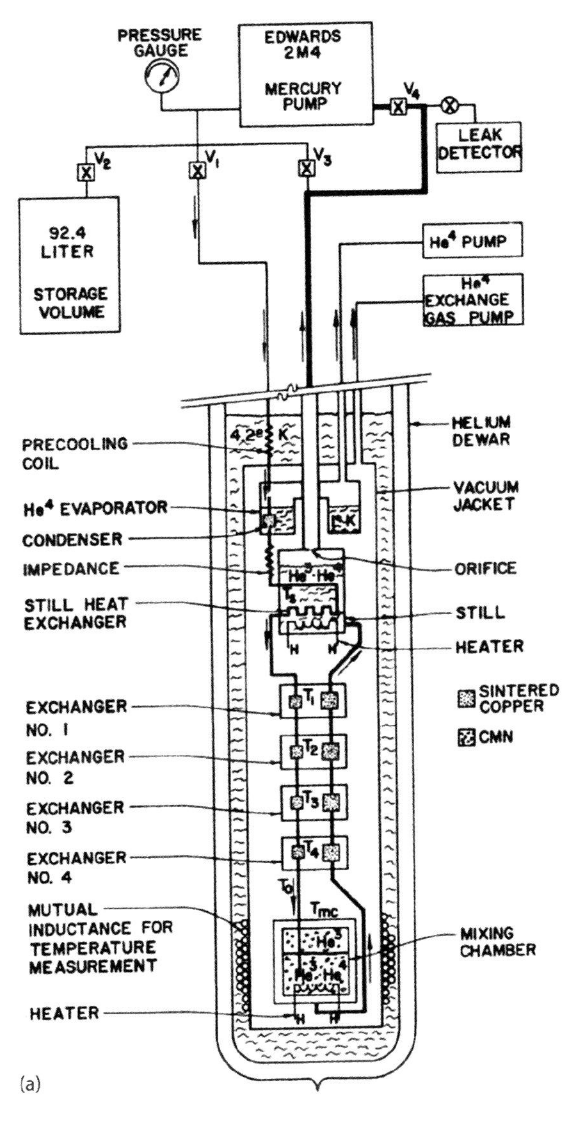

FIGURE 11.1 (a) Wheatley's four exchangers version dilution refrigerator design. (*Continued*)

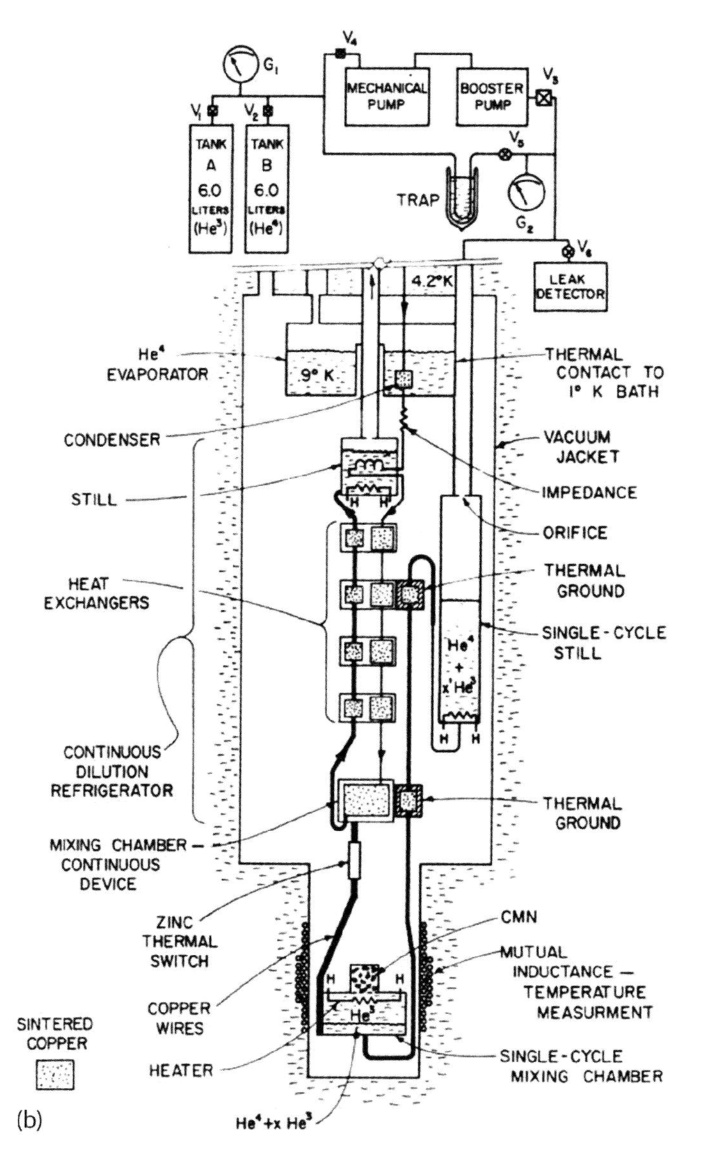

FIGURE 11.1 (Continued) (b) Design of the *one-cycle dilution refrigerator* attached to the four exchangers version of the continuous one reached 4.5 mK. (*Continued*)

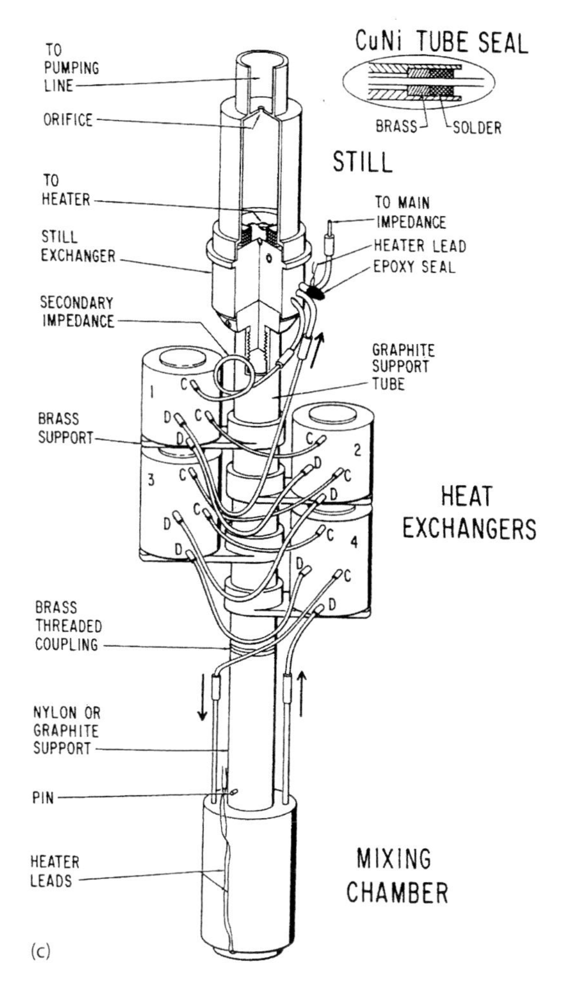

FIGURE 11.1 (Continued) (c) Wheatley's second four exchangers version dilution refrigerator design.

11.1.3 2000s–2018: Cryogen-Free Systems

The cryogen-free DR system manifests a new milestone of DR development in the twenty-first century. Vibration reduction on the cryogen-free DR system has become a new technical challenge for the users in certain applications.

11.2 3He–4He Mixture: Its Phase Diagram and Properties [16,17]

A mixture of 3He and 4He, usually called ***3He–4He mixture,*** is the coolant of DR systems. Illustrated in Figure 11.2 is the phase diagram of (liquid) mixture in ***temperature–concentration plane*** in zero magnetic field under its saturated vapor pressure (SVP).

The *X*-axis represents the 3He "***atomic concentration***" referred to as χ_3.

$$\chi_3 = \mathbf{N}_3/(\mathbf{N}_3 + \mathbf{N}_4) \tag{11.1}$$

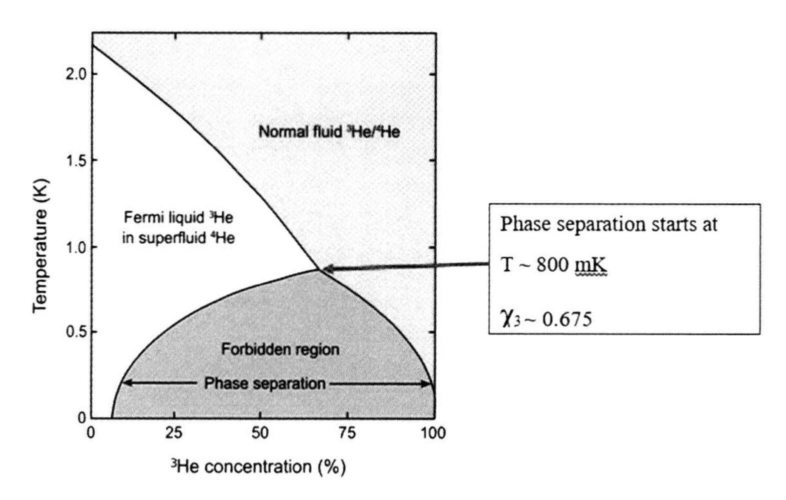

FIGURE 11.2 Phase diagram of 3He–4He mixture under saturated vapor pressure. (From Cryocourse 2016, School and Workshop, in *Cryogenics and Quantum Engineering*, Aalto University, Espoo, Finland, September 26–October 3, 2016.)

Therefore, the 4He concentration should be

$$\chi_4 = \mathbf{N_4/(N_3 + N_4)} \tag{11.2}$$

where N_3 and N_4 are the numbers of 3He atoms and 4He atoms in the mixture, respectively, often referred to as the molar numbers.

The *Y*-axis represents the absolute temperature (K).

For an example, $\chi_3 = 0$ when the mixture is pure 4He, which is represented by *Y*-axis. The 4He superfluid transition takes place at 2.17 K.

Figure 11.2 is also called *T ~ χ phase diagram of the 3He–4He mixture.*

The 3He–4He mixture has several unique properties:

- The λ-point temperature where normal liquid 4He becomes superfluid is depressed when 3He is added to 4He, that is, the superfluid transition temperature decreases when the 3He concentration, χ_3, increases;
- The mixture undergoes spontaneous ***phase separation*** into a 3He-rich phase (referred to as the concentrated phase) and a 3He-poor phase (referred to as the dilute phase, or 4He-rich phase) around 0.8 K when $\chi_3 = 0.67$;
- Two phase lines (or "branches") exist below the transition temperature as shown in Figure 11.2. Both 3He and 4He concentrations will change along these two lines respectively when the temperature decreases. The region under these two phase separation lines is non-stable for the mixture;
- **3He concentrated phase:** The branch with $\chi_3 > 0.67$, denoted as χ_{3c}, is ***3He concentrate phase***, where
  - ***4He concentration*** in this phase, denoted as χÇ$_{4d}$, decreases exponentially when the temperature decreases, as expressed by Eq. 11.3:

$$\chi_{4d} = \mathbf{0.85T^{1.5}e^{-0.56/T}} \text{ [19]} \tag{11.3}$$

- The concentration of 3He is $\chi_{3c} = 1 - \chi_{4d}$, and it reaches one (1) at T = 0;
- The liquid 4He in the concentrated 3He solution is ***normal liquid***;

- **3He dilute phase:** The branch with $\chi_3 < 0.67$, denoted as χ_{3d}, is ***3He dilute phase,*** where
  - The 3He concentration and its finite solubility in 4He in this phase at low temperatures are of great importance. Edwards et al. [20] measured the 3He concentration along the solubility curve as shown in Figure 11.3a; and
  - ***3He concentration*** in this dilute phase decreases with temperature and fits the empirical Eq. 11.4 below 0.15 K.

$$\chi_{3d} = \chi_0 (1 + \beta T^2) \text{ where } \beta = 10.4 \text{ K}^{-2} \tag{11.4}$$

$At\ T = 0$

$$\chi_0 = \chi(\mathbf{6.40 \pm 0.07})\% \ [20,21]$$

Both χ_0 and β vary with temperature and pressure [22]. For example, the solubility of 3He in 4He can be increased to almost 9.5% by raising the pressure to 10 bars [22,23].

The behavior of 4He in the 3He dilute phase is of great interest.

The binding energy of a single 3He atom to liquid 4He is larger than the binding energy to liquid 3He to other atoms of 3He at T~0 K. Therefore, 3He will dissolve in liquid 4He even at absolute zero temperature. The finite solubility of 3He in liquid 4He is a crucial factor to make the refrigeration happen.

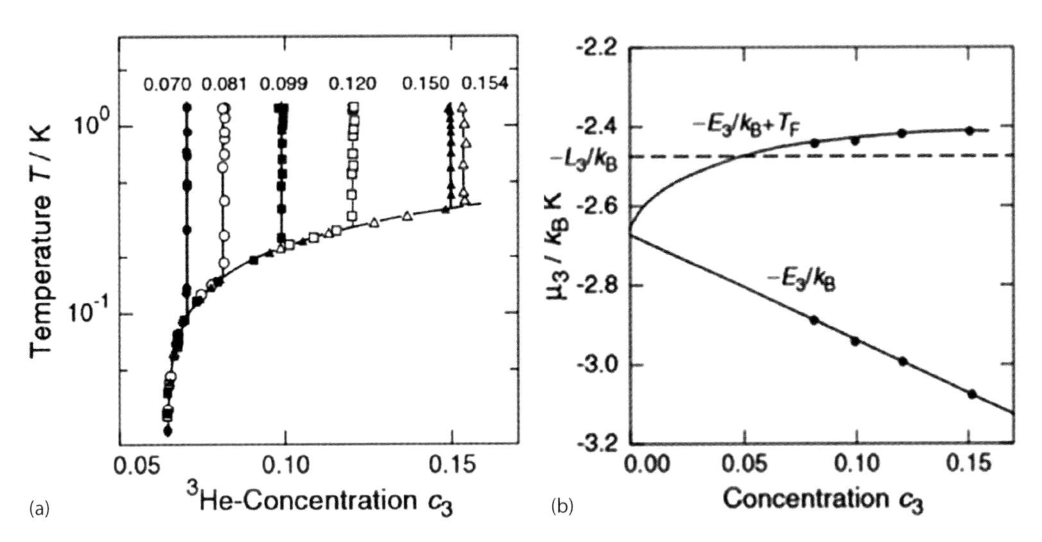

FIGURE 11.3 (a) The solubility cure for 3He in 4He at the SVP (Edwards et al. 1969). (b) Determination of the equilibrium concentration of 3He in 4He at the absolute zero. *X*-axis: 3He concentrate in dilute phase. *Y*-axis: Chemical potentials.

The liquid 4He in the diluted 3He solution is a ***superfluid*** [19]. 3He atoms located inside the DR can be driven by the osmotic pressure gradient and move up in the superfluid 4He along the proper channel during the circulation (refer to Section 11.3.2).

- The λ-line intercepts the phase separation line that generates a three-phase boundary. The point of interception is called the tricritical point, TCP;
- 3He and 4He mixture is the only system that experiences a phase separation due to the mass difference only; and
- After phase separation, the 3He-rich liquid will "float" on top of the 3He-diluted liquid due to its smaller density.

Understanding the finite, non-zero 3He solubility in 4He is an important step to investigate the concentration of 3He in 4He.

The thermodynamic properties of the 3He–4He mixture are a function of temperature and pressure, just like that of pure 3He and 4He. The following discussion will be focused on the saturated liquid at or very close to 0 K.

Assume $\boldsymbol{\mu}_{3c}(\boldsymbol{T})$ and $\boldsymbol{\mu}_{3d}(\boldsymbol{T})$ are the chemical potentials of 3He at temperature T in the concentrated phase and dilute phase, respectively. Then $\boldsymbol{\mu}_{3c}(\boldsymbol{T})$ is a constant determined by the latent heat, $-\boldsymbol{L}_{3He}$, of liquid 3He under that temperature T. $\boldsymbol{\mu}_{3d}(\boldsymbol{T})$ is a function of 3He concentration, $\boldsymbol{\chi}_{3d}$. At the same time, it is enhanced by the Fermi energy, that is, $\mu_{3d}(T) = -\mathrm{E}_3 + \mathrm{k_B T_F}$ due to the Pauli Exclusion principle, where $-\boldsymbol{E}_3$ is its binding energy, approximately a linear function of temperature as shown in Figure 11.3b.

The chemical potential of 3He in the concentrate phase is equal to the chemical potential of 3He in the dilute phase at the phase separation boundary when this two-phase system is under thermodynamic equilibrium, that is, $\mu_{3c}(T)$ intercepts $\mu_{3d}(T)$ at $\chi_{3d} = 6.6\%$ at $T = 0$ K under the mixture SVP (Figure 11.3b).

The preceding conclusion implies that a reduction in the concentricity of the dilute phase, $\boldsymbol{\chi}_{3d}$, will unbalance the chemical potential at the phase boundary. In that case, the 3He from the concentrated phase will cross the phase boundary and enter the dilute phase to rebalance the chemical potential.

We now focus on the thermodynamic process of the DR and explain why cooling occurs. **The discussion is limited to "zero" temperature under the mixture SVP, where $\boldsymbol{\chi}_{3d} = 6.6\%$.**

Take:

$\boldsymbol{C}_{3c}$ $(\boldsymbol{T})$: Heat capacity of concentrated 3He at temperature T
$\boldsymbol{C}_{3d}$ $(\boldsymbol{T})$: Heat capacity of dilute 3He at temperature T
$\boldsymbol{H}_{3c}$ $(\boldsymbol{T})$: Enthalpy of concentrated 3He at temperature T
$\boldsymbol{H}_{3d}$ $(\boldsymbol{T})$: Enthalpy of dilute 3He at temperature T
$\boldsymbol{T}_c$: Temperature of concentrated 3He
$\boldsymbol{T}_d$: Temperature of dilute 3He
$\boldsymbol{T}_B$: Temperature at the phase separation boundary
$\boldsymbol{S}_{3c}$ $(\boldsymbol{T})$: Entropy of concentrated 3He at temperature T
$\boldsymbol{S}_{3d}$ $(\boldsymbol{T})$: Entropy of dilute 3He at temperature T

Take the measured values:

$$C_{3c}(T) = 22T[\text{Joule/mole} \cdot \text{K}] \quad (11.5)$$

$$C_{3d}(T) = 106T[\text{Joule/mole} \cdot \text{K}] \quad (11.6)$$

$$H_{3c}(T=0) = 0,$$

and

we get the following enthalpies of the concentrated liquid 3He:

$$H_{3c}(T) = \int C_{3c}(T)\, dT = 11.0T_c^2[\text{Joule/mole}] \quad (11.7)$$

H_{3d} (T) calculation takes a few more steps.

Start from $\mu_{3c}(T) = \mu_{3d}(T)$ at the phase separation boundary, we have:

$$H_{3c}(T) - TS_{3c}(T) = H_{3d}(T) - TS_{3d}(T)$$

$$H_{3d}(T) = H_{3c}(T) + T_{3d}S_{3d}(T) - T_{3c}S_{3c}(T)$$

$$= 11.0T_{3c}^2 + T_{3d}\int C_{3d}(T')/T'\, dT' - T_{3c}\int C3c(T')/T'\, dT'$$

$$= 106T_{3d}^2 - 11.0T_{3c}^2 \quad (11.8)$$

The enthalpy change of 3He at the phase separation boundary = Equation 11.8 – Equation 11.7.

Take the temperature at the phase separation boundary, T_B,

$$H_{3d}(T) = 95.0T_B^2[\text{Joule/mole}] \quad (11.9)$$

$$H_{3c}(T) = 11.0T_B^2[\text{Joule/mole}] \quad (11.10)$$

The enthalpy change of 3He at the phase separation boundary becomes

$$H_{3d}(T) - H_{3c}(T) = 84T_B^2[\text{Joule/mole}] > 0 \quad (11.11)$$

Conclusions:

1. The enthalpy of 3He atoms in the dilute phase is greater than that in the concentrated phase. Therefore, the enthalpy of 3He atoms in the concentrated phase increases when they pass the phase boundary into the dilute phase.
2. The above result implies that an endothermic process occurs when 3He atoms pass the phase boundary. Energy will be taken from the "environment" during that process and cooling occurs.

3. The enthalpy difference of 3He at the phase separation boundary is proportional to the square of the temperature at the phase separation boundary per Equation 11.11 and can be used to estimate the cooling power of the DR. Pragmatically, the *mixing chamber* (see **Section 11.3**) temperature, T_M, is used to estimate the effective cooling power of the DR:

$$\boldsymbol{Q} = \boldsymbol{n} 84 T_M{}^2 - Q_{ext} \tag{11.12}$$

where $\boldsymbol{n}$ is the number of 3He atom passing (often using mole numbers) across the phase separation boundary per unit time (also called 3He molar circulation rate), and T_M is the mixing-chamber temperature.
Q_{ext} is the total parasitic heat load imposed on the mixture at the phase separation boundary.

4. Temperature limit for the mixing chamber assuming $Q_{ext} = 0$
We rewrite **Eq. 11.12** with **Eq. 11.7**, **Eq. 11.8**, and put $Q_{ext} = 0$

$$\boldsymbol{Q} = n[106 T_{3d}{}^2 - 22 T_{3c}{}^2] \tag{11.13}$$

The temperature of the concentrated 3He, T_{3c}, in **Equation 11.13** is the 3He liquid temperature after exiting the final heat exchanger and before entering the mixing chamber. The temperature of the concentrate 3He, T_{3c} is always higher than the temperature of the dilute 3He liquid, T_{3d}.
Therefore, the minimum possible temperature that the mixing chamber can reach is

$$\mathrm{T}_{\mathrm{Mmin.}} \sim 0.1\,(Q/n)^{1/2} \tag{11.14}$$

when $T_{3d} = T_{3c} = T_M$ and $Q_{ext} = 0$ **[24]**
Minimizing T_c before the concentrated 3He enters the mixing chamber is crucial for the DR design and fabrication. An excellent analysis on the cooling power of the DR was developed [25].

11.3 Realization of the Cryogen-Free Dilution Refrigerator System

11.3.1 What Is Dilution Refrigerator?

A DR is a recirculating system using 3He–4He mixture as the coolant.

Dilution refrigerator operation can be represented with a so-called ***U-tube model*** as illustrated in Figure 11.4. The concentrated (or 3He-rich) phase stays in one side of the U-tube and the dilute (or 4He-rich) phase stays in the other side. The ***Mixture Condensing Unit*** (MCU) is located at the top of the concentrated side, and the ***Still*** is located at the top of the dilute side. The ***mixing chamber*** is located at the bottom of the ***U-tube model***. The ***phase-separation boundary*** is located inside the mixing chamber. Thermal exchange occurs between the two sides of the U-tube. After the warmer incoming mixture (concentrated 3He) gets liquefied by the MCU, it is further cooled by the outgoing mixture in the dilute side. This thermal exchange process is accomplished by a component called the ***Heat Exchanger***.

Refer to Figure 11.4a and assume a DR with $\chi_3 = 28\%$. The DR cools with the following process:

- The fridge is precooled for mixture condensation.
- The DR is further cooled down by pumping on the Still, and the mixture phase separation occurs (**P** and **Q**).
- 3He concentration in the dilute phase decreases when the temperature gets lower (**P --> R**).
- 3He concentration in the concentrated phase increases when the temperature gets lower (**Q --> S**).
- Continue pumping on 3He from the Still.

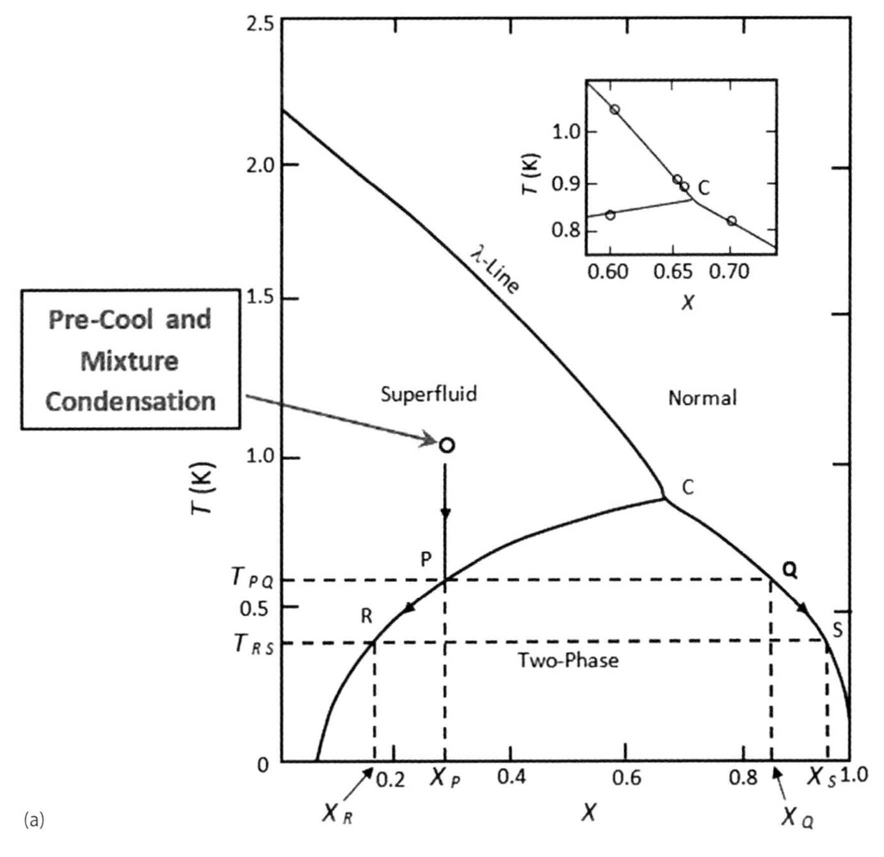

(a)

FIGURE 11.4 (a) Dilution refrigerator cool-down procedures. *(Continued)*

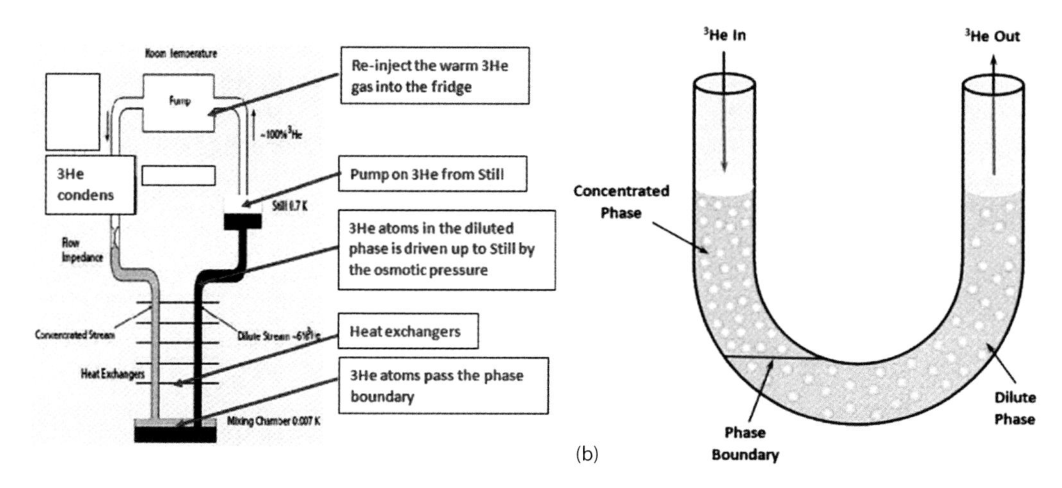

FIGURE 11.4 (Continued) (b) U-tube model of the dilution refrigerator operation.

- 3He atoms pass the phase separation boundary into the dilute phase and cooling occurs; 3He atoms in the dilute phase are driven up to the Still by osmotic pressure gradients (see more detailed discussion in **Section 11.3.2**).
- Re-inject the (warm) 3He gas into the refrigerator for condensation.
- Heat exchange between the warm returning 3He (gas and liquid) and the cold exiting 3He (vapor or liquid) takes places inside heat exchangers throughout the entire process.

Most of the above processes take places in the so-called ***Dilution Refrigerator Core*** (referred to as DR-core in future discussions). Illustrated in Figure 11.5 are the main components of the DR-core, followed by detailed descriptions.

11.3.1.1 Mixture Condensing Unit

The MCU is located on top of the concentrated side of the U-tube. Its function is to condense the returning gaseous mixture (typically greater than 95% 3He) into liquid.

There are two alternative approaches:

1. Use the Joule–Thomson mixture condensing unit (JTMCU).
   This approach is referred to as "JT condensing mode" or "***JT mode***."
   The JTMCU mainly includes the JT-heat recuperative exchanger (JTHE) made of stainless-steel convoluted hose and the JT-valve (or impedance).
   The mixture is cooled to approximately 2.2–2.5 K and gets partially condensed inside the JT-heat exchanger. The mixture is then further cooled to lower temperature through the JT valve before it enters the Still.

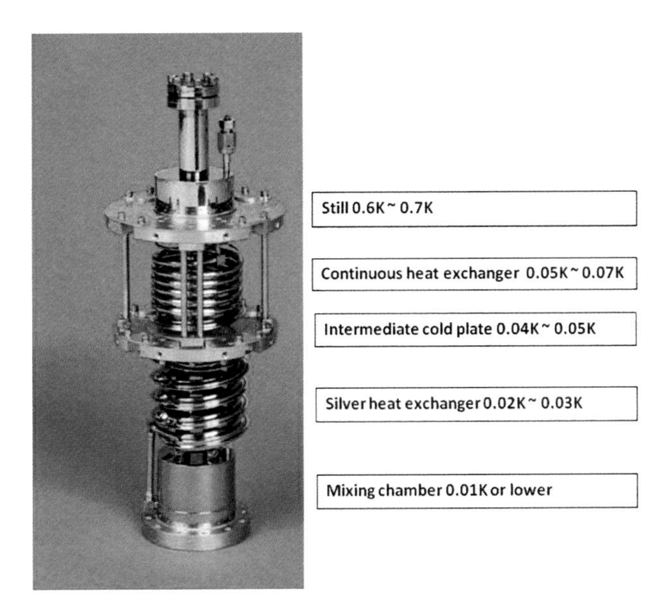

FIGURE 11.5 Main components of DR-core and their typical operating temperatures. (Courtesy of Janis Research Company, LLC.)

Since the JTMCU has less cooling power than the 1 K pot, the mixture condensation with JTMCU is expected to be slower at the beginning. The condensation will become smoother at lower temperatures.

A compressor is needed keep a proper head pressure for effective 3He condensation when ***JT mode*** is applied to a "wet" DR system. However, a compressor is not necessary for the cryogen-free DR since the returning mixture can be precooled to a lower temperature by the cryocooler in cryogen-free systems. More details will be discussed in Section 11.3.3.

A gold-plated copper flange can be soldered to the middle of the JTHE as illustrated in Figure 11.6a, which serves an independent 2.5 K thermal anchor stage. **Typical values** of an example of JTMCU are listed below for readers' reference. All values may vary for different systems [26].

Outer flexible hose for the diluted mixture:

Material: Stainless steel
Diameter: ~ 0.6–0.75″
Length: ~ 7.0–10.0″

Inner capillary for the concentrated mixture:

Material: Stainless steel
Diameter: 0.030″/0.035″ with 0.004″/0.006″ wall thickness
Length: approximately 10 m

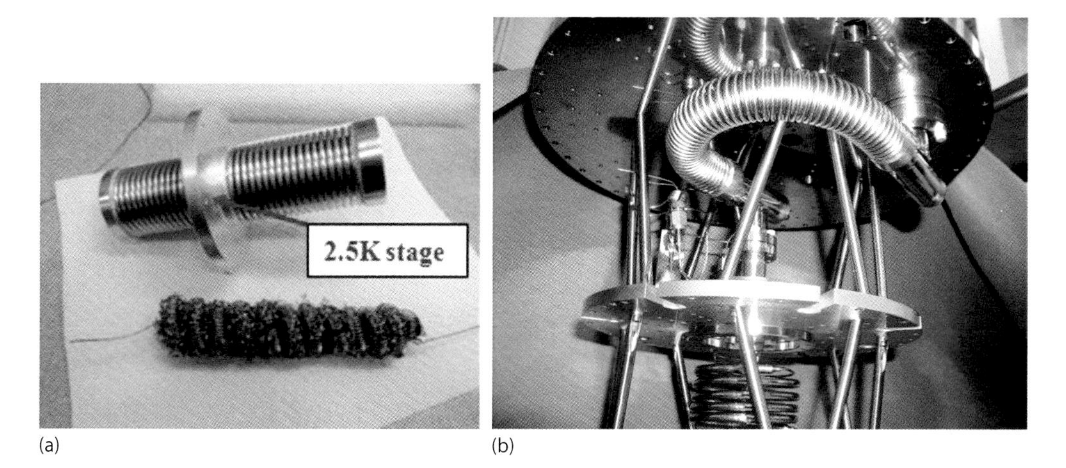

(a) (b)

FIGURE 11.6 (a) Examples of Joule-Thomson heat exchanger. (b) Joule-Thomson heat exchanger installed inside a cryogen-free DR. (Courtesy of Janis Research Company, LLC.)

2. Use the **recirculation mixture condenser** for mixture condensation.

   An alternative approach is to equip a *recirculation mixture condenser* (also referred to as ***4He-1 K-stage***) in a cryogen-free DR. This 1 K-loop is similar to the recirculating 1 K system as discussed in **Section 8.2.3**, and the mixture can be condensed with this 1 K-loop, as illustrated in Figure 11.7 [27].

   The cooling power of the 1 K stage will be enhanced when separate continuous heat exchanger is installed [27]. The 3He of the DR is condensed by this 1 K stage with the continuous heat exchanger with high efficiency due to the large cooling power of the recirculating 1 K mixture condenser.

   This approach is referred to as "***1 K operation mode***." The 1 K operation mode is not common practice for the commercial cryogen-free DR systems due to its higher cost and technical complexities. However, it provides faster initial condensation of the 3He/4He gas mixture due to its higher cooling power provided by the 1 K stage than that by the JT mixture condenser. In addition, the 1 K operation mode approach provides a user-friendly 1 K stage for these measurements that do not need millikelvin temperatures.

   The refrigeration power of the 1 K stage with and without continuous heat exchanger are illustrated by the upper and lower curve, respectively.

3. Comparison between ***1 K operation mode*** and ***JT operation mode***

   The ***1 K operation mode*** provides higher cooling power. The ***JT operation mode*** is obviously more user friendly. It also generates lower acoustic noise [28].

11.3.1.2 Main Flow Impedance

Flow impedance has been discussed in detail in **Section 6.2.4.2**. It is located below the MCU and maintains a proper mixture *condensing pressure* for efficient 3He condensation.

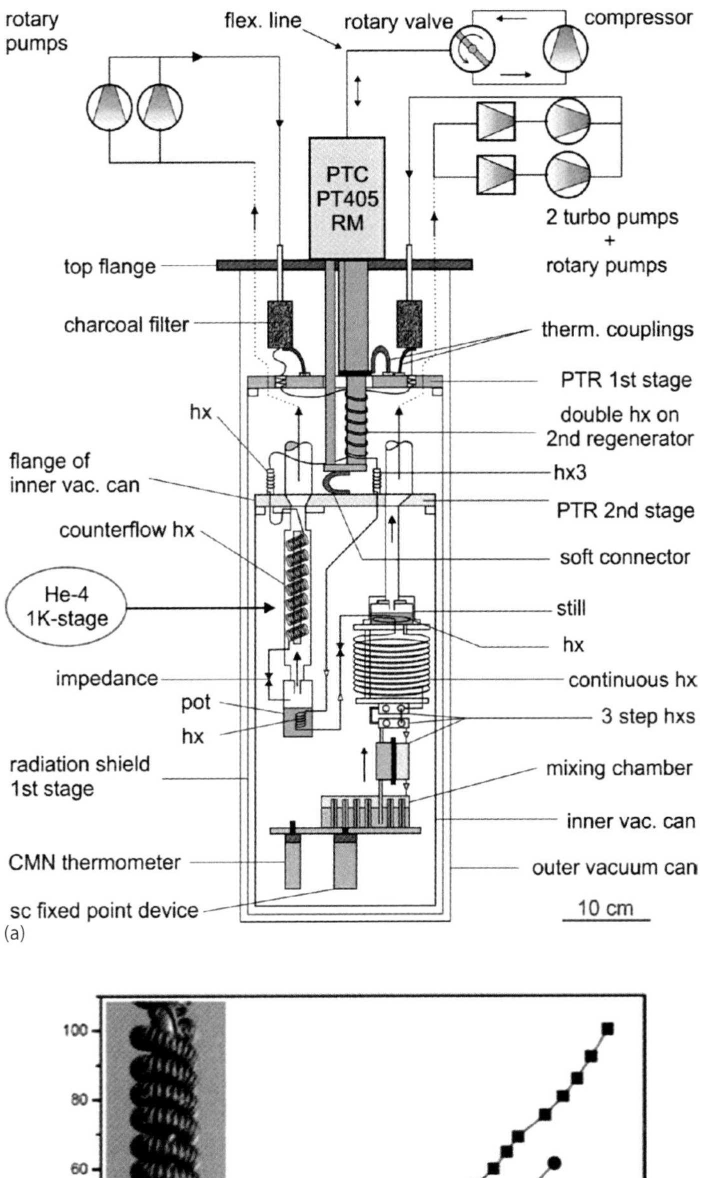

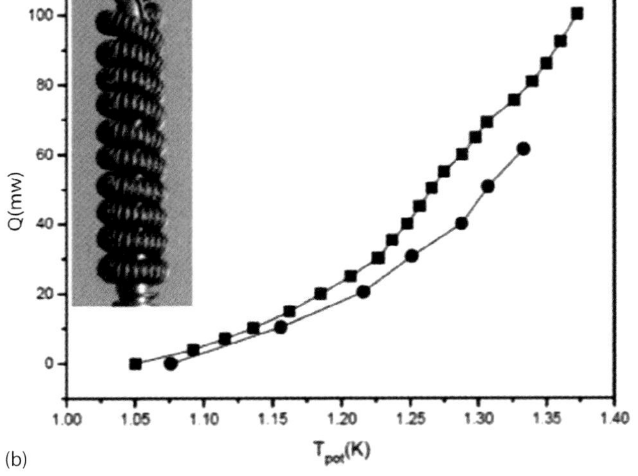

FIGURE 11.7 (a) The 1 K stage equipped with a continuous heat exchanger. (b) Image of continuous heat exchanger installed with the 1 K stage.

Flow impedance also controls the mixture circulation rate within the proper range. If the impedance is too low, poor JT expansion and overly high circulation rate will introduce too much heat to the Still. If the impedance is too high, the system will not be able to achieve the desired circulation rate. The value of the main condensing impedance should match the maximum mixture circulation rate for each system.

Typical values of the main condensing impedance are of the order of approximately 10^{10}–10^{11} cm^{-3} depending on the circulation rate needed.

11.3.1.3 Still and the Second Flow Impedance

The still is usually a cylindrical chamber made of stainless steel. It located at the top of the dilute phase side with a typical design as illustrated in Figure 11.8.

The Still is a partially filled "reservoir" for the dilute liquid mixture. The rest of the Still is mainly filled with 3He vapor due to its much higher partial vapor pressure than that of 4He at the Still temperature (0.6~0.7 K). The 3He vapor is being continuously removed during normal operation.

In order to keep the Still at the proper temperature (with the proper vapor pressure inside it) and maintain a higher circulation rate for the needed cooling power, heat needs to be applied to the Still. Care must be taken not to overheat the Still and cause too much unwanted 4He flow. High 4He circulation will introduce much higher heat load to the DR due to its higher heat of condensation.

A knife edge (or an orifice) or a heater (so-called film-burner) is usually installed on top of Still to suppress the flow of superfluid 4He.

The still has an active cooling power due to the evaporative cooling process. A coiled capillary referred to as "Still Heat Exchanger" is installed inside the Still. The returning 3He flows inside this capillary and gets cooled by the Still.

A second flow impedance is installed below the Still. Its main function is to prevent the mixture in the heat exchangers below the Still from backflow, therefore maintaining a stable mixture circulation.

Typical values of the second condensing impedance are of the order of $10^{9}cm^{-3}$ and vary from system to system.

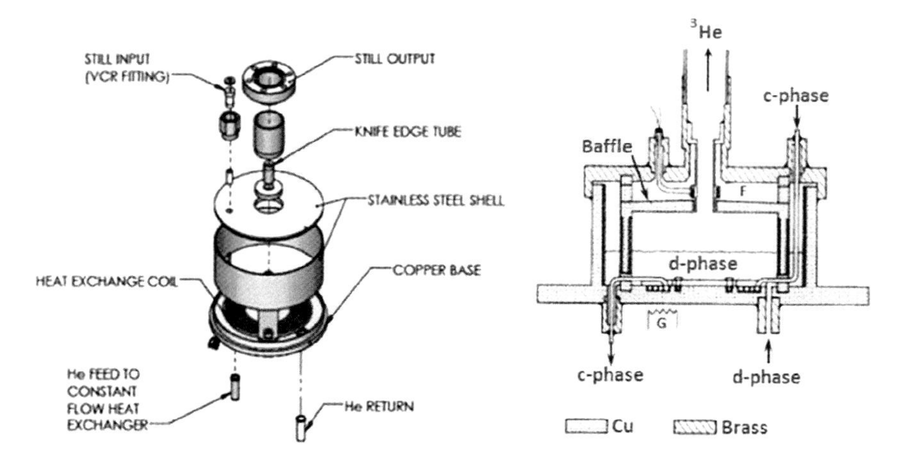

FIGURE 11.8 Typical Still deign. (Courtesy of Janis Research Company, LLC.)

11.3.1.4 Heat Exchangers and Intermediate Cold Plate (ICP) [17,29]

In order for the DR to reach the desired base temperatures, the cold outgoing 3He vapor in the **dilute phase** is employed to gradually cool down the returning warm 3He step-by-step to a temperature near that of the phase boundary. A mechanism needs to be built to achieve effective heat exchange in between the two sides of the U-tube (DR-core). This mechanism is implemented by the ***heat exchangers***, and it is one of the most important elements determining the performance of the DR.

The mixtures in the two phases do not have a direct contact because they are flowing in separate channels inside the U-tube (DR-core). The heat exchange can only occur through a metal separator between the two sides of the U-tube (DR-core). The mechanism is highly inefficient due to the Kapitza resistance [30], $\boldsymbol{R}_K$, at the interface between the liquid mixture and the solid walls of the heat exchangers at low temperatures, which is proportional to T^{-3} at temperature below 0.10 K.

A common approach to improve the heat exchange between the mixtures in the two phases is to increase the contact surface area between the liquid and the solid. This approach is realized by so-called Heat Exchangers. Two types of heat exchangers are commonly used in the dilution refrigerator systems. They are referred to as **continuous heat exchanger** and **discrete heat exchanger** respectively.

1. *Continuous heat exchanger*: After the returning mixture is cooled to the Still temperature, the retuning mixture enters the continuous heat exchanger (**CHE**) located below the Still and the second flow impedance.

   The CHE is made of two tubes that are arranged one inside the other as illustrated in Figure 11.9.

   The outer tube is made of a metal with low thermal conductivity, such as stainless steel, brass, CuNi, etc.

   A few meter-long CuNi or stainless-steel capillary is installed inside the outer tube after it is coiled into a spiral with many turns. The lengths and the diameters of the capillary depends on the design and performance of the individual DR.

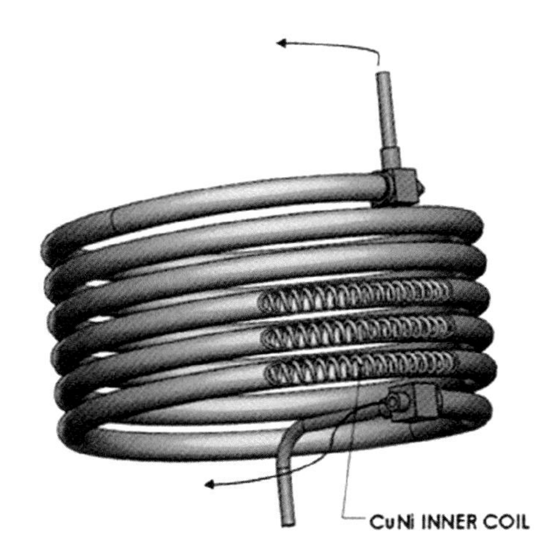

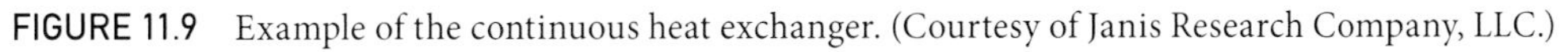
FIGURE 11.9 Example of the continuous heat exchanger. (Courtesy of Janis Research Company, LLC.)

The concentrated 3He at higher temperatures flows through this capillary, while the colder 3He from the dilute phase flows in the opposite direction in the outer tube. Heat exchange in between these two liquid occurs. Typically, the concentrated 3He can be cooled down to approximately 50–70 mK by the CHE.

Typical values of CHEs are listed as below for readers reference. All values may vary for different systems.

Outer tube of the CHE
Material: stainless steel, brass, Cu(70%)Ni(30%),
Diameter: ~ 3.5 mm ~ 6.0 mm
Wall: ~ 0.2mm ~ 0.25 mm
Length: ~ 35" ~ 50″

Inner capillary of the CHE
Material: stainless steel, Cu(70%)Ni(30%)
Diameter of: ~ 2.0mm ~ 3.5 mm
Wall: 0.1 mm
Length: ~7.5 ft. ~ 10 ft

Readers can estimate the surface area as well as volume of the both phases.

2. *Discrete (silver) heat exchanger*: The discrete heat exchanger used to employ copper powder. Modern DRs typically use silver powder. Therefore, the discrete heat exchanger is commonly known as the "silver heat exchanger" now (Figure 11.10).

A much larger contact surface area is needed when a much lower temperature (e.g., 30 mK or less) is required because the Kapitza resistance, $\boldsymbol{R}_K$, becomes enormously high in that temperature range. Lower temperatures are realized by implementing the silver heat exchanger (**SHE**) installed between the ICP and the mixing chamber.

The SHE starts with a thin copper-nickel foil (approximately 0.25-mm thick) with a layer of a few grams of sintered silver attached on both sides.

The surface area of each gram of sintered silver is typically 1.0–3.0 m^2. This enormous area reduces the Kapitza resistance by one or two orders of magnitudes.

The CuNi foil with the sintered silver powder is then sandwiched by two stainless-steel covers, that is, each heat exchanger consists of two chambers with flow channels in them, one for the incoming concentrated 3He, and the other for the outgoing dilute 3He.

The high viscosity of liquid 3He below 0.1 K as described in **Section 9.2.6** explains the reason why the cross-sectional area of the channels inside the **SHE** is much larger than that of the capillary inside the **CHE**.

Since the returning mixture needs to be cooled step-by-step before it can reach a temperature near that at the phase boundary, multiple "steps" are usually needed in order to cool the returning 3He to the desired low temperatures. This is why the SHE is also referred to as "***step heat exchanger***." More ***step heat exchangers*** are generally needed when the fridge needs to reach lower temperatures.

Typical values of the channel diameters of the SHE examples are listed as follows for reference. They vary for different systems.

Channel diameter of the diluted phase: ~ 4.0mm ~ 8.0 mm
Channel diameter of the concentrated phase: ~ 3.0mm ~ 5.0 mm

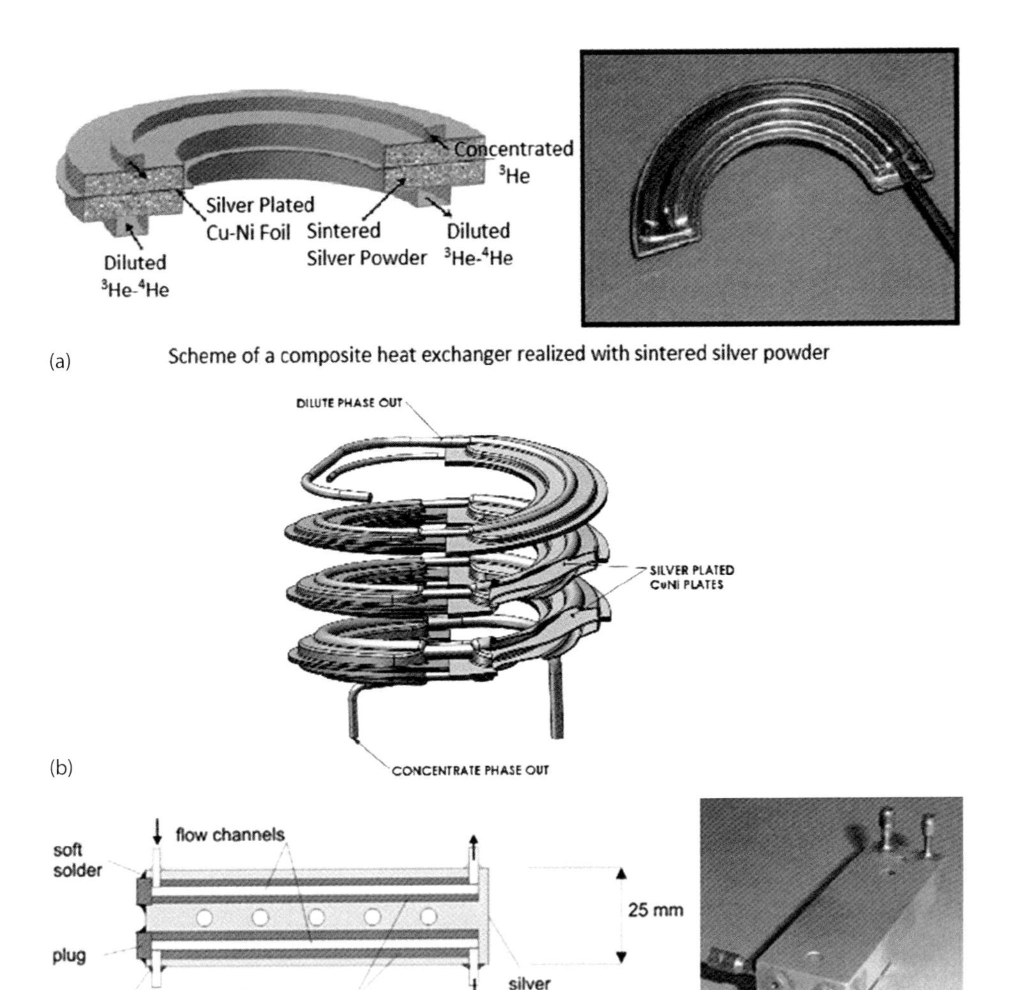

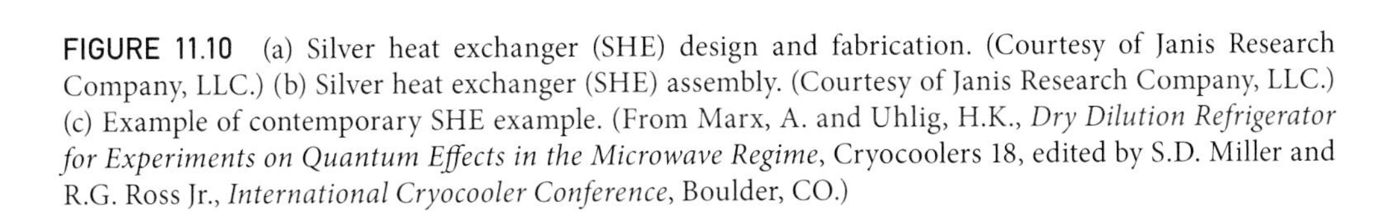

FIGURE 11.10 (a) Silver heat exchanger (SHE) design and fabrication. (Courtesy of Janis Research Company, LLC.) (b) Silver heat exchanger (SHE) assembly. (Courtesy of Janis Research Company, LLC.) (c) Example of contemporary SHE example. (From Marx, A. and Uhlig, H.K., *Dry Dilution Refrigerator for Experiments on Quantum Effects in the Microwave Regime*, Cryocoolers 18, edited by S.D. Miller and R.G. Ross Jr., *International Cryocooler Conference*, Boulder, CO.)

Amount of silver power in ***each*** step heat exchanger: ~ 3.0g ~ 10g
Surface area of each gram of sintered silver: ~ 1.0 m^2 ~ 3.0 m^2 (usually not more than 4.0 m^2)

3. *Intermediate cold plate* (*ICP*): Because a fairly large temperature difference exists between the CHE and SHE, most dilution refrigerator cores include an ICP between the CHE and SHE.

   The ICP is usually a short cylindrical container with some sintered silver installed inside it. The retuning 3He will be further cooled inside the ICP before entering the SHE.

The ICP also provides an additional thermal stage for wires and supports, which is very helpful to minimize the heat load to the mixing chamber.

Typical ICP temperature is around 50 mK or lower.

11.3.1.5 Mixing Chamber

The mixing chamber (MC) is usually a cylindrical chamber made of stainless steel. The bottom of this chamber is normally made of gold-plated oxygen-free high-conductivity (OFHC) copper.

The MC is located at the bottom of the **DR-core** and contains both phases of the 3He–4He mixture. Ideally, the phase separation boundary should be located near the middle of the MC. Multiple OFHC copper pins with silver sintered on their surfaces are used as the heat exchangers. They are bolted onto the inner surface of the MC bottom plate as shown in Figure 11.11b.

More silver-sintered pins are needed if the fridge needs higher cooling power.

A 3He pump-out tube is installed near the bottom of the MC (in the 4He-rich phase) below the phase boundary. This tube is connected to the dilute channels of all heat exchangers and terminated at the bottom of the Still. The dilute mixture typically fills the dilute side up to the middle of the Still.

When 3He is extracted from the Still by a hermetic pump, the dilute 3He in the mixing chamber is driven along the channel all the way up to the Still. 3He in the condensed phase will cross the phase boundary and enter the dilute phase to balance the chemical potential at the phase boundary per our previous discussion. The 3He extracted from the Still will be sent back to the MCU for recondensation. The active cooling occurs at the phase separation boundary inside the mixing chamber, which is the coldest place in the whole dilution refrigerator.

Typical 3He concentration of DR systems is ~20–30%.

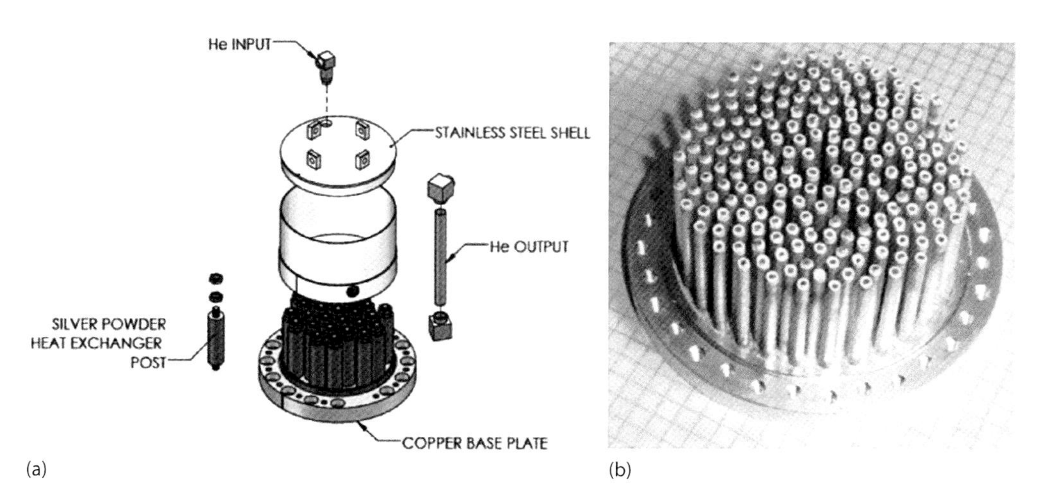

FIGURE 11.11 (a) Mixing chamber design. (Courtesy of Janis Research Company, LLC.) (b) Silver-sintered heat exchanger inside the mixing chamber [32].

11.3.2 Additional Details

1. DR users often have the following questions.
   a. **What drives the 3He atoms in the dilute phase up from the MC to the Still?**
   The answer is the osmotic pressure gradient between the MC and the Still.

The **osmotic pressure** (π) is defined as the minimum pressure that needs to be applied to a solution to prevent the inward flow of its pure solvent across a semipermeable membrane.

Consider a U-tube as shown in Figure 11.12a. The left-hand side has pure solvent, while the right-hand side has its solution. A semipermeable membrane is installed in between them at the bottom of the U-tube. Only the pure solvent can penetrate the membrane and enter the solution, but the solution cannot pass through the membrane. An **osmotic pressure** (π) will build up in the solution and prevent the pure solvent from entering the solution. Reduction of the solvent concentration inside the solution will reduce the osmotic pressure in the right side of the tube. Solvent from the left tube will enter the solution in the right tube and rebalance the pressures between the two tubes.

Take the superfluid 4He in the diluted phase and the dilute 3He in superfluid 4He as the "*solution*" and the "*solvent*," respectively. The osmotic pressure, π, of the diluted 3He–4He mixture *at very low temperature* is a function of ~ $T_d\,X_{3d}$, where T_d and X_{3d} (also denoted as $X_{\mathbf{3d\ liquid}}$) are the temperature of the diluted mixture and the 3He concentration in the diluted mixture, respectively.

When the DR is circulating in a dynamic steady state, the Still temperature is much higher than that of the MC. The diluted 3He concentration in the Still

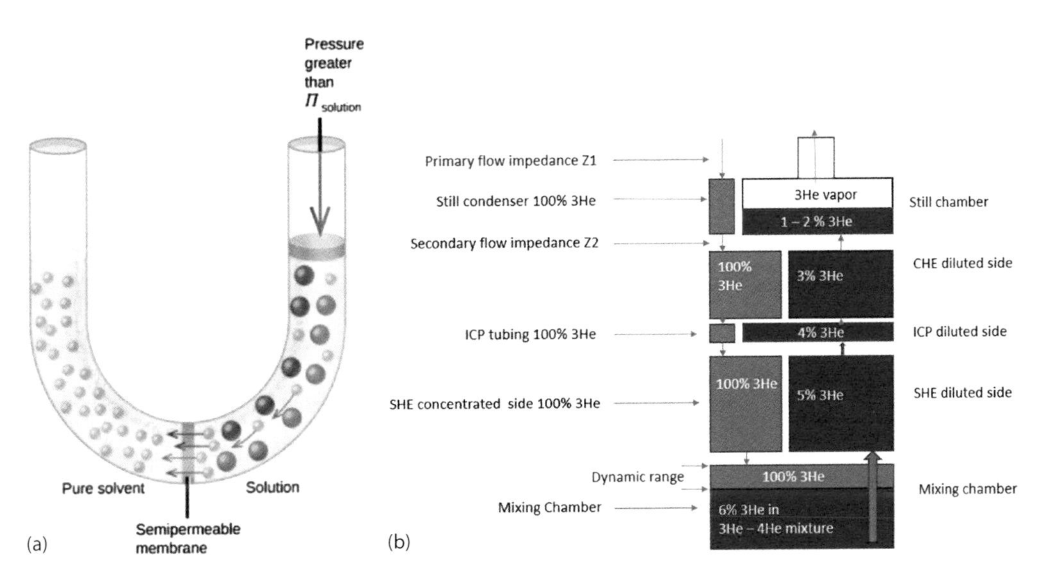

FIGURE 11.12 (a) Osmotic pressure. (b) 3He concentration when the DR is circulating in a dynamic steady state with Still at 0.7 K. (Private discussion with Vladimir Shvarts, April 1, 2019.)

is expected to be lower than that in the diluted phase inside the MC. Listed in Table 11.1 are some values of the vapor pressures and 3He concentration inside Still as function of the Still temperature [24].

For an example, Figure 11.12b depicts the distribution of 3He concentration inside the DR core when the Still and the MC is at 0.7 K and 10 mK, respectively.

Reduction of 3He concentration in the Still by pumping 3He out for circulation will cause a decrease of the osmotic pressure in the Still, with the proper impedance of the connecting channel.

3He atoms will be driven up along the channel all the way up from the MC to the Still by the osmotic pressure gradient inside the DR-core. The same amount of 3He from the concentrated phase in the MC will pass the phase boundary and enter the dilute phase. 3He circulation continues.

The maximum osmotic pressure difference between the MC and the Still is between 1.6 and 2.2 kPa [11,16]. A hydraulic pressure in the mixture exists between the Still and the MC due to their different vertical locations. The DR will not work if the hydraulic pressure exceeds that maximum possible osmotic pressure.

Table 11.1 Vapor Pressures and 3He Concentration Inside Still as Function of the Still Temperature

T (K)	P3 (mbar)	P4 (mbar)	P3 + P4 (mbar)	$\chi_{3ds\ vapor}$ (%)	$\chi_{3ds\ liquid}$ (%)
0.4	0.00286	7×10^{-6}	0.00286	99.99	1.5
0.5	0.0105	1.1×10^{-5}	0.0105	99.8	1.4
0.55	0.0173	5.6×10^{-5}	0.0175	99.6	1.4
0.6	0.0255	2.25×10^{-4}	0.0258	99.1	1.2
0.65	0.0361	6.92×10^{-4}	0.0368	98.1	1.1
0.7	0.0481	0.0018	0.0499	96.4	1.0
0.75	0.0624	0.0043	0.0667	93.6	0.9
0.8	0.077	0.0086	0.0861	90.7	0.8
0.85	0.095	0.015	0.110	86.3	0.8
0.9	0.103	0.030	0.143	78.9	0.7
1.0	0.150	0.0752	0.226	66.6	0.6
1.2	0.203	0.203	0.419	50	0.5

where:

T: Still temperature

P3: Partial vapor pressure of 3He inside the Still

P4: Partial vapor pressure of 4He inside the Still

P3 + P4: Total vapor pressure inside the Still

χ_{3ds} *vapor:* 3He concentration (%) in vapor

χ_{3ds} *liquid:* 3He concentration (%) in liquid

b. Why does the DR circulate with 3He in majority (say, above 95%) while the 3He concentricity in the Still is very low?

If the 3He concentration in the Still, χ_{3dS}, is very low per Table 11.1 (i.e., less than 1% above 700 mK), why is more than 95% of the circulating mixture 3He?

The answer is due to the much higher partial vapor pressure of 3He at ~0.6–0.7 K than that of 4He in that temperature range (Figure 11.13). More than 95% of the vapor inside the Still is occupied by 3He vapor per Table 11.1. This is why 3He is predominantly evaporated from the liquid even with lower concentration.

In order to achieve the desired cooling power during the actual operation, heat needs to be applied to the Still in order to drive more 3He out and increase the circulation rate.

Having the right quantity and 3He–4He concentration ratio for the mixture is critical so that the phase boundary is located inside the mixing chamber, and the top surface of the diluted liquid 3He is located inside the still. Otherwise, the fridge will not perform properly.

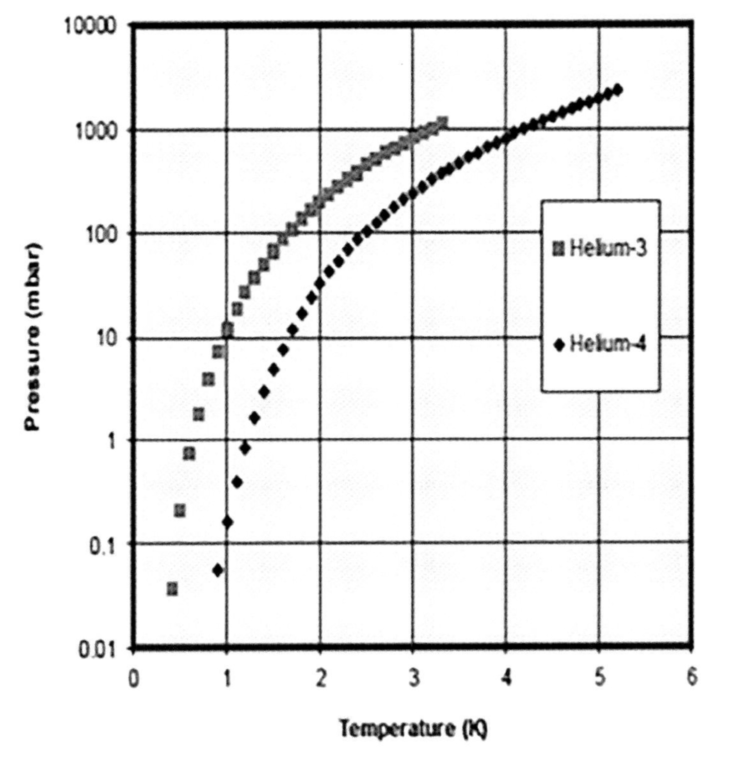

FIGURE 11.13 Partial vapor pressure of 3He and 4He vs. temperatures in Still.

2. Accessories for a complete dilution refrigerator system.

The following auxiliary items are necessary to make a complete DR system:

a. Hermetic pumping station: The main function of the pumping station is to extract 3He out from the Still and send it back to the system for mixture circulation.

   Refer to Figures 9.25 and 9.26 for candidates of the pumps.

b. Gas handling system (GHS) (Figure 11.14): The GHS including valves, gauges, connecting lines, cold trap, etc., is a multifunction element to implement the system operations

c. Cold traps: The most frequently used cold trap is the "*liquid nitrogen temperature trap.*" It is filled with activated charcoal or zeolite and then connected in series with the 3He returning loop. The trap is submerged in liquid nitrogen, and the charcoal will adsorb air and other impurities with critical temperatures higher than 77 K. A different trap is needed if the mixture contains hydrogen impurities.

   Internal charcoal traps are employed inside dry DRs.

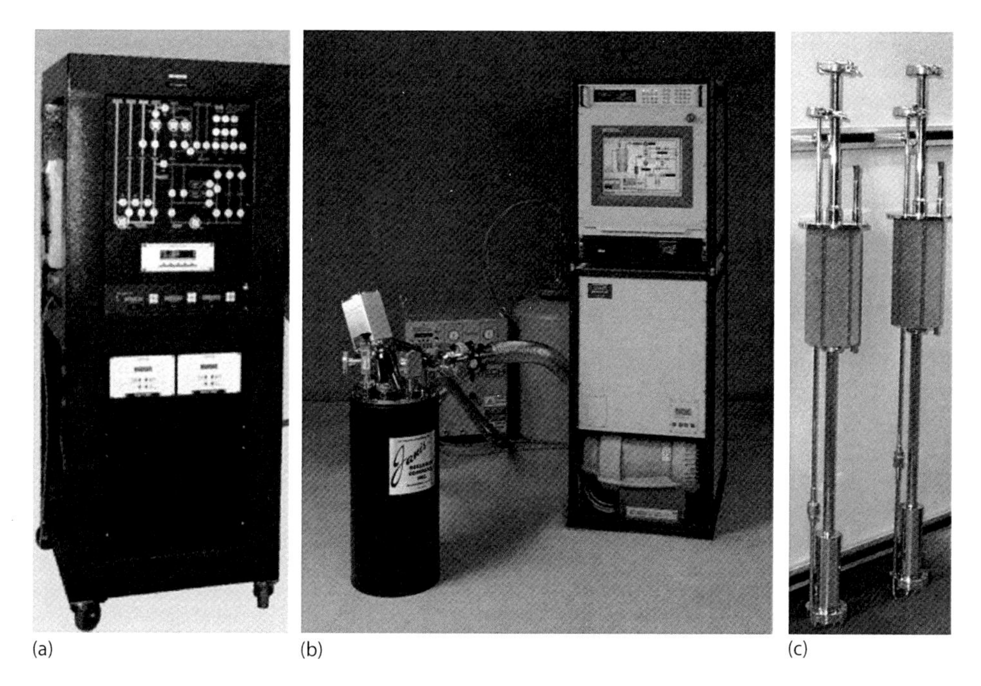

FIGURE 11.14 (a) Example of gas handling system for dilution refrigerator. (Courtesy of Leiden Cryogenics.) (b) Example of gas handling system for dilution refrigerator (c) Example of liquid nitrogen temperature cold trap. ([b and c] Courtesy of Janis Research Company, LLC).

11.3.3 Cryogen-Free Dilution Refrigerator

The cryogen-free DR, often referred to as "***dry DR***," was first developed in early 1990s [31]. It has gradually evolved into the indispensable cryogenic workhorse in the millikelvin temperature regime. Uhlig has played a leading role directing the ***dry DR*** development with his tireless and creative work during past quarter century [32].

Instead of using liquid cryogens for precooling, the cryogen-free DR employees a closed-cycle refrigerator for mixture precooling and condensation. Illustrated in Figure 11.15 are some typical configurations of ***dry DR*** systems.

11.3.3.1 Special Features of the Cryogen-Free Dilution Refrigerator

1. *No cryogen (liquid helium) is needed*: A cryogen-free DR does not need any external supply of cryogenic liquids but employs the cryocooler for "precool" and mixture condensation.
2. *Compact, convenient to operate, and less maintenance*: The cryocooler-based cryogen-free DR is much more compact than the conventional system. The dry DR also has fewer cold seals than the conventional wet DR system. Limited maintenance is needed for a dry DR as compared to a wet DR.
3. *Pulse-tube refrigerator (PTR)*: Although different types of commercial cryocoolers are available, the two-stage PTR with remote motor, for example, model PT-400RM series pulse-tube cryocooler fabricated by Cryomech, Inc., is the preferable choice for the dry DR. The PTR does not have cold moving parts refrigerator and thus it generates less vibration.

   The first stage of that series of PTR can reach approximately 40 K, which is much lower than the liquid nitrogen temperature in wet DR systems. The second stage of that series of PTR can reach 3 K or lower, which is below the normal 3He condensing temperature.
4. *Second-stage regenerator*: The second-stage regenerator (as well as the second-stage pulse tube) has considerable amount of cooling power [33], which should be fully utilized.
   a. "10K" thermal stage: An additional thermal stage made of OFHC copper can be mounted at approximately the middle of the second-stage pulse tube and that of the second stage regenerator as illustrated in Figure 11.15.

      This intermediate thermal stage is referred to as a *10 K cold plate* since the temperature of this thermal stage is around 10 K.

      This cold plate intercepts the conductive heat load from the first stage of the cryocooler and provides a colder thermal anchoring stage for wires and supporting tubes.

      The new model PT420RM PTR has approximately 0.5-watt cooling power at this intermediate 10 K stage. This amount of cooling power is highly desirable for many applications including quantum computer development where a large heat load from a few hundreds of coaxial cables needs to be intercepted.

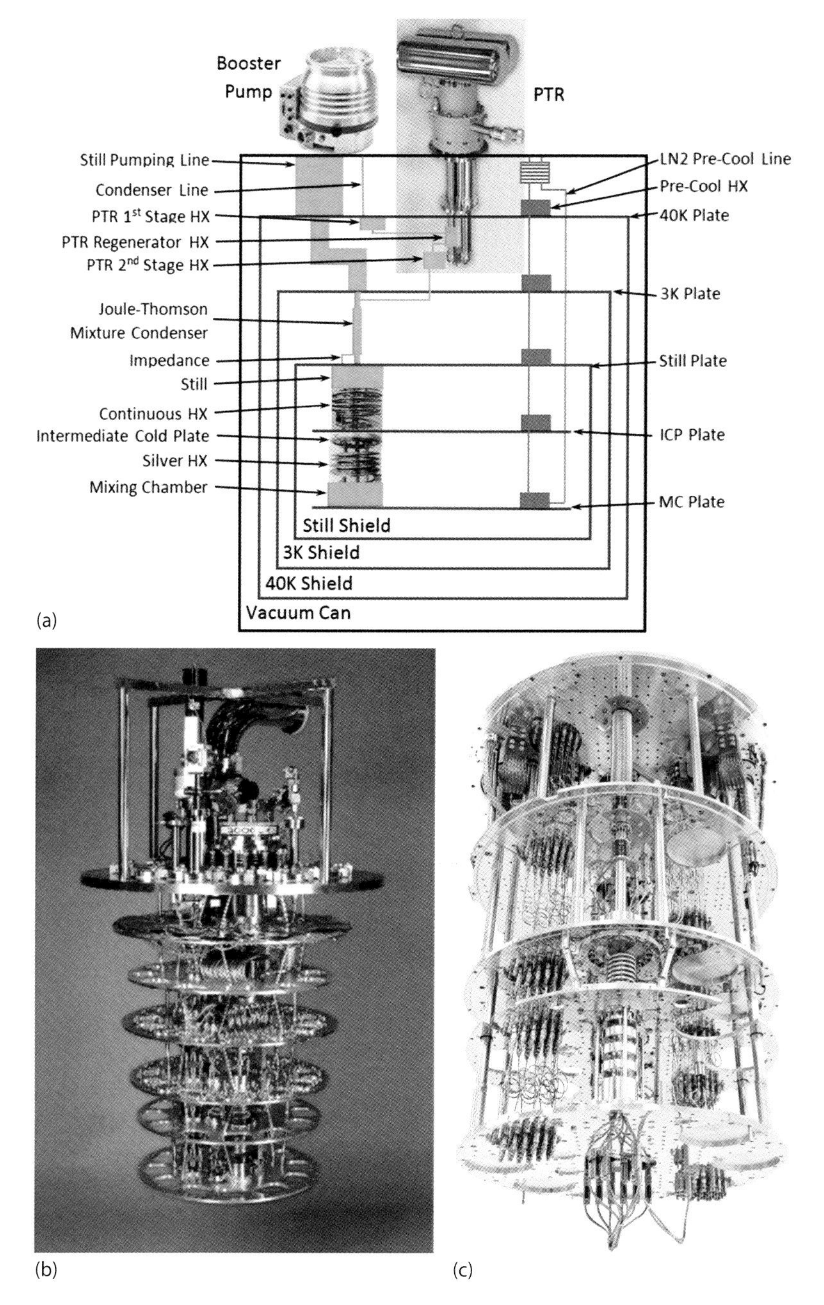

FIGURE 11.15 (a) Conceptual sketch of the cryogen-free dilution refrigerator. (b and c) Example of the cryogen-free dilution refrigerator. ([b] Courtesy of Janis Research Company, LLC; [c] BlueFors Cryogenics Oy.)

b. Second-stage heat exchanger: Model PT400RM series PTR takes the U-shape design as described in **Chapter 3: 4 K Regenerative Cryocooler** with the accessible second-stage regenerator. A capillary for the returning 3He gas can be coiled around and then soldered onto the second-stage regenerator as a heat exchanger. After being precooled at the first stage, the returning 3He gas can be cooled to near 4 K or lower by this heat exchanger. This approach greatly reduces the heat load to the second stage of the PTR.

5. *Joule-Thomson mixture condensing unit* (*JTMCU*): JTMCU is the key device for mixture condensation in a dry DR. Figure 11.16a shows the 3He enthalpy as a function of pressure representing the cooling process of JTMCU.

   3He in ***region I*** is in the gaseous state. Gas and liquid coexist in ***region II***. 3He in ***region III*** is in the liquid phase.

   After the returning 3He is cooled below 3 K (region I in Figure 11.16a) by the second stage of the PT cryocooler, it will partially condense inside the inside the JTHE (regions I–II in Figure 11.16a). The JT expansion takes place at the impedance (regions II–III in Figure 11.16a) where the enthalpy of 3He does not change since it is an isenthalpic process. Both the temperature and pressure drop dramatically.

   The 3He gets fully condensed inside the Still heat exchanger (regions III–IV in Figure 11.16a) and cooled down to the Still temperature.

   The precooling and condensation process of the returning 3He inside the cryogen-free DR can be summarized as follows:

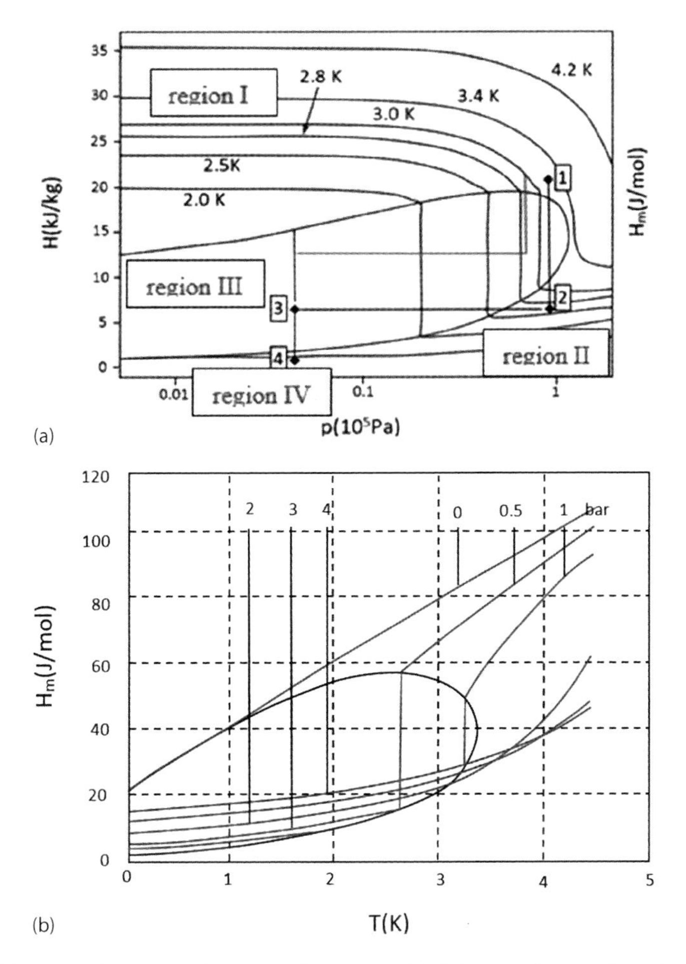

FIGURE 11.16 (a) 3He enthalpy as a function of pressure. (b) 3He enthalpy as a function of temperature.

Reinjected into the dry DR (300 K) → precooled by the first stage (40K) → cooled by the second-stage regenerator heat exchanger (10–4 K) → further cooled down by the second stage (3 K) and condensation may start depending on the head pressure – continuous cooling in the JT heat exchanger (~2.2–2.5 K) and condensation usually starts there → partially condensed after JT expansion → fully condensed inside the Still heat exchanger (~0.5–0.7 K).

According to Figure 11.16b, 3He will start condensing at pressures much lower then 1.0 bar when its temperature is near 2.5 K. In this case, there is no need to boost the mixture returning pressure with an external compressor (*not the cryocooler "compressor"*) in the dry DR system when the previously described approach is taken. This important feature has eliminated a major source of noise and vibration.

6. *Heat switch and liquid nitrogen precooling lines* (Refer to Section 6.3): Most of the dry DR systems are precooled to 3 K with a heat switch. The heat switch can be either a mechanical heat switch, gas-gap heat switch, or heat pipe, instead of exchange gas since dry DR systems normally do not have a separate vacuum space inside the cryostat other than the overall dewar vacuum.

   Aluminum block heat exchangers connected by stainless-steel flexible hoses can be installed at stages with different temperatures to form a so-called pre-cooling line as shown in Figure 11.17. Flowing liquid nitrogen through the precooling line will shorten the precooling time substantially.
7. *Vibration, internal pressure pulsation, and thermal contraction*: Vibration from the cryocooler is the main drawback of a dry DR, in particular when highly sensitive measurements at very low temperatures are needed. A typical example is the atomic resolution measurements with STM in the millikelvin temperature range.

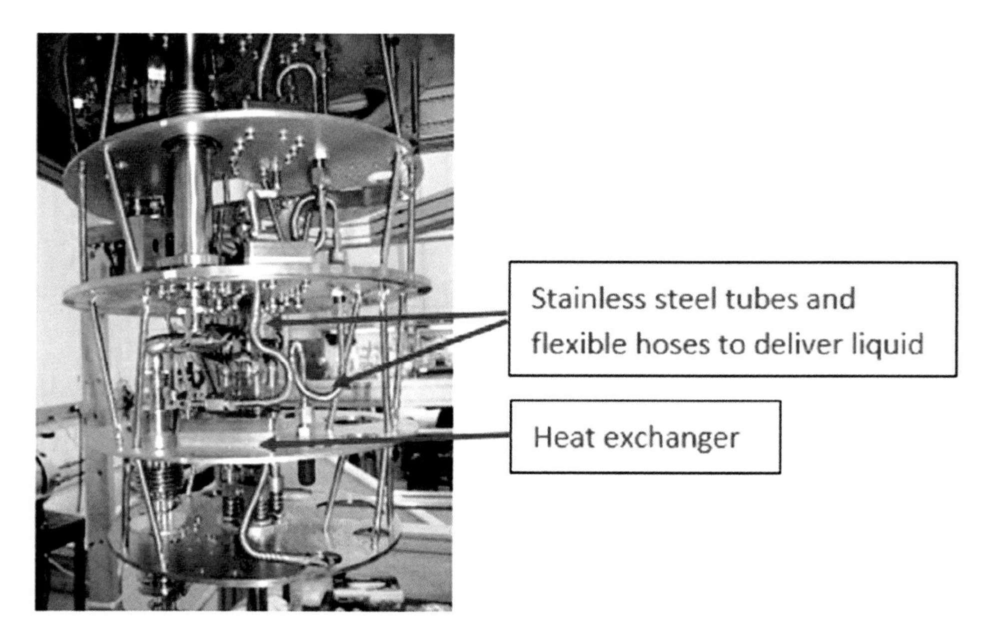

FIGURE 11.17 LN precooling line.

Vibration measurements with accelerometer on a cryogen-free DR has been performed [34]. The measurements were focused on the DR top plate at room temperature and the MC plate being kept at 4 K. The dominant frequencies of the vibration generated by the tested cryogen-free DR system are between 7 and 28 kHz.

Effective methods of suppressing vibration in DR systems include, but are not limited to, using wall-mounted solid pipes to connect the pulse tube motor to the compressor, moving the pulse tube motor away from the cryostat frame, etc.

A common practice has been to place weights, such as sand bags or lead bricks, on the system for vibration reduction. However, measurements have shown that this approach has very little effect other than redistribution of the vibration in the frequency domain.

Some commonly used approaches for vibration reduction are listed, but not limited, below:

- Using two-stage pulse-tube cryocooler with the ***remote motor option***
- Installation of spring-loaded bellows stage between the PTR and cryostat
- Employing more flexible high-pressure connecting hoses
- Connecting the cryocooler cold stages to the cryostat with flexible copper braided straps
- Designing rigid supports of different parts of the cryostat
- Rigid support of the Still pipe
- Suspending the pumps with a separate frame
- A short flexible section of bellows tubing installed between the DR Still pumping port and the rest of the Still pumping line
- Seating the cryostat on a vibration suspension table
- Installing Gimbals in between stages with different levels of vibration.
- Employing eddy current damping
- Design buildings with separate concrete foundation

More sophisticated approaches are needed for special application. For example, a mechanical low-pass filter vibration isolation was developed by Wit et al. [35]. This setup includes a thermal link with high thermal conductance, and it is operated at low millikelvin temperature range for STM applications in a cryogen-free DR.

The following photos (Figure 11.18) show some examples to reduce the vibrations for reader's references.

11.3.3.2 Normal Operation of a Cryogen-Free DR and Comparison between DR and the Continuously Circulating 3He Refrigerator

The operation of a cryogen-free DR is very similar, if not identical, to that of the continuously circulating 3He refrigerator discussed in Section 9.3.2. The key process includes: *system precooling, refrigerant condensation,* and *refrigerant continuous circulation*. This section will focus on the differences of these two systems to avoid redundant discussion.

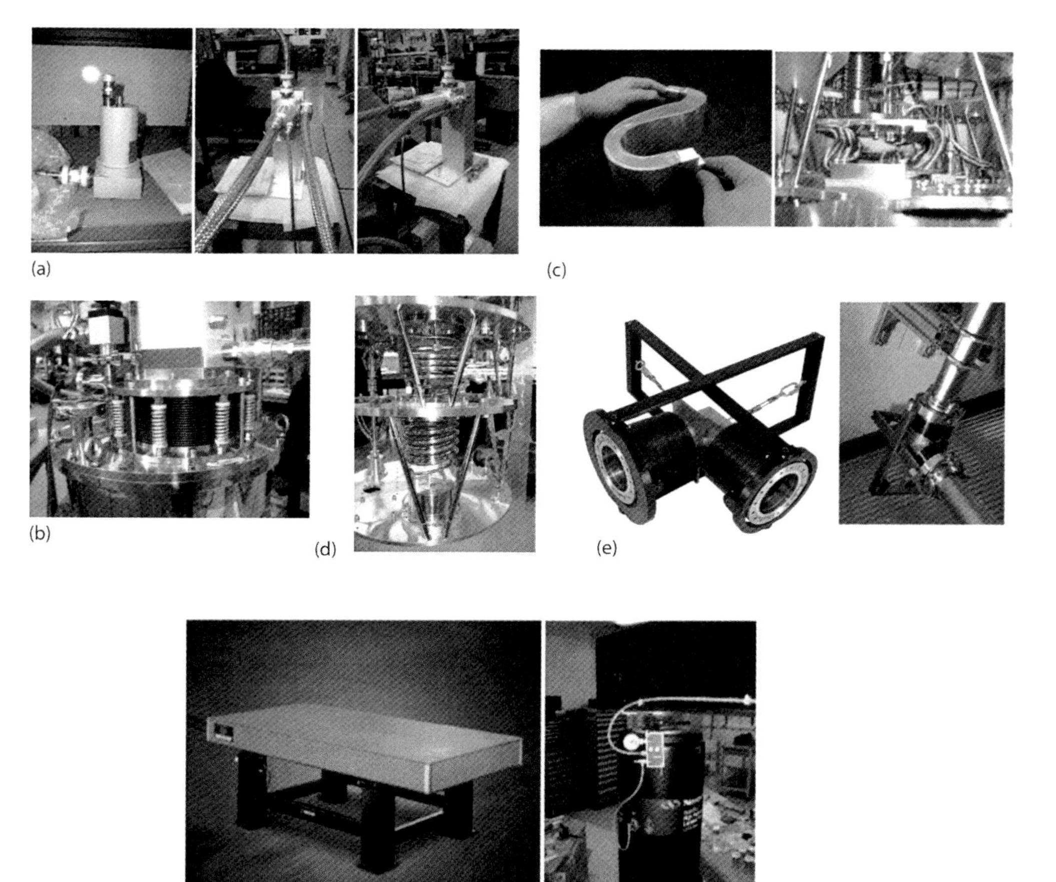

FIGURE 11.18 Vibration reduction approaches for reader's references. (a) PTR remote motor, (b) spring loaded bellows stage for PTR cold head, (c) flexible thermal links, (d) rigid support design, (e) gimbals, and (f) vibration suspension.

1. *Refrigerant:* The coolant in the 3He cryostat is pure 3He, while that in the DR is 3He and 4He mixture.
2. *Circulation rate control*: Unless heating (including the heat load from the experiments) is applied to the 3He pot, the circulation rate of the 3He cryostat depends on the pumping station only.

   The circulation rate of a DR is usually controlled by the Still temperature. Higher Still temperatures with more heat applied to the Still results in a higher circulation rate, which usually generates higher cooling power.

   However, a higher circulation rate is actually a double-edged sword because it also brings more warm 3He gas back to the system per unit time. The higher heat load to the refrigerator may cause a higher temperature at the mixing chamber.

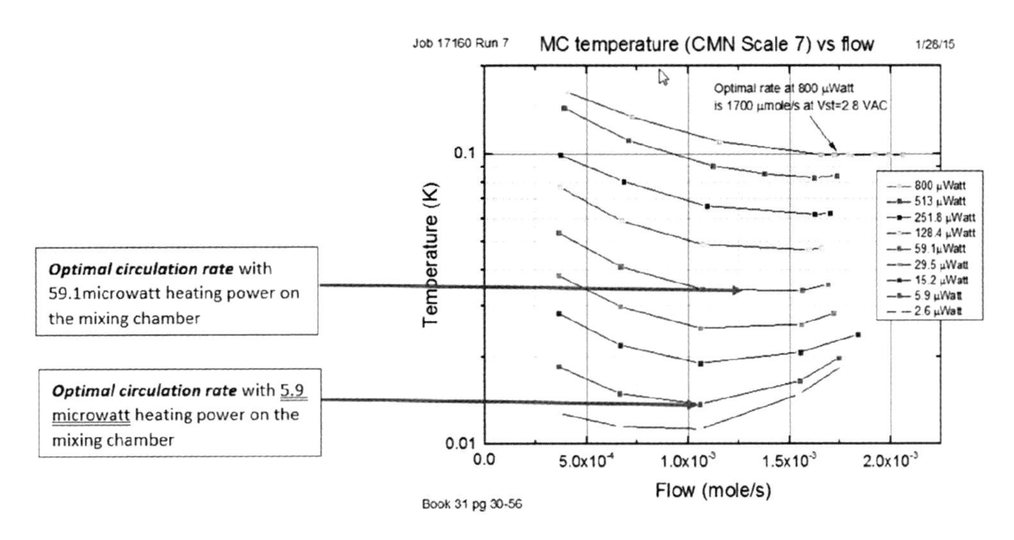

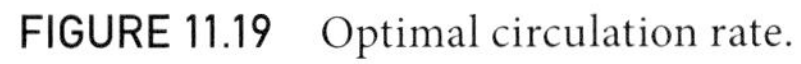

FIGURE 11.19 Optimal circulation rate.

If the lowest possible temperature is desired by the user, a balance between the circulation rate and the heat load from the returning warm 3He gas is needed. The circulation rate when the mixing chamber reaches the lowest base temperature (or called "temperature minimum") is called the "***optimal circulation rate***." Figure 11.19 shows a plot of the MC temperature versus the circulation rate with different amounts of heat applied to the MC. The temperature minimum moves to lower circulation rates when less heating is applied to the MC. This is because lower circulation rate is needed to cover lower heating. It also brings less heat load to the refrigerator from the returning 3He gas, which helps to reach the lowest possible temperature at the MC.

3. *Cooling mechanism and cooling power*: The most important difference between the DR and 3He cryostat is their cooling mechanism (Figure 11.20).

   A 3He cryostat makes use of the heat of evaporation of 3He when liquid 3He evaporates. 3He evaporation is a normal "***evaporative cooling***" process.

   The DR cooling originates from the enthalpy difference of the two phases of liquid 3He. The concentrated 3He "***evaporates inversely***" into the dilute 3He where 4He acts as a "***vacuum space***."

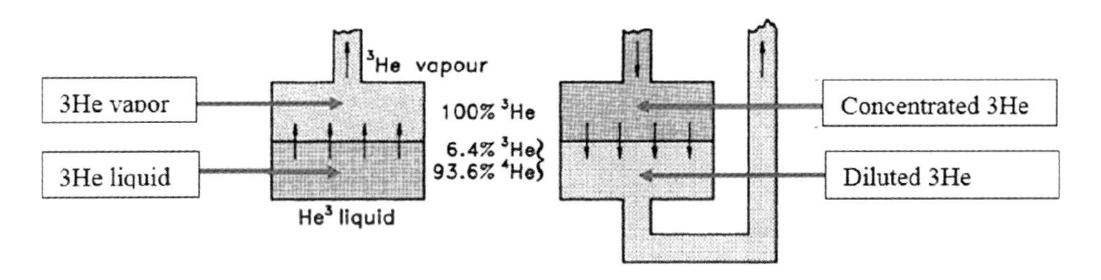

FIGURE 11.20 Cooling mechanism of 3He refrigerator and dilution refrigerator.

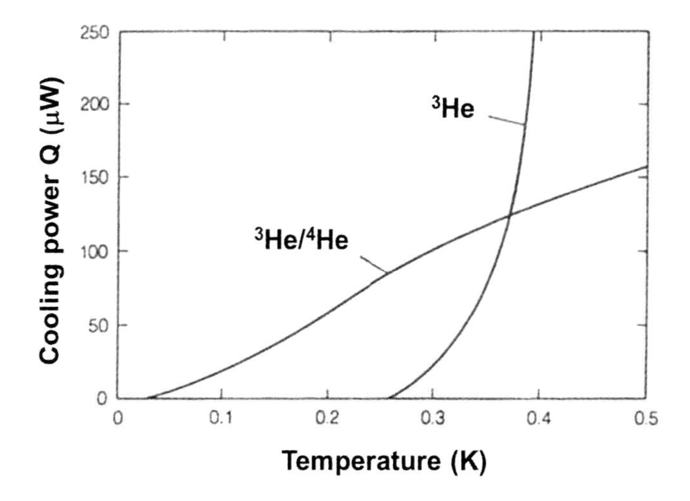

FIGURE 11.21 Cooling power comparison between a DR and a 3He evaporation cryostat as a function of temperature, if both systems use the same pump with a helium gas circulation rate of 5 L/s.

Although the cooling powers of both systems are proportional to their circulation rates, the base temperatures and the cooling powers are quite different from each other.
Figure 11.21 shows the cooling power comparison as a function of temperature between a DR and a 3He cryostat.

We can conclude from the Figure 11.21 that the lowest temperature a typical 3He cryostat can reach is limited to approximately 250–300 mK.

A 3He refrigerator should be considered when base temperatures above 300 mK are needed.

A DR is needed for temperatures below 300 mK.

Under the same circulation rate, a DR system has higher cooling power than that of 3He refrigerator below approximately 400 mK, while a 3He refrigerator has much higher cooling power than that of a DR above 400 mK. The cooling power of the 3He system increases exponentially with temperature.

11.3.4 Push for Lower Temperatures

While the cryogen-free DR became a mature commercial product, an effort to achieve lower temperatures, such as the cryogen-free Nuclear Adiabatic Demagnetization Refrigerator (NADR) systems to reach sub-millikelvin, is of great interest. A few cryogen-free NADR cryostats have emerged from the horizon during the past few years [36,37].

The cooling mechanism of NADR is similar to the ADR discussed in Chapter 10. Instead of using the electronic spin magnetic moment for magnetic refrigeration under the adiabatic environment, refrigeration by NADR is achieved by the adiabatic demagnetization of nuclear spin magnetic moment [38]. The paramagnetic salt pills in ADR are replaced by a **nuclear stage** in the NADR system.

Since the magnetic moments of nuclear spin is three orders of magnitude smaller than the electronic magnetic moments (i.e., Bohr magneton), the ordering temperature due to nuclear dipole–nuclear dipole interaction in the nuclear stage is much lower, typically the order of 0.1 μK Therefore, NADR is able to reach microkelvin range.

Listed below, but not being limited to, are features generally required for the good material candidate of the **nuclear stage** of the NADR system:

- Good thermal conductivity,
- Reasonably large nuclear magnetic moment,
- Non-ferromagnetic,
- Non-superconductor,
- Good coupling between the nuclear spin and the conduction electrons, and
- Easy to be machined.

Copper (with 2.47 nuclear magnetons) is the commonly used material for the nuclear stage.

An alternative candidate for the nuclear stage is the intermetallic compound $PrNi_5$.

The nuclear stage needs to be precooled (typically around 10 mK) by a DR in an applied magnetic field through a superconducting heat switch discussed in **Section 6.3.4**. The superconducting heat switch should be in the normal state during precooling. After the nuclear stage is precooled properly, magnetic refrigeration starts when the heat switch is disengaged and the magnetic field is reduced.

Vibration isolation between the nuclear stage and the magnet is crucial in NADR systems since their coupling will induce eddy current heating during the demagnetization in the nuclear stage. Furthermore, slots needed to be added on the copper nuclear stage to minimize the eddy current heating.

Detailed discussion of design, fabrication, and operation of NADR cryostat is beyond the scope of this book. Interested readers can refer to references [16] and [38].

Figure 11.22a illustrates a newly set-up cryogen-free NADR system with aluminum heat switch and copper nuclear stage at the International Center for Quantum Material, Peking University, China. The NADR system is integrated with a cryogen-free double magnet including a 9 tesla solenoid with 5.0″ clear bore for the nuclear demagnetization, and a 12 tesla solenoid with 3.0″ clear bore for the experiments. Both solenoids are cooled by a model PT415RM pulse-tube cryocoolers as shown in Figure 11.22b.

Figure 11.23 illustrates the nuclear stage including the heat switch assembly and the low eddy current copper nuclear stage.

This system is under the final performance test and we look forward to the performance report.

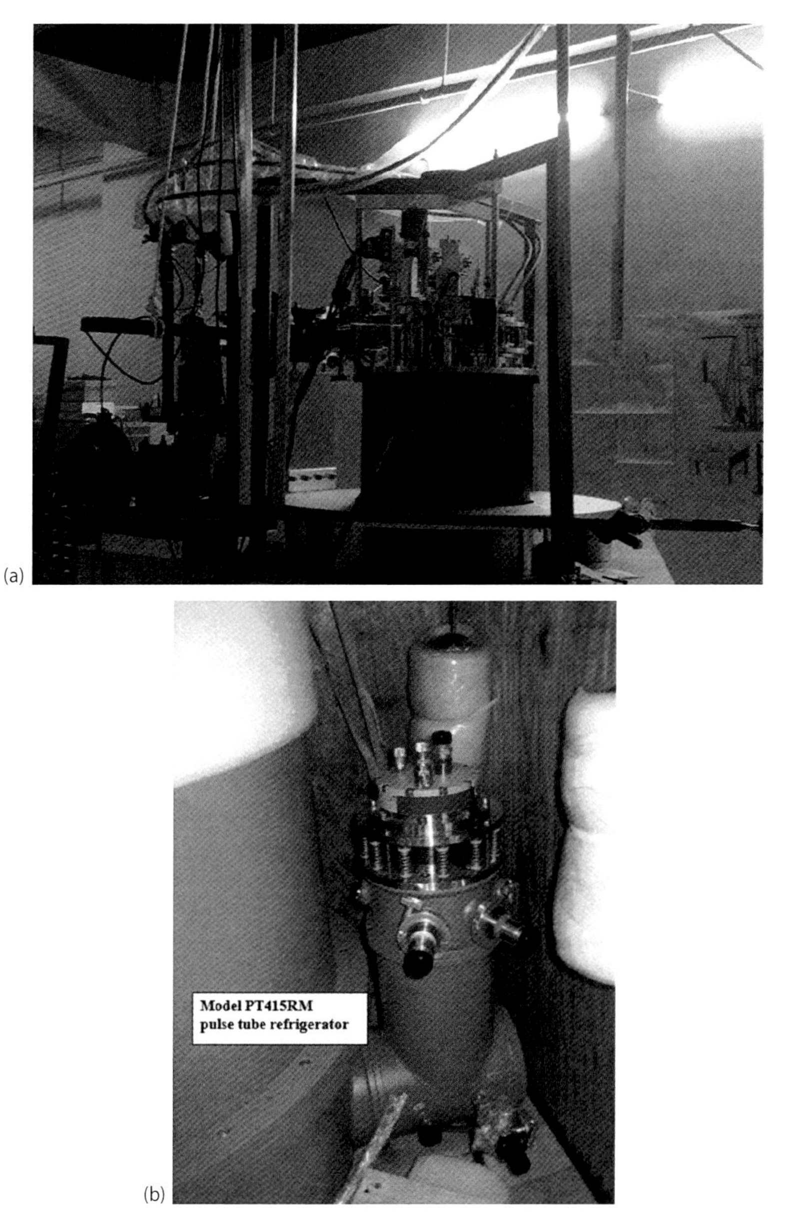

FIGURE 11.22 (a) Cryogen-free NADR system at the International Center for Quantum Materials, Peking University, China. (b) Double cryogen-free magnet cooled by pulse-tube cryocooler. (Private communication with Rui-rui Du and Xi Lin; Courtesy of Prof. R.R. Du and Prof. X. Lin at the International Center for Quantum Materials, Peking University, China.)

FIGURE 11.23 (a) Heat switch surrounded by an aluminum solenoid. (b) Low eddy current copper nuclear stage. (c) Whole nuclear stage. (Private communication with Rui-rui Du and Xi Lin; Courtesy from Prof. R.R. Du and Prof. X. Lin at the International Center for Quantum, Peking University, China.)

Acknowledgments

The contributor acknowledges Munir Jirmanus for reviewing this chapter for its scientific integrity. Sincere acknowledgment goes to Yong-hua Huang for his helpful discussion on 3He properties.

The contributor acknowledges his colleagues at Janis Research Company for their help:
Thanks to Choung Diep for making drawings; thanks to Dan Logan for his comments; thanks to Jeonghoon Ha for his final proofreading. Last but not the least, thanks to Ann Carroll for her constant help with software difficulties, reference searches, and copyright requests.

References

1. London, H.: *Proceedings of the International Conference on Low Temperature Physics*, Oxford, Clarendon Laboratory (1951), p. 157.
2. London, H., Clarke, G.R., and Mendoza E.: Osmotic Pressure of He3 in Liquid He4, with Proposals for a Refrigerator to Work below 1°K, *Physical Review* **128**, 1992 (1962).
3. Wheatley, J.C., Viches, O.E., and Abel, W.R.: Principles and Methods of Dilution Refrigeration, *Physics* **4**, 1 (1968).
4. Wheatley, J.C., Rapp R.E., and Johnson, R.T.: Principles and Methods of Dilution Refrigerator II, *Journal of Low Temperature Physics* **4**(1) (1971).
5. Ventura, G., and Risegari, L.: *The Art of Cryogenics Low-Temperature Experimental Techniques*, Elsevier Ltd., 2008.
6. Das, P., de Bruyn Ouboter, R., and Taconis, K.W., *Proceedings of the 9th International Conference on Low Temperature Physics*, (1965), p. 1253.
7. Hall, H.E., Ford, P.J., and Thompson, K: A Helium-3 Dilution Refrigerator, *Cryogenics* **6**, 80 (1966).
8. Nagano, B.S., Borisov, N.S., and Liburg, M.: A Method of Producing Very Low Temperatures by Dissolving He^3 in He^4, *JETP*, **50** (1966), 1445.
9. Vilches, O.E.: *John Wheatley and His First Dilution Refrigerators, Fifty Years of Dilution Refrigerators*, Manchester, UK, 2015.
10. Radebaugh, R. Siegwarth, J.D., and Hoslte, J.C.: *Proceedings of the of 5th International Cryogenic Engineering Conference,* Kyoto, K. Mendelssohn (ed.). IPC Science and Technology Press (1974a) 242.
11. Radebaugh, R., Siegwarth, J.D., Oda, Y., and Nagano, H.: *Proceedings of the of 5th International Cryogenic Engineering Conference* (Kyoto), K. Mendelssohn (ed.). IPC Science and Technology Press (1974b) 235.
    School of Physics and Materials, Lancaster University, Lancaster L41 4YB, United Kingdom, European Union.
12. Wheatley, J.C.: *Proceedings of the 15th Scottish Summer School in Physics*, St. Andrews (1975).
13. Frossati, G.: Obtaining Ultralow Temperatures by Dilution of 3He into 4He, *Journal de Physique Colloques*, **39** (C6) (1978), pp. C6-1578-C6-1589.
14. Ebner, C., and Edwards, D.O.: The Low Temperature Thermodynamic Properties of Superfluid Solutions of ^3He in ^4He, *Physics Reports* **2**(2), (1971) 77–154.
15. Graham, B.: *50 Years of Dilution Refrigeration Oxford Instruments*, 50 Years of Dilution Refrigeration, 2015.
16. Lounasmaa, O.V.: *Experimental Principles and Methods Below 1K*, Academic Press: London, UK, 1974.
17. Pobell, F.: *Material and Methods at Low Temperatures*, 3rd edition, Springer-Verlag, Berlin, Germany, 2007.
18. Cryocourse 2016: School and Workshop, in *Cryogenics and Quantum Engineering*, Aalto University, Espoo, Finland, September 26–October 3, 2016.

19. Laheurte, J.P., and Keyston, J.R.G.: Behaviour of He 4 in Dilute Liquid Mixtures of He 4 in He 3, *Cryogenics* 485 (1971).
20. Edwards, D.O., Ifft, E.M., and Sarwinski, R.E.: Number Density and Phase Diagram of Dilute He3-He4 Mixture at Low Temperatures, *Physical Review* **177**(1) (1969).
21. Ghozlan, A., and Varoquaux, E.: *Régulateur de température pour réfrigérateur à dissolution, Revue de Physique Appliquée* **10**(6) (1975).
22. Dobbs, E.R.: *Helium Three*, Oxford University Press, 2000. Stahl, C., Hunklinger, S.: *Low-Temperature Physics*, Springer, 2005.
23. Enns, C., and Hunklinger, S.: *Low-Temperature Physics*, Springer, 2005.
24. Frossati, G.: Experimental Techniques: Methods for Cooling Below 300 mK, *Journal of Low Temperature Physics* **87**, 604 (1992).
25. Takano, Y.: Cooling Power of the Dilution Refrigerator with a Perfect Continuous Flow Heat Exchanger, *Review of Scientific Instruments* **65**(5) (1994).
26. Uhlig, K.: "Dry" Dilution Refrigerator with Pulse Tube Precooling, *Cryogenics* **44**, 53–57 (2004).
27. Uhlig, K.: *"Cryogen-free" Dilution Refrigerators*, 50 Years of Dilution Refrigeration, Manchester, invited talk, September 16, 2015.
28. Song, Y.J., Otte, A.F., Shvarts, V., Zhao, Z., Kuk, Y., Blankenship, S.R., Band, A., Hess, F.M., and Stroscio, J.: A 10mK Scanning Probe Microscopy Facility, *Review of Scientific Instruments* **81**, 1211–1211 (2008).
29. Oda, Y., Fujii, G., Ono, T., and Nagano, H.: Practical Design of Heat Exchanger for Dilution Refrigerators: Part 2, *Cryogenics*, **139** (1983).
30. Cousins, D.J., Fisher S.N., Guenault, A.N., Pickett, G.R., Smith, E.N., and R. P. Turner, R.P.: T 3 Temperature Dependence and a Length Scale for the Thermal Boundary Resistance between a Saturated Dilute 3He-4He Solution and Sintered Silver, *Physical Review Letters* **73**(19) (1994).
31. Pari, P.: Dilution Refrigerator with No Liquid Helium Supply, Fast, R.W. (ed.), *Advances in Cryogenic Engineering*, Vol. 35, 1079 (1990).
32. Uhlig, K., and Hehn, W.: 3He/4He Dilution Refrigerator with Gifford-McMahon Precooling, *Cryogenics* **33**, 1028 (1993).
Uhlig, K., and Hehn, W.: 3He/4He Dilution Refrigerator Combined with Gifford-McMahon cooler, *Cryogenics* **34**, 587 (1994).
Uhlig, K., and Hehn, W.: 3He/4He Dilution Refrigerator Precooled by Gifford-McMahon refrigerator, *Cryogenics* **37**, 1028 (1997).
Uhlig, K.,: 3He/4He Dilution Refrigerator with Pulse-tube Refrigerator Precooling, *Cryogenics* **42**, 73–77 (2002).
Uhlig, K.: 3He/4He Dilution Refrigerator Precooled by Gifford-McMahon Cooler II. Measurements of the Vibrational Heat Leak, *Cryogenics* **42**, 569–575 (2002).
Uhlig K., and Wang C.: Cryogen-free Dilution Refrigerator Precooled by a Pulse-tube Refrigerator with Non-magnetic Regenerator, *Advances in Cryogenic Engineering* **51A**, 939–945 (2005).
Uhlig, K.: Dry Dilution Refrigerator with High Cooling Power, *Advances in Cryogenic Engineering* **53B**, 1287–1291 (2007).
Uhlig, K,: Condensation Stage of a Pulse Tube Precooled Dilution Refrigerator, *Cryogenics* **48**, 138–141 (2008).
Uhlig, K.: Dilution Refrigerator with Direct Pulse Tube Precooling, *Cryocoolers* **15**, 491–495 (2008).
Uhlig, K.: 3He/4He Dilution Refrigerator with High Cooling Capacity and Direct Pulse Tube Precooling, *Cryogenics* **48**, 511–514 (2008).
Uhlig, K.: Improved Design of the Intermediate Stage of a Dry Dilution Refrigerator, *Advances in Cryogenic Engineering* **55A**, 641–647 (2009).
Uhlig, K.: Concept of a Powerful Cryogen-free Dilution Refrigerator with Separate 1K Stage, *Cryocoolers* **16**, 509–513 (2010).
Uhlig, K.: Cryogen-free Dilution Refrigerator with Separate 1K Cooling Circuit, *Advances in Cryogenic Engineering* **57B**, 1823–1829 (2011).
Uhlig, K.: Cryogen-free Dilution Refrigerator with 1K-stage, *Cryocoolers* **17**, 471–477 (2012).
Uhlig, K.: Cryogen-free Dilution Refrigerators, LT26, invited talk, *Journal of Physics*: Conf. Ser. 400 052039 (2012).

Marx, A., and Uhlig, H.K., Dry Dilution Refrigerator for Experiments on Quantum Effects in the Microwave Regime, *Cryocoolers* 18, edited by S.D. Miller and R.G. Ross Jr., *International Cryocooler Conference*, Boulder, CO, 2014.
Uhlig, K.: Dry Dilution Refrigerator with He-4 1K-Loop, *Cryogenics* 66 (2015).
Uhlig, K.: Alumina Shunt for Precooling a Cryogen-free He-4 or He-3 Refrigerator, *Cryogenics* **79**, 35–37 (2016).
Uhlig, K.: Concepts for a Low-vibration and Cryogen-free Tabletop Dilution Refrigerator, *Cryogenics* **87**, 29–34 (2017).

33. Wang, C.: Extracting Cooling from Pulse Tube and Regenerator in a 4 K Pulse Tube Cryocooler, *Cryocoolers* **15**, 177–184 (2009).
34. Schmoranzer, D., Luck, A., Collin, E., and Fefferman, A.: Cryogenic Broadbank Vibration Measurement on a Cryogen-Free Dilution Refrigerator, *Cryogenics* (2019).
35. de Wit, M., Welker, G., Heeck, K., Buters, F. M., Eerkens, H. J., Koning, G., van der Meer, H., Bouwmeester, D., and Oosterkamp, T. H.: Vibration Isolation with High Thermal Conductance for a Cryogen-Free Dilution Refrigerator, *Review of Scientific Instruments* **90**, 015112 (2019).
36. Batey, G., Casey, A., Cuthbert, M. N., Matthews, A. J., Saunders, J., and Shibahara, A.: A Microkelvin Cryogen-free Experimental Platform with Integrated Noise Thermometry, *New Journal of Physics* **15** (2013). *New Journal of Physics* **15**, 113034 (2013).
37. Todoshchenko, I., Kaikkonen, J.-P., Blaauwgeers, R., Hakonen, P. J., and Savin, A.: Dry Demagnetization Cryostat for Sub-millikelvin Helium Experiments: Refrigeration and Thermometry, *Review of Scientific Instruments* **85**, 085106 (2014).
38. Andres, K., and Lounasmaa, O.V.: in *Progress in Low Temperature Physics*, Vol. 8, ed. by D.F. Brewer, North-Holland, Amsterdam, the Netherlands (1982) p. 221.

Appendix A

Patent of the Improved Process for the Artificial Production of Ice

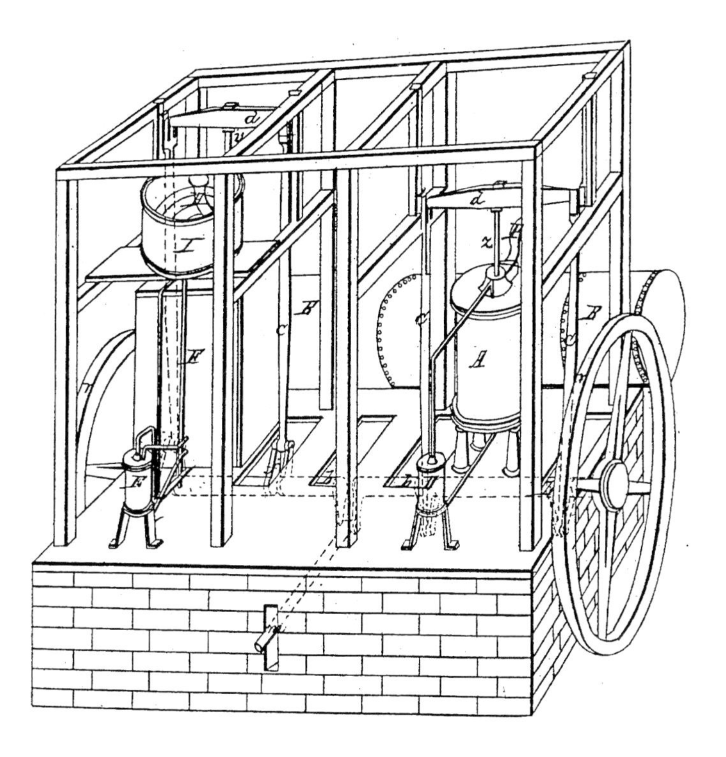

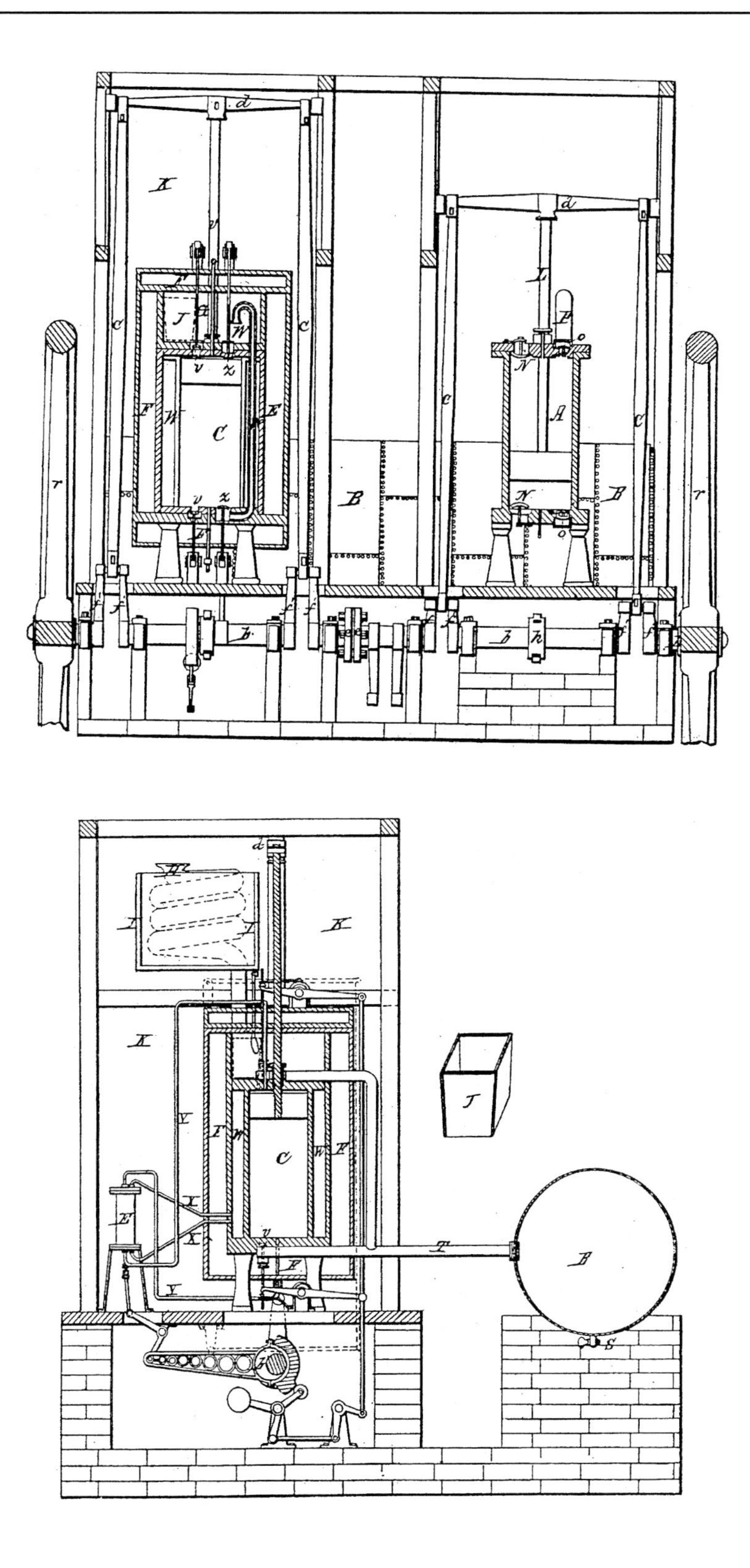
d
K
C
B
A
L
r
c
T
S

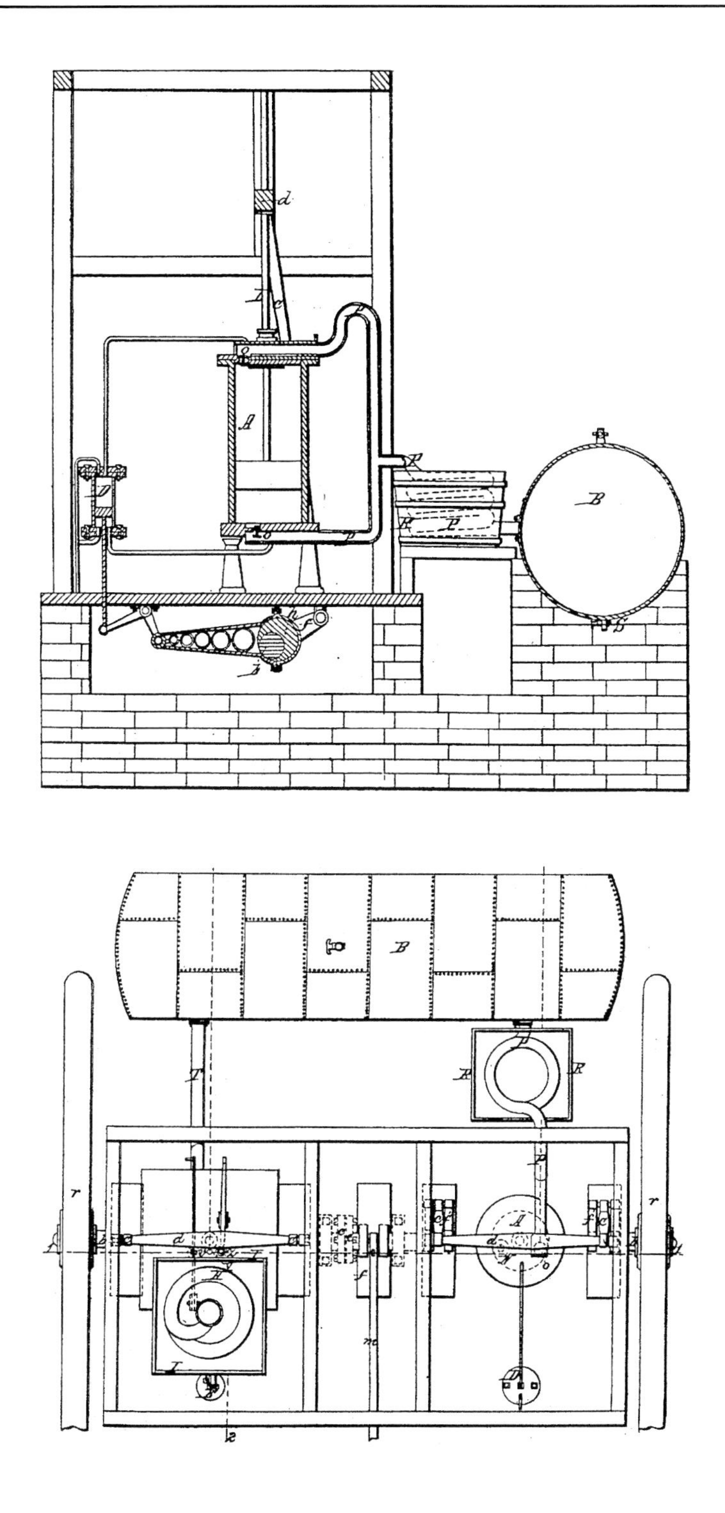
d
B
r
r

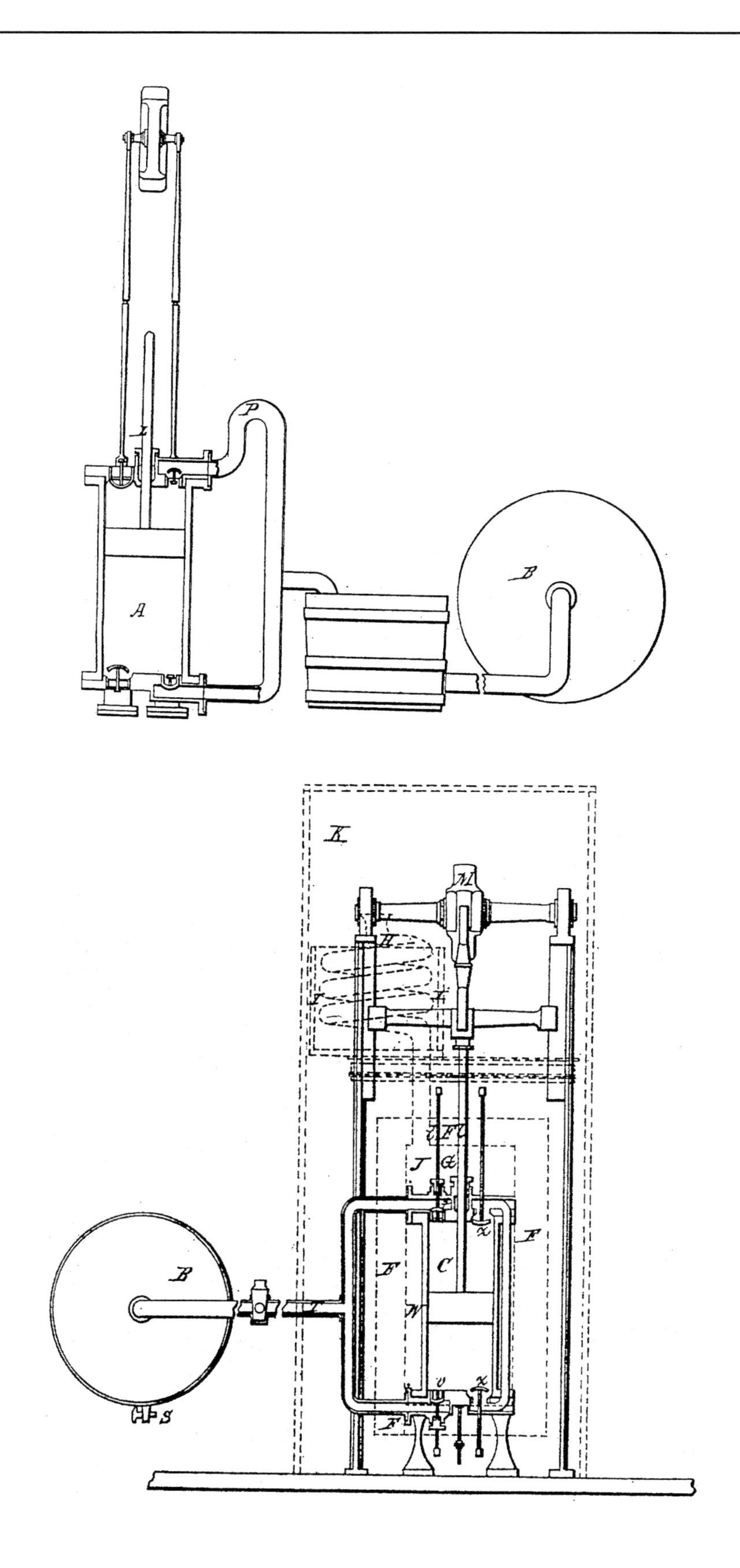
P
B
A
K
M
C
W
B
S

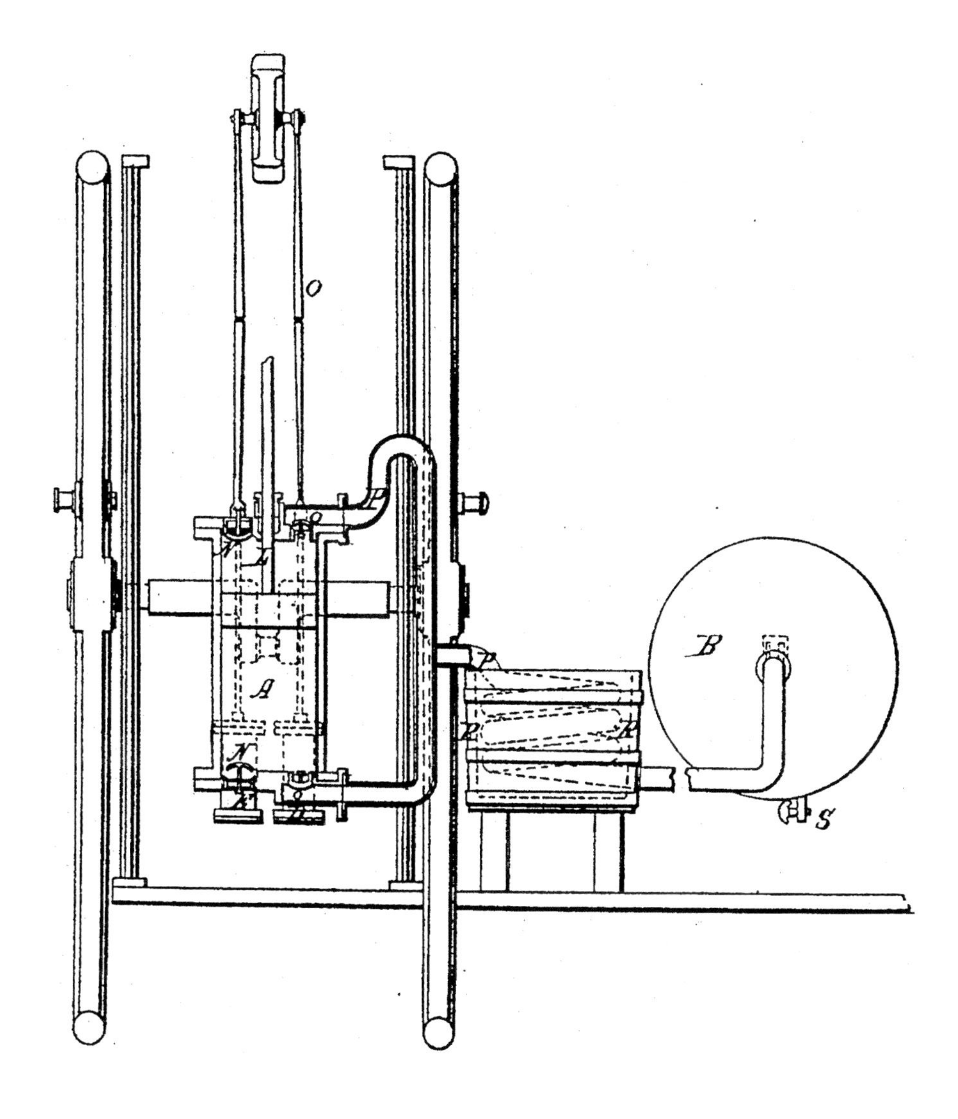
O
A
B
N
S

United States Patent Office.

JOHN GORRIE, OF NEW ORLEANS, LOUISIANA.

IMPROVED PROCESS FOR THE ARTIFICIAL PRODUCTION OF ICE.

Specification forming part of Letters Patent No. 8,080, dated May 6, 1851.

To all whom it may concern:

Be it known that I, JOHN GORRIE, of the city of New Orleans, in the parish of Orleans and State of Louisiana, have invented a new and useful Machine for the Artificial Production of Ice and for General Refrigeratory Purposes, of which the following is a full, clear, and exact description, reference being had to the annexed drawings of the same, making part of this specification, in which—

Figure 1 is a perspective view of the machine. Fig. 2 is a vertical longitudinal section, taken through the condensing-pump and expanding-engine, for the purpose of showing more clearly their internal structure and arrangements. Fig. 3 is a vertical transverse section through the expanding-engine and its appendages, showing, also, the section of the reservoir. Fig. 4 is a similar section, taken through the condensing-pump. Fig. 4½ is a perspective view of the vessel in which the ice is formed, removed from the refrigerating-chamber. Fig. 5 is a top view of the apparatus. Fig. 6 is a vertical longitudinal section of the machine modified in the structure and arrangement of some of its parts; and Figs. 7 and 8 are, respectively, vertical transverse sections of the same through the expanding-engine and condensing-pump.

The same letters indicate the same parts in all the figures.

It is a well-known law of nature that the condensation of air by compression is accompanied by the development of heat, while the absorption of heat from surrounding bodies, or the manifestation of the sensible effect, commonly called "cold," uniformly attends the expansion of air, and this is particularly marked when it is liberated from compression.

The nature of my invention consists in taking advantage of this law to convert water into ice artificially by absorbing its heat of liquefaction with expanding air. To obtain this effect in the most advantageous manner it is necessary to compress atmospheric air into a reservoir by means of a force-pump to one-eighth, one-tenth, or other convenient and suitable proportion of its ordinary volume. The power thus consumed in condensing the air is, to a considerable extent, recovered at the same time that the desired frigorific effect is produced by allowing the air to act with its expansive force upon the piston of an engine, which, by a connection with a beam or other contrivance common to both, helps to work the condensing-pump. This engine is constructed and arranged in the manner of a high-pressure steam-engine having cut-offs and working the steam expansively. When the air, cooled by its expansion, escapes from the engine, it is made to pass round a vessel containing the water to be converted into ice, or through a pipe for effecting refrigeration otherwise, the air while expanding in the engine being supplied with an uncongealable liquid whose heat it will absorb, and which can in turn be used to absorb heat from water to be congealed. By this arrangement I accomplish my object with the least possible expenditure of mechanical force, and produce artificial refrigeration in greater quantity from atmospheric air than can be done by any known means.

The apparatus for producing the refrigeratory effects before stated consists, essentially, of a large double-acting force-pump, A, with its jet-pump D, Figs. 1 and 4, condensing-tub R, and worm P, as represented in the drawing No. 4, a reservoir, B, made of such metal in the manner of a steam-boiler, a double-acting expanding-engine, C, provided with cut-offs, a jet-pump, E, a tub, I, and worm H, for cooling water, the engine C and the chamber G above it being inclosed in an insulating-box, F, which box, together with the worm and tub H, are inclosed in a second insulating room or chamber, K. The pumps, engine, and other moving parts are provided with the necessary mechanical appliances for putting and keeping them in motion and connecting them with the prime mover, which may be either a steam-engine or other available power.

It is believed that the precise nature of my invention and discovery, and the manner in which the refrigerating effects are produced, can be more clearly and fully set forth by describing the construction and operation of the apparatus in connection. I shall therefore adopt that plan in the remainder of the specification.

In the apparatus represented in Figs. 1, 2, 3, 4, and 5 the piston-rods of the pump A and engine C are attached to cross-heads *d d*, which are connected by rods *c c c c* with the cranks

$f'f'$ of a revolving shaft, b. This main shaft b is divided into two parts, which are connected by the flanges e, secured to their adjacent ends and bolted together in such a manner that the cranks on that section of the shaft which drives the pump may be placed at any angle of inclination to those which are attached to the opposite section of the shaft and worked by the engine. This angle of inclination should be such that the maximum force of the engine C may be exerted about the same time that the maximum resistance is offered by the pump A—or, in other words, that the engine, supposing that it works under a tension of eight atmospheres, may commence its stroke about the same time that the pump has completed six-sevenths of its stroke. The power of the prime mover is applied to the crank f by a connecting-rod, m, Fig. 1, and the motion is regulated by fly-wheels r r.

The pump A is constructed like the ordinary double-acting force-pump for air or water. It receives atmospheric air through valves N N, placed in its opposite ends, which open inward, and after compressing it to the degree required forces it through the eduction-valves o o, which open outward into a pipe, P, through which it passes into the reservoir B. A part of the pipe P is bent into a worm, which is immersed in a tub of cold water, R, to cool the air as it passes through the same.

Air, while being compressed, evolves sensible heat, which, if not absorbed or extinguished as fast as given out, will increase its elastic force greatly beyond what is due to the degree of its compression, thus requiring a greater degree of mechanical power than is absolutely necessary to compress it in the reservoir. To obviate this difficulty as far as practicable, I immerse the pump in a cistern of cold water and inject into its interior a jet of water, also cold, and in a finely-divided or other state, by means of a small pump, D, which bears a suitable relation in its size to the pump A, being in the machine I have built in the proportion I have represented in the drawings, and capable of discharging twenty-eight cubic inches of water for every cubic foot of air compressed by the latter. The pump D is double-acting, receives its motion from the eccentric h on the shaft b, and forces the water through tubes and perforated plates inserted in the lids of the pump A, into the body of air in the cylinder A. This pump D is so constructed and arranged as to force the jet into the pump A at the beginning, and continue it to the end of the stroke of its piston, so as to meet and absorb the heat as fast as evolved.

It is contemplated, particularly when the apparatus is made upon a large scale, to return the water of injection through the pump D under the pressure of the air in the reservoir, so as to recover some of the power consumed in making the injection, and also to supersede the necessity of employing a greater quantity of water than is required to insure a proper temperature. In this use of the injection-water it is obvious that the valves of the pump D cannot be self-acting, but must be operated mechanically. Air, being a bad conductor of heat, cools very slowly when in large masses, and as it is essential to the success of this process that most of the heat set free from the air by compression should be absorbed as soon as possible after it is given out, and as the immersion of the cylinder and the injection of the jet of cold water have proved inadequate to effect this result, I have superadded a worm, P, immersed in a tub of water, R, through which the air in its transit from the pump A to the reservoir B must pass, and this effectually absorbs the remainder of the heat.

Instead of the worm, some other form of refrigerating apparatus may be used.

The reservoir B is made of twenty or thirty times the capacity of the condensing-pump A, and is provided with a stop-cock, S, inserted into its under side, through which the water of the injection, precipitated from the air, may be discharged. It is also furnished with a gage for measuring with precision the pressure of the air within it. This consists of a glass tube, closed at the upper end, filled with dry atmospheric air, inserted into and in communication with the air in the reservoir through the intervention of a cup of mercury.

The engine C is supplied with the condensed air, which works it from the reservoir B through the pipe T. Its valves V are so arranged as to cut off the supply of dense air from the reservoir when the cylinder has received a quantity which is equal to the quantity condensed by one stroke of the pump A. This air received into the engine tends to dilate with a force corresponding to the degree of its compression, which force acts alternately upon the opposite sides of the piston, communicating to it a reciprocating motion which aids in working the condensing-pump, and thus a part of the force expended in condensing the air is here reclaimed.

The refrigeratory effects of air dilating from the removal of pressure can be rendered available to the fullest extent only when the expansion is gradual, because time is required to enable it to absorb heat, and therefore the employment of the compressed air as a mechanic agent retards its expansion sufficiently to allow it to absorb the greatest amount o heat from the liquid of the jet and the walls o the cylinder, thus being an advantage to the freezing process.

The jet for the engine C is furnished by the pump E. The capacity of this pump bears the same relation to that of the engine that the pump D does to the pump A. It is arranged and operated in precisely the same manner as the pump D, except that its supply of liquid is obtained through the pipes X from the cistern W, into which, after it has performed the office of intermixture with the expanding air in the cylinder, it is returned through the eduction-valves Z. As the temperature of the engine C, the cistern which surrounds it, and the expanding air must be kept considerably below the freezing-point of water to make ice advantageously, it follows that the liquid which fills the cistern W must be uncongealable at the low temperature at which this portion of the apparatus is required to be kept. As the waste of this liquid in a properly-constructed machine is only that which arises from evaporation, and as this latter is small at the low temperature maintained, its first cost is not a consideration of much importance, so that any liquid uncongealable at the low temperature required may be employed—as proof-spirit a solution of common salt, or of the nitrate or carbonate of potash, &c. As the air when expanded in the engine C is at a much lower temperature than that at which it is received into the pump A, and as its volume is directly as its temperature, it follows that the capacity of the engine C must be less than that of the pump A, in order that the engine may be filled with air fully expanded at a tension not less than the atmospheric, or that it may not be consumed faster than it is compressed into the reservoir. The difference between the capacity of the pump A and the engine C should be directly proportioned to the intensity of cold required to be produced, and hence the mechanical force applied by the engine will in all cases be less than that consumed by the pump, in the proportion that the heat of the air escaping from the engine is less than that entering the pump. This excess in the consumption of force by the pump over its production by the engine, together with the amount of force necessary to keep the other parts of the apparatus in motion, must be supplied, as already mentioned, from some extraneous source. The air, after its expansion in the cylinder C, passes out intimately mixed with the liquid of injection at the eduction-valves Z, and through tubes connected therewith into the cistern W, which is filled with fluid nearly to the top of the vessels J, containing the water to be converted into ice. In this way the mass of fluid in the cistern is constantly having its heat absorbed by the expanding air and cooled fluid of the jet, while at the same time it is constantly absorbing the heat of liquefaction from the water in the vessels J, which is thereby congealed. It will thus be seen that the uncongealable liquid of the cistern W merely acts as a medium to transmit the heat of the water to be congealed to the expanding air in the engine.

Experience has shown that if the vessels J be filled with water previous to the commencement of the process of refrigeration, the operation will be greatly retarded by the formation of an insulating coating of ice on its surface and on the interior of the sides and bottom of the vessels. This difficulty has, however, been obviated by furnishing the water in a small stream, and only as fast as it is frozen, by means of a flexible tube, *t*, from the cistern I. The congelation is further hastened by rocking or jarring the vessels slightly by connecting them with some of the moving parts of the machinery. Ice is specifically lighter than water, and will therefore float on it; hence it follows that if water be admitted into the vessels J so gradually as to be frozen in films, as every film is formed it must be displaced by a film of water, which, in virtue of its superior gravity, passes under it, to be in like manner frozen to the under side of the first film of ice, and thus by successive increments of ice formed in this manner a solid block filling the vessel J may be produced. By widening the vessel J from the bottom upward the removal of the block of ice is not only rendered more easy, but the formation of it is assisted. The smaller part of the block, by gradually ascending into the larger part of the vessel, keeps a passage open around it, through which the water to be frozen runs down to the bottom. In this manner the important advantage is gained of freezing water from its under surface instead of its upper, thereby exposing every particle of it to the frigorific action of the cistern-liquid without subjecting it to the intervention of an insulating coating of ice.

To further facilitate the removal of the ice from the vessels J, they are not only made a little smaller at the bottom than at the top, but are lubricated with a thin coating of oil or grease by means of a sponge.

The cover of the refrigerating-chamber G is removable for the purpose of introducing and withdrawing the freezing-vessels J, which are suffered to remain in it such length of time as experience may determine is most advantageous. These vessels should be made of good conductors of heat, and may be of any suitable and convenient size.

In order that the capacity of the expanded air for the absorption of heat may be rendered more fully available, and also for the purpose of more effectually separating from it the liquid of the jet, it is conducted from the chamber G through the pipe H, which is bent into the form of a worm and surrounded by water in the tub I. The air in its passage through this pipe deprives the water in the tub of a portion of its heat, and as this cistern-water is the source of supply for the freezing-vessels J, its incidental refrigeration by the escaping air is so much gain to the process.

In the modified form of the apparatus (represented in Figs. 6, 7, and 8) the force-pump A and expanding-engine C are connected with the opposite ends of a lever-beam, M, by means of links, jointed to their respective piston-rods L and U. This connection insures uniformity of action between the pump and engine, and enables the latter to act directly upon the pump, to aid in working it. As an inspection of the drawing of this modified arrangement will render it fully understood, if made in connection with the description of the process as performed by the machine, (represented in Figs. 1, 2, 3, 4, and 5,) I have deemed an explanation of the same in detail to be unnecessary.

The several parts of the foregoing apparatus may be made of such materials as it may be deemed advisable by the constructor to employ in reference to the efficiency of the machine, economy in its cost, or other considerations that may influence him, and the form and arrangement of the several parts may be varied indefinitely without essentially changing the character of the invention.

It will have been seen that a great object aimed at in the construction of the machine is as perfect a system of compensations (chemical and mechanical) as possible. Thus the heat evolved and carried off in the condensation of air is replaced in the expanding-engine by an abstraction of heat from the water to be frozen through the intervention of the liquid in the cistern. In the consumption and production of mechanical force these compensating equivalents are more general and more marked. It has already been intimated that the power consumed in compressing air is nearly all recovered in the force exerted by its subsequent dilatation, and it has been shown in what way the force required to inject the water for receiving the heat of the condensed air may be, in a great measure, derived from the pressure of the air in the reservoir. It is evident that a mechanical apparatus admitting of such a system of compensations must operate, in theory at least, without the consumption of any power other than that required to overcome its friction, and to supply the loss arising from the difference of temperature, and consequently of bulk, between the air as it exists before condensation and after expansion; and, practically, the working of the machine is found not to differ materially from this result, and thus it presents by far the most comprehensive application of natural laws to the economical production of cold that it is believed has ever been devised.

Having thus fully made known my improved process of manufacturing ice and explained and exemplified suitable machinery for carrying the same practically into operation, I wish it to be understood that I do not claim as my invention any of the several parts of the apparatus in themselves; but

What I do claim as my invention, and desire to secure by Letters Patent, is—

1. The employment of a liquid uncongealable at the low temperature at which it is required to keep the engine, to receive the heat of the water to be congealed and give it out to the expanding air.

2. The employment of an engine for the purpose of rendering the expansion of the condensed air gradual, in order to obtain its full refrigeratory effects, and at the same time render available the mechanical force with which it tends to dilate, to aid in working the condensing-pump irrespective of the manner in which the several parts are made, arranged, and operated.

3. Supplying the water gradually and slowly to the freezing-vessels and congealing it by abstracting the heat from its under surface, substantially as herein set forth.

4. The process of cooling or freezing liquids by compressing air into a reservoir, abstracting the heat evolved in the compression by means of a jet of water, allowing the compressed air to expand in an engine surrounded by a cistern of an unfreezable liquid, which is continually injected into the engine and returned to the cistern, and which serves as a medium to absorb the heat from the liquid to be cooled or frozen and give it out to the expanding air.

JOHN GORRIE.

Witnesses:

JOHN G. RUAN,

J. R. POTTS.

Appendix B

Critical Temperatures and Critical Fields of Selected Superconductors

- Changchun Institute of Optics, Fine Mechanics and Physics, Chinese Academy of Sciences, Changchun, Jilin 130033, China
Received September 18, 2016, Revised November 12, 2016, Accepted November 13, 2016, Available online November 17, 2016.

Mullard Space Science Laboratory, UCL, Holmbury St. Mary, Dorking, Surrey RH5 6NT, UK.
Received June 24, 2009, Revised February 15, 2010, Accepted February 22, 2010, Available online February 26, 2010.

Material	Critical Temperature (K)	Critical Field (Gauss)
Aluminum	1.19	102
Cadmium	0.55	28.8
Gallium	1.09	55
Indium	3.4	285
Iridium	0.14	19
Lanthanum (β)	6.10	1,600
Lead	7.19	803
Mercury (α)	4.15	415
Molybdenum	0.92	98
Niobium	9.2	1,390 (H_{c1})
		2,680 (H_{c2})

(Continued)

Material	Critical Temperature (K)	Critical Field (Gauss)
Osmium	0.66	65
Rhenium	1.70	193
Ruthenium	0.49	66
Tantalum	4.48	805
Technetium	7.77	1,410
Thallium	2.39	170
Thorium	0.68	150
Tin (α)	3.72	305
Titanium	0.39	100
Uranium	0.68	
Vanadium	5.41	430 (H_{c1})
		820 (H_{c2})
Zink	0.88	53
Zirconium	0.55	47

Reference

Brookhaven National Laboratory Selected Cryogenic Data Notebook Compiled and edited by J.E. Jensen, W.A. Tuttle, R.B. Stewart, H. Brechna, A.G. Prodell, Revised August 1980.

Appendix C

Preparation for Aluminum Heat Switch Fabrications

Step 1: Wash the aluminum foils in alkaline (22 g/L of $Na_3PO4{\cdot}12H_2O$ and 22 g/L of Na_2CO_3) cleaner at 75°C for 60 s.
Step 2: Acid bath (equal volumes of concentrated HNO_3 and water) for 15 s.
Step 3: Zincate solution (e.g., 1.0 g/L $FeCl_3{\cdot}6H_2O$, 100 g/L ZnO, 525 g/L NaOH, 10 g/L $C_4H_4KCO_6{\cdot}4H_2O$) at (22 +/ 2)°C for 60 s.
Step 4: Acid bath (equal volumes of concentrated HNO_3 and water) for 30 s.
Step 5: Repeat steps 3 and 4 until the aluminum is slightly but uniformly etched. The aluminum may be briefly dried to apply a protective lacquer on parts not to be plated, but do not touch the parts to be plated. The foils must be kept wet all times following this step until completion of the process.
Step 6: Zincate solution at (22 +/ 2)°C for 10 s.
Step #7: Copper strike: connect the electrodes and turn on the power before immersing the item into the solution (41.3 g/L CuCN, 50.8 g/L NaCN, 30 g/L Na_2CO_3, 60 g/L $C_4H_4KNO_6{\cdot}4H_2O$), so that plating begins immediately; use copper anode and plate at room temperature for the first 2 min at 26 mA/cm^2, and at 13 mA/cm^2 for 2 more min.

References

Gloos, K., Smeibidl, P., Kennedy, C., Singsass, A., Sekovski, P., Mueller, R.M., Pobell, F. The Bayreuth Nuclear Demagnetization Refrigerator, *J. Low Temp. Phys.* 73, 101 (1988).
Yao, W., Knuuttila, T.A., Nummila, K.K., Martikainen, J.E., Oja, A.S., and Lounasmaa, O.V. A Versatile Nuclear Demagnetization Cryostat for Ultralow Temperature Research, *J. Low Temp. Phys.*, 120, 121–150 (2000).

Appendix D

Selected Ultra-High-Vacuum-Compatible Items

This appendix is to share the author's limited personal experience with readers on some selected UHV-compatible items often used in cryogenic systems.

Many more UHV components are not listed in this appendix. New products will also be provided by venders continuously.

Venders have provided some UHV-compatible products. Many equally qualified venders are not listed in this appendix.

Readers should refer to the "Buyer's Guide" edited by *Physics Today* for additional information.

All listed information is for reader's reference only. It is the reader's own responsibility to select the right components for their specific applications.

1. UHV-compatible machining lubricant and coolant:
   Blasocut BC 35 SW
   This is a water-miscible, chlorine-free, mineral-oil-based metalworking fluid especially designed for use in soft water where good foam control is required.
   Dilute it with water before use.
   http://www.cuttingfluids.com.mx/es/wp-content/uploads/2014/09/Blasocut_BC35SW.pdf
   Vender:
   Blasser Swiss Lube
   31 Hatfield Goshen, New York 10924
   USA
   Tel: (845) 294-3200
   Fax: (845) 294-3102
   Website: http://www.mfgday.com/events/2017/blaser-swisslube-inc-3

2. UHV-compatible solder and flux
   1. **604 silver solder**

Brand Name:	Handy & Harmon 604	
Components:	60%	Ag
	30%	Cu
	10%	Sn
Flux:	Harris Stay Silv black Hi-T flux	
Melting temperature:	1,115 F (600°C)	
Flowing temperature:	1,325 F (720°C)	

   2. **430 solder**

Brand Name:	Stay-Brite	
Components:	96%	Sn
	4%	Ag
Flux:	Supersafe #30 water based “blue” flux	
Melting temperature:	430 F	
Application:	Copper to copper; stainless to copper with special flux	

   Notes:

   Theoretically, this solder is not truly UHV compatible, but it is acceptable in small quantities, such as soldering wires and small parts.

   Use Eutector 157 flux for stainless-steel solder.

   3. **Indium solder**
   4. **Supersafe #30 water-based “blue” flux**

      The solder flux must be thoroughly cleaned and removed.

      Clean the soldering joints with warm water (preferably in ultrasonic), and finally wipe with Propanol.

3. UHV-compatible cleaning solution from Alfa Aesar.
   1. **Deionized (DI) water with item number: 36645**
   2. **Grade acetone**
   3. **Grade propanol**

      Vender:
      Alfa Aesar
      Sales
      Tel: 1-800-343-0660 or 1-978-521-6300
      Fax: 1-800-322-4757
      or 1-978-521-6350
      email: ecommerce@alfa.com
      Technical Service
      Tel: 1-800-343-7276 or 1-978-521-6405
      (M-F, 8AM-5PM EST)
      Fax: 1-978-521-6350
      email: tech@alfa.com
      Website: https://www.alfa.com/en/

4. UHV compatible epoxy
   Vender:
   Epoxy Technology Inc
   14 Fortune Dr., Billerica, MA 01821
   USA
   Phone: (978) 667-3805
   Website: http://www.epotek.com/site/
   1. **EPO-TEK E4110**
      This is a silver-based conductive epoxy for thermal anchor.
   2. **EPO-TEK T7110**
      This is a non-conductive UHV-compatible epoxy for adhesions, thermal contacting, and electrical isolation.
   3. **EPO-TEK H74F**
   4. **EPO-TEK H70E**
      This is a non-conductive epoxy for adhesive.
      Some users do not accept this epoxy being UHV compatible.
5. UHV-compatible wires, wire feedthrough, high-frequency coaxial cables, and high-frequency feedthrough and connectors.
   Note:
   Few venders declare UHV compatibilities on wires, coaxial cables, feedthroughs, and connectors.
   The following products have been used in UHV systems and have proven records.
   1. **Wires from New England Wire Technology**
      New England Wire Technologies
      130 S Main St., Lisbon, NH 03585
      USA
      Phone: (603) 838-6624
      Website: https://www.newenglandwire.com
      a. **Model N13-38E-103 (shielded manganin wires in twisted pair)**
         Wires: 2 × 38 AWG heavy polyurethane insulated manganin wires in twisted pairs
         Jacket: 0.005″ ETFE to 0.020″ ± 0.001″ o.d.
         Shielding: 44 AWG **Cu-Ni** braided shield, 90% minimum coverage
      b. **Shielded manganin wires in twisted pair (shielded manganin wires in twisted pair model name TBD)**
         Wires: 2 × 38 AWG heavy polyurethane-insulated manganin wires in twisted pairs
         Jacket: 0.005″ ETFE to 0.020″± 0.001″ o.d.
         Shielding: 44 AWG **316 Stainless-steel** braided shield, 90% minimum coverage
      c. **Shielded single manganin wires (shielded manganin wires in twisted pair model name TBD)**
         Wire: 1 × 38 AWG heavy polyurethane-insulated manganin wire
         Jacket: 0.006″ EFP to 0.017″ ± 0.002″ o.d.
         Shielding: 44 AWG **316 Stainless-steel** braided shield, 90% minimum coverage

d. **Model N13-36M-131 (shielded S/C wires)**
   Wires: 2 × 36 AWG NbTi in 70% Cu—30% Ni clad insulated with .006″ FEP and cabled into twisted pair
   Jacket: 0.005″ clear FEP
   Shielding: 44 AWG **316 Stainless-steel** braided shield, 90% minimum coverage
e. **Model N13-38M-100 (shielded S/C wires)**
   Wires: 2 × 38 AWG NbTi in 70% Cu—30% Ni clad insulated with .003″ FEP and cabled into twisted pair
   Jacket: 0.005″ clear FEP
   Shielding: 44 AWG **316 Stainless-steel** braided shield, 90% minimum coverage
f. **Model N23-44M-100 (shielded S/C wires)**
   Wires: 2 × 44 AWG NbTi in 70% Cu—30% Ni clad insulated with .003″ FEP cabled in twisted pairs
   Jacket: 0.005″ clear FEP
   Shielding: 44 AWG **316 Stainless-steel** braided shield, 90% minimum coverage
g. **Model N13-30E-100 (shielded copper wires)**
   Wires: 2 × 30 AWG heavy polyurethane-nylon-insulated copper wires in twisted pairs
   Jacket: 0.006″ translucent ETFE
   Shielding: 44 AWG type **304 Stainless-steel** braided shield
h. **Model N12-36M-131 (shielded s/c "coax")**
   Wires: 1 × 36 AWG NbTi superconductor with 70Cu-30Ni cladding, 0.006″ FEP to 0.017″ ± 002″ o.d.—Translucent
   (Single wire inside the shield, sometimes called "coaxial cable)
   Shielding: 44 AWG type **316L Stainless steel** 90.0% min. coverage
i. **Model N12-38R+00002-0 (shielded ph-br "coax")**
   Wire: 1 × 38 AWG hard bare NewAloy 75, 0.0045″PFA to 0.013″ ± 0.001″ o.d.—Translucent
   (Since it is single wire inside the shield, sometimes people also called it "coaxial" cable)
   Shielding: 44 AWG type **316L Stainless steel** 90.0% min. coverage to 0.022″ ± 0.001″ OD
j. **Model N12-36M-133 (50-ohm coax with low noise coating)**
   Wire: 36 AWG NbTi in 70% Cu—30% Ni clad insulated with .006″ FEP and cabled into—Translucent
   Shielding: 44 AWG type **316L Stainless steel** 90.0% min coverage to 0.022″ ± 0.001″ OD

2. **Wires from California Fine Wires**
   California Fine Wire Co.
   338 S 4th St., Grover Beach, CA 93433
   USA
   Phone: (805) 489-5144
   Website: http://www.calfinewire.com/index.html

- Insulated manganin wires with Polyimide insulation (UHV specially rated)
- Insulated ph-bronze wires with Polyimide insulation (UHV specially rated)
- California Fine Wire
- Insulated copper wires with Polyimide insulation (UHV specially rated)
- Silver plated copper wires with Kapton film
- Vacuum range: 1×10^{-11} torr (really UHV rated)

3. **Sub-D wire feedthrough and PEEK mating connector in the vacuum side**
   MDC Vacuum Products, LLC
   30962 Santana St, Hayward, CA 94544
   USA
   Phone: (510) 265-3500
   Website: https://www.mdcvacuum.com/MDCMain.aspx
   Double-ended 25-pin sub-D feedthrough (with male pins on both sides) welded on CF-flange
   Vender: MDC Vacuum Products, LLC
   Item number: 1163601
   Mating PEEK connector in the vacuum side
   Item number: 1269501
4. **PEEK mini connector**
   Reichenbach International Inc.
   Model: MTS04-FPC (PEEK insert with 4 × 3A female contacts)
   MTS04-MPC (PEEK insert with 4 × 3A male contacts)
   Alternative vender:
   CMR Direct
   Website: info@cmr.direct.com
5. **Model N12-50F-257-0 flexible coaxial cables from New England Wire Technology**
   New England Wire Technologies
   130 S Main St., Lisbon, NH 03585
   USA
   Phone: (603) 838-6624
   Website: https://www.newenglandwire.com
   Outer conductor: 304 stainless steel
   Inner conductor: 304 stainless steel
   FEP dielectric
   FEP jacket
   Impedance: 51+/5 ohms
   Operating voltage (Vrms): 200V
   Breakdown Voltage (Vpeak): 8,000V
   Operating attenuation: db/100 feet at 20°C

1 MHz:	230 dB
10 MHz:	318 dB
100 MHz:	396 dB
500 MHz:	588 dB

Note:

a. By rights, the FEP jacket should to be removed. Practically no negative impact on vacuum has been reported from it.
b. The user must bake the system really well if any coaxial cables are installed because a lot of dirty stuff are trapped in between the outer conductor and the inner conductor.

6. **Model UT-85 semi-rigid coaxial cables from Micro-coax**
   Example #1: Model UT-85-SS-SS semi-rigid coaxial cable
   Outer conductor: 304 stainless steel
   Inner conductor: 304 stainless steel
   PTFE dielectric
   Example #2: Model UT-85-BSS semi-rigid coaxial cable
   Outer conductor: 304 stainless steel
   Inner conductor: Be-Cu inner
   PTFE dielectric
   Note:
   The user must bake the system really well if any coaxial cables are installed because water might be trapped in between the outer conductor and the inner conductor.
7. **SMA feedthrough**
   Example #1: Double-ended SMA feedthrough welded on CF-flange
   MDC Vacuum Products, LLC
   Item number: 1221001
   Note:
   Electrically floating feedthrough is also available, and it can also be welded on a CF-flange.
   Alternative vender:
   Ceramaseal
   1033 State Route 20
   New Lebanon,
   NU 12125, USA
   Tel: (518) 794-7800
   (800) 752-seal
   Fax: (518) 794-8080
   Example #2: Double-ended SMA feedthrough installed in vacuum
   (This item is also highly non-magnetic)
   East Coast Microwave Distributors, Inc.
   70 Tower Office Park
   Woburn, MA 01801
   USA
   Tel: 1-800-786-2576
   1-781-279-0900
   Website: http://sourceesb.com/eastcoastmicrowavedist/linecard
   This is UHV compatible, non-magnetic double-ended SMA feedthrough

8. **SMA connector**
   Use the **brass-free** SMA connector from Microstock, Inc.
   Remove the O-ring inside the connector before use.
   Example: UHV SMA male connector
   Accu-glass
   Item number: 111027
   Alternative vender:
   Microstock, Inc.
   P.O. Box 91
   West Point, PA 19486-0091
   USA
   Tel: (215) 699-0355
   Fax: (215) 699-0286
   Email: micrstok@ix.netcom.com
   www.microstock-inc.com
9. **SSMC bulkhead and plugs**
   East Coast Microwave Distributors, Inc.
   70 Tower Office Park
   Woburn, MA 01801
   USA
   Tel: 1-800-786-2576
   1-781-279-0900
   Website: http://sourceesb.com/eastcoastmicrowavedist/linecard
   Example #1: SSMC right-angle bulkhead (mounted cable jack)
   Crimp type for RG178/196
   Gold plated
   Vender: East Coast Microwave
   Item number: 7106-1521-002
   More details:
   This item includes two parts:
   Part A: A piece with a right-angle piece with a male straight section and a bulkhead with a cap.
   Part B: A gold-plated cylindrical "sleeve."
   The right-angle piece can be mounted onto a thin gold-plated copper plate with spring washer and nut as shown in the above photos.
   The bulkhead is located behind the copper plate in the above photos (so you can barely see them), and the exposed parts are the male straight sections.
   The cap will be removed before the wire is soldered to the bulkhead.
   Slide the cylindrical "sleeve" onto the wire.
   Solder the wire to the central pin inside the bulkhead and then check for shorting to ground after the wire is
   Add a dab of T7110 epoxy inside the bulkhead and "fix" the wire (as well as provides some insulation).
   Install and epoxy the cap is onto the bulkhead.

Spray the shield around the short (~ 0/125″ long) "neck" on the bulkhead.
Slide the cylindrical "sleeve" around the bulkhead "neck."
Crimp the cylindrical "sleeve" onto the bulkhead "neck" but do not crush the "neck."
Add a dab of T7110 epoxy to the end of the "sleeve."
The male straight section matches the mating female SSMC right angle cable plug. See below for some details.
Example #2: SSMC right-angle cable plug (used as mating connector to SSMC right-angle bulkhead)
Crimp type
Gold plated
Item number: 7105-1521-002

10. **Superconducting semi-rigid coaxial cables from COAX**
COAX provides superconducting coaxial cables with connector preinstalled.
The superconducting coaxial cables have extremely small thermal conductivities, and it is the only candidate to be applied in ultra-low temperature (in millikelvin temperature range) UHV system at high frequencies with low loss.
Vender:
COAX Co., LTD
461-1 Edacho, Aoba-ku, Yokohama-shi
Kanagawa-ken, 225-0013
Japan
Tel: +81-15-572-3300
Fax: +81-15-572-3858
Email: stanabe@coax.co.jp
Website: http://www.coax.co.jp

6. UHV compatible heater and thermometers for low temperatures
   1. **UHV compatible heater**
   The best candidate is the non-magnetic cartridge heater.
   Vender:
   Hingham Bay Corporation
   32 Scotland Blvd # 8,
   Bridgewater, MA 02324
   USA
   Phone: (508) 697-3164
   Website: http://www.hinghambay.com/
   2. **UHV-compatible thermometers**
   Lakeshore SD-type silicon diode, RuO, or Cernox
   The user can start with SD-type thermometers and make different packages for mounting.
   UHV-compatible epoxy needs to be used to thermally anchor the wires.
   Vender:
   Lake Shore Cryotronics, Inc.
   585 McCorkle Blvd, Westerville, OH 43082
   USA

Main: (614) 891-2243
Fax: (614) 818-1600
Website: www.lakeshore.com

7. Kapton-based multilayer superinsulation
   Multek Flexible Circuits Inc.
   1150 Sheldahl Road, Northfield MN 55057-9444
   USA
   Tel: 507-663-8564
   Fax: 507-663-8300
   www.sheldahl.com
8. NASA website for outgas rate of various commercial materials
   https://outgassing.nasa.gov/

Index

Note: Page numbers in italic and bold refer to figures and tables, respectively.

D

E

F

G

H

I

T

U